Handbook of
PARTICLE
PHYSICS

CRC SERIES *in* PURE *and* APPLIED PHYSICS

Dipak Basu
Editor-in-Chief

FORTHCOMING TITLES

Introduction to Molecular Biophysics
Jack Tuszynski

Physics of Semiconductor Electron Devices
Pratul K. Ajmera

Principles and Applications of Ultrasonics
David Cheeke

High Field Electrodynamics
Frederic V. Hartemann

Handbook of
PARTICLE
PHYSICS

M. K. Sundaresan

CRC Press
Taylor & Francis Group
Boca Raton London New York

CRC Press is an imprint of the
Taylor & Francis Group, an **informa** business

CRC Press
Taylor & Francis Group
6000 Broken Sound Parkway NW, Suite 300
Boca Raton, FL 33487-2742

First issued in paperback 2019

© 2001 by Taylor & Francis Group, LLC
CRC Press is an imprint of Taylor & Francis Group, an Informa business

No claim to original U.S. Government works

ISBN-13: 978-0-8493-0215-2 (hbk)
ISBN-13: 978-0-367-39736-4 (pbk)
Library of Congress Card Number 2001025210

Library of Congress Cataloging-in-Publication Data

Sundaresan, M. K.
    Handbook of particle physics / by M. K. Sundaresan
        p. cm. — (Pure and applied physics)
    Includes bibliographical references.
    ISBN 0-8493-0215-3 (alk. paper)
    1. Particles (Nuclear physics)—Handbooks, manuals, etc. I. Title. II. Pure and applied physics (Boca Raton, Fla.)

QC783 .S86 2001
539.7′2—d21

                                                          2001025210

Visit the Taylor & Francis Web site at
http://www.taylorandfrancis.com

and the CRC Press Web site at
http://www.crcpress.com

Preface

There has been very impressive progress in elementary particle physics during the last half a century. Literally thousands of particles have been discovered. Their physical properties have been measured by ingenious experiments, the inter-relationships between the different particles have been systematized in terms of fundamental interactions, and efforts have been made to understand them in terms of cleverly constructed theories. This field has attracted some of the most able scientists and their creativity has been recognized by the award of many Nobel Prizes for the discoveries in this field. Many good books have been written in the field, some of which serve as text books. However, there is no readily available handbook for the physicists (working in other areas of physics, and who are nonexperts in this field), from which they may obtain quickly the gist of some idea or jargon, explained in more than lay terms. It is with the intention of providing this service that this handbook has been written. It contains explanations in sufficient detail, including theoretical formulations, which will provide understanding and significance of the concepts in this field.

The field of particle physics is a vast one, and to do proper justice to this field, many volumes will have to be written. That would be a very ambitious project. Our aim is a modest one. This is not a book from which one can learn particle physics and certainly not one for experts in the field. Its targets are the nonspecialists. What we have attempted, is to provide the essential information in a handbook of limited size, restricted to fewer than five hundred pages. The three chapters of the book are devoted to brief descriptions of (1) a historical overview of developments in particle physics as a whole, (2) the historical development of accelerators to reach high energies, and (3) the historical development of multi-purpose detectors for making experimental observations in particle physics. These are followed by a Glossary, explaining items of interest in particle physics (including its peculiar jargons), arranged in alphabetical order for ease of retrieval, and addressed to the general body of scientists with relevant mathematical and technical training. It is hoped that it can serve as a source of information on particle physics, not only to the general body of physicists, but also to graduate students in particle physics, who want to get to the heart of the matter quickly, without wading through large amounts of material in text books on the subject. It is hoped that it will also serve as a handy reference for others. The level of presentation is at an advanced stage, assuming basic knowledge of quantum mechanics and relativity, as is available in most graduate curricula. The items in the Glossary are fairly self-contained with cross references to other material of a similar nature in other parts of the Glossary. Some items are repeated under different headings, intending to bring out different perspectives in the topics. For those interested in more details than are provided in this handbook, references to other sources are provided liberally, throughout the Glossary. An effort has been made to include as many of the significant key words and jargon as possible, which are currently in use in particle physics. It is hoped that there are no glaring omissions. With the realization that developments in astrophysics and cosmology have a significant bearing on particle physics and vice versa, some sections are devoted to these topics and included in the Glossary.

Acknowledgements

- The "Review of Particle Physics", *Eur. J. Phys.* C15, 1-878, 2000 (and some of the previous editions) have been an invaluable resource in getting some of the material that is presented here.

- LEP experimental collaborations, ALEPH, DELPHI, L3, OPAL, have all been generous in giving permissions to download pictures of their respective detectors from their Web pages. CERN gave permission to use the picture of the CERN site that appears in their Web pages. I am grateful to all these groups for their permissions.

- Materials on SLC and SLD detectors were obtained from the Web pages of SLAC and are included here with permission from SLAC. For this I am grateful to SLAC.

- The book *Particle Detectors*, by C. Grupen, published by Cambridge University Press has been extremely helpful in providing material on detectors and detector developments. I thank Cambridge University Press for giving me permission to reproduce some of the figures appearing in this book. I also thank Dr. C. Fabjan for permission to reproduce a figure from one of his publications, explaining the workings of the Ring Imaging Cherenkov Detector. Specific acknowledgements may be found in the location where the figures appear.

- I am also grateful to Professor S. Gasiorowicz and John Wiley and Sons, to Addison-Wesley Longman Publishers Inc., and to the *Annual Reviews of Nuclear Science* for permission to include here some of their copyright material. Specific acknowledgements may be found in the location where the materials appear.

- I would also like to acknowledge permissions received from the American Institute of Physics and the American Physical Society to reproduce material from the *Physical Review* and *Physics Today*. Specific acknowledgements relating to these will be found in the locations where they appear.

- I am grateful to the SNO Collaboration for permission to use a picture depicting an event in the SNO detector.

The material for this book was produced using PCTEX32 and I have received a great deal of assistance from Jean-Guy Boutin in coping with many problems of a technical nature peculiar to LaTeX and in the final production of the book. Page headers and chapter heads were set using LaTeX packages `fancyhdr` by Piet van Oostrum and `FncyChap` by Ulf A. Lindgren, respectively. I would also like to acknowledge help received from I. S. Batkin during the course of preparation of this book. I am grateful to the Department of Physics, Carleton University, Ottawa, for the generous use of the facilities of the department during the course of the preparation of this book. CRC PRESS editor Carole Gustafson deserves special thanks for her constructive and congenial advice.

Finally, I would like to thank my wife, Bharathy, for help with some aspects of the work on the bibliography and for being patient with me for my spending many long hours on this book every day for nearly eighteen months.

M. K. Sundaresan
Ottawa, December 2000

Natural Units

In this book we have used units in which $\hbar$ (Planck's constant divided by 2π) and c (the velocity of light) are both unity ($\hbar = c = 1$). Such a system of units has been called *Natural Units*. If the value of the product $\hbar c$ is worked out, it has the value 197.3 MeV fermi, where 1 fermi $= 10^{-13}$ cm. In this sytem, energy and momentum have dimensions of inverse length, and length and time have the same dimensions:

$$\text{Energy} = \text{momentum} = \text{Length}^{-1} = \text{Time}^{-1}.$$

The mass of a particle, m, stands for mc^2, and is given in MeV. Thus, the mass of the electron is 9.1×10^{-28} g, or 0.511 MeV, or $1/(3.86 \times 10^{-11}$ cm).

Angular momentum is measured in units of $\hbar$, and in these units, it is dimensionless. The fine structure constant $\alpha = e^2/(4\pi)$ is dimensionless and has the value (1/137.04). The Fermi weak interaction constant G_F has the dimension of GeV^{-2} (or length^2) and has the value $G_F = 1.166 \times 10^{-5}$ GeV^{-2}. The unit of cross section is the *barn* and equals 10^{-24} cm^2.

Occasionally, where necessary, constants such as Planck's constant and the velocity of light are explicitly indicated. In most of the book, however, the natural units are used.

Contents

Contents

Contents

List of Figures

List of Tables

CHAPTER 1

Historical Overview

The search for the elementary constituents of all matter has occupied generations of human beings since the speculations of the early Greek philosophers and other philosophers from different parts of the world. Customarily, Democritus is associated with the hypothesis that all elements are made up of minute constituents called atoms. Yet real progress in the field started being made only in the sixteenth century. The formulation of the laws of motion by Galileo and Newton paved the way for a quantitative study of the motions of particles endowed with an inertial property called *mass*. Experiments were developed to test how well the hypotheses in the formulation of the laws of motion fared. The "scientific era" in which science develops by progress in theory and experiment may be said to have its origin then.

The first steps in understanding the properties of gases in terms of the mechanical motions of the constituents of the gas as material particles was undertaken by Daniel Bernoulli in the early part of the eighteenth century (1738). This date may be said to mark the origin of the kinetic theory of gases. However, significant progress in understanding the structure of matter came from studies in chemistry. James Dalton, very early in the nineteenth century (1803), took the atomic hypothesis a number of steps further in understanding, both qualitatively and quantitatively, many of the observed facts in chemistry. This was the period in which the table of atomic weights of elements was first constructed. In the same century, other major achievements were the

1

discovery of Avogadro's law (1811) and the formulation of laws of electrolysis (1833) from experimental studies by Faraday. Also, in the later part of the century, Mendeleev, a Russian chemist, found that when the elements were arranged according to their atomic weights, their chemical properties showed a periodic behavior; the chart containing this finding goes under the name of the *Periodic Table of Elements*. These studies established the atomic hypothesis on a firmer footing. In spite of these achievements, there was considerable scepticism in accepting the fact of atoms as real constituents of matter until the early twentieth century.

The impressive progress in chemistry, although achieved on the basis of the atomic hypothesis, did not depend on the detailed properties of atoms, such as their mass, size, or shape. Knowledge of these further properties of atoms had to await studies in the electrical discharges in gases undertaken toward the end of the nineteenth century and in the beginning of the twentieth century. These studies had two aspects to them. On the one hand, one could study what happens to atoms of gases when they are subjected to electrical discharges through them, that is, study whether the atoms break up, and if so, what the products of the breakup are. On the other hand, one could study the properties of the light emitted from the atoms in the discharge. Both these types of studies developed in parallel. The former studies led eventually to unraveling the properties of the products of the breakup of atoms in the electrical discharge, in particular the electron. The latter studies spawned the development of optical spectroscopy as a tool for the study of atomic structure. The turn of the century was also the period when quantum concepts were introduced by Planck (1900) to understand the thermodynamics of radiation. Discoveries of X-rays by Röntgen in 1895 and of natural radioactivity by Becquerel in 1896 were other elements which played a substantial role in leading to our understanding of the structure of atoms.

Electron Discovered (1897)

In studying the discharge of electricity through gases, J. J. Thomson, Crookes, and others studied cathode rays and, from the deflections they suffered in electric and magnetic fields, established that charged particles of both signs existed in the discharge. From their motions in the fields they obtained information on the ratio of charge to mass for these particles. They found that many of the negatively charged particles had small masses compared to the atomic mass, while the positively charged particles, called *ions*, had masses of the size of atomic masses. The negatively charged particles having masses much less than atomic masses were given the name *electrons*. This was the first elementary particle to be discovered (1897).

Photon—Quantum of Radiation (1905)

Based on Planck's idea of the quantum, Einstein in 1905 extended it further and assumed that light exists as quanta (called *photons*), with the energy of a quantum being given by $h\nu$, where, h is the constant introduced by Planck, and ν the frequency of the light. This led to a complete understanding of the features observed experimentally by Hertz, Lenard, and Millikan on the phenomenon of photoelectric emission of electrons from metals. Thus was born the second elementary particle, the *photon*, in 1905.

Natural Radioactivity

The nature of the radioactive emanations discovered by Becquerel was clarified soon after their discovery. It was established that they consisted of three components, called α, β, and γ radiation. Rutherford and collaborators made a detailed study of the α and β emanations. They measured the penetrating power of these emanations in materials. It was found that the α emanations were absorbed in a few centimeters of air, while the β emanations were absorbed in an aluminum foil about one millimeter thick. By a variety of means, it was established that the α emanations consisted of positive doubly charged helium ions (1903), while the β emanations had negative electric charge. Comparison of the absorption of cathode ray electrons and of the β emanations in thin foils showed that it was possible to identify them as the same. It was thus established that β emanations consist of electrons. The γ emanations were found not to be affected by electric or magnetic fields and, hence, were electrically neutral. By studying the penetration properties of the γ emanations through materials, for example, lead, and comparing with the penetration properties of X-rays through the same material, it was established that these two could be the same. Definitive understanding that γ emanations are electromagnetic radiation of high energy came later. Marie Curie, and Rutherford and collaborators, found that in radioactive transmutations, the chemical identities of the atoms changed with time. Thus, in α emission, a substance changes its chemical nature and ends up as a substance with chemical properties two units down in the periodic table of elements.

Thomson Model of Atom

On the basis of all these phenomena, it was concluded that electrons formed an integral part of all matter. Because matter on the whole is electrically neutral, in an atomic view of matter, each atom must be electrically neutral. Thus, the view emerged that an atom is a neutral combination of a positive charge compensated by the negative charge of the electrons. Having settled this, one had to know the size of the atoms and how much mass the positive and negative charge components contributed to the mass of the atom. From studies in chemistry and other fields, it was generally concluded at this time that the size of atoms must be about 10^{-8} cm. Also, the periodic table of atomic elements existed. Based on these ideas and on the periodic table, Thomson proposed a model in which an atom consists of a uniformly positively charged sphere of radius 10^{-8} cm in which was distributed an equal negative charge in the form of electrons.

Rutherford Model of Atom

The model proposed by Thomson had to be abandoned in 1911 as it could not explain the large angle scattering (almost backward scattering) of α particles by atoms observed by Geiger and Marsden, associates of Rutherford. The Thomson model predicted an extremely small amount of backward scattering, in disagreement with observations. The observed data on the α scattering were well accommodated in a model of the atom that Rutherford proposed, in which the atom has a massive positively charged nucleus of radius about 10^{-13} cm, carrying the bulk of the mass of the atom, surrounded by electrons moving in orbits whose radii are of the order of 10^{-8} cm.

Measurement of Electron Charge by Millikan, X-ray Diffraction, Isotopes

The charge on the electron was measured by Millikan in 1911 using an ingenious method. Combining this with Thomson's measurement of the charge to mass ratio for the electron, the mass of the electron could be determined. It was found to be much smaller than the mass of the lightest atoms, confirming earlier indications to the same. The discovery that X-rays could be diffracted was made by von Laue in 1912, and very soon after that, Moseley introduced the concept of atomic number Z to classify elements. Thus every element had an atomic number and atomic weight associated with it. Developments of mass spectrographs by Aston in England and Dempster, Bainbridge, and Nier in USA, measured atomic masses more and more accurately. These measurements showed that chemical species came with different masses, called *isotopes*, and these masses were very nearly integral. In Rutherford's model, the atomic number Z was interpreted as the number of units of positive charge carried by the nucleus, the unit of charge having the same magnitude as that carried by the electron. The neutral atom will have Z electrons revolving around a nucleus with atomic number Z.

Bohr Model of the Atom and the Beginnings of Quantum Mechanics

In 1913, based on Rutherford's ideas, Bohr produced the theory for the structure of the simplest of atoms, namely hydrogen, an atom with a nucleus of $Z = 1$, around which one electron revolves. In doing this, he showed that application of classical physics to this system poses an immediate problem. Classical electromagnetism predicts that accelerated charges radiate energy in the form of electromagnetic radiation. Since the electron in the Rutherford model revolves around the nucleus, it is undergoing acceleration and, hence, must radiate. If the electron radiates away its energy, it will spiral in to the nucleus in short order. He abandoned classical physics to discuss the motion and introduced his now-famous quantum condition on the electron's orbital angular momentum to describe the motion. With that he was able to describe the quantized energy levels of the hydrogen atom and the spectrum it would exhibit. The results were in extremely good agreement with measurements from the experimental spectra of atomic hydrogen. This provided the starting point for many remarkable developments in atomic physics and, subsequently, in the late 1920's, led to the formulation of quantum mechanics by Heisenberg, Schrödinger, de Broglie, Born, Dirac, Pauli, and others. Quantum mechanics was rapidly developed and applied to problems of atomic structure, emission and absorption of radiation by atoms, etc. Quantum statistics dealing with assemblies of identical particles were developed.

Chemistry, Nuclear Physics as Separate Disciplines

At about this time, studies of the structure of matter started recognizing two distinct areas. Chemical reactions between atoms of elements, leading to formation of compounds etc., involved only the electrons, with no changes to the nuclei of the atoms, and energy changes of the order of a few electron-volts (eV). Then there were the radioactive transformations in which, the nucleus itself underwent a change, leading to a different chemical element, and the energy change was a million times larger (MeV). A clear distinction emerged between studies in these two areas, the former could be classified as chemistry and the latter as nuclear physics. The study of the nucleus and the exploration of its constituents became the goal of these latter studies.

Proton Discovered (1919)

Among the first nuclear physics experiments performed, mention must be made of the work of Rutherford in 1919. Using a natural radioactive source to provide energetic α particles, he broke up nitrogen nuclei and showed that hydrogen nuclei were produced as a result of the bombardment. The name given to the hydrogen nucleus was *proton*. This represents the third elementary particle to be discovered.

Studies of many nuclear disintegrations revealed the existence of protons as products. Thus, one could envisage a model of the nucleus of an atom of atomic weight A to be made up of A protons and $A - Z$ electrons, which would give the nucleus a charge $+Z$. However, there are problems with this model. Quantum mechanical considerations showed that if electrons were confined inside a region of size 10^{-13} cm, their kinetic energies would increase to the point where they would not stay bound inside such a nucleus. An alternative model for the composition had to be found.

Need for Neutral Component in the Nucleus

Around 1920, Rutherford drew attention to an important conclusion from all the work on the scattering of alpha particles by various elements. If the element had an atomic weight A, these experiments showed that the nucleus carried a charge Ze, with $Z = A/2$, where e is the fundamental unit of charge. The numbers of extra nuclear electrons in the atoms were also determined to be very close to $Z = A/2$ from studies of Thomson scattering of X-rays from atoms. If the positive charge on the nucleus is due to $A/2$ protons, then the question arises as to what holds these together against their electrical repulsion. Further, the nucleus with $Z = A/2$ protons would only account for half the atomic mass. These facts could be reconciled if there were another component in the nucleus, which is electrically neutral and contributes the rest of the mass of the nucleus. This was the first strongest hint for the existence of neutrons, which were found somewhat later. In this same work, Rutherford also suggested that the atomic number, or the equivalent nuclear charge, is the more natural variable in terms of which to classify the periodic table of elements rather than the atomic weight.

Strong Interactions as Distinct from Electromagnetic

In 1921, Chadwick and Bieler, studying the scattering of α particles in hydrogen, found that the characteristics they were observing in the scattering could not be accounted for solely in terms of the Coulomb interactions between the alpha particle and the proton in hydrogen. This was the first evidence for the strong nuclear force as distinct from the electromagnetic force.

Intrinsic Angular Momentum—Electron Spin Introduced

By 1925 much experimental work on atomic spectra had been carried out and compared with Bohr's theory. Among other problems, one feature observed in the spectra was a fine structure in the spectral lines. It was found that many spectral lines, which were predicted in Bohr's theory to be single lines, were actually closely spaced doublets. The solution for this problem came from two sources. First, Pauli suggested in 1924 that the electron in the atom was described, in addition to the quantum numbers of Bohr's theory, by another quantum number which could take on two possible values. Goudsmit and Uhlenbeck, in 1925, went a step further than Pauli, and suggested that this observed fine structure could be accommodated in Bohr's theory, if the electron carried, in addition to the orbital angular momentum, an intrinsic angular momentum, called *spin*. For the spin they proposed a two-valued variable, which could orient itself either parallel or antiparallel to the orbital angular momentum vector. If there were a small energy difference between the states corresponding to these orientations, this would explain the observed doublet fine structure. Thus, the notion that an elementary particle could carry an intrinsic angular momentum, called *spin*, was introduced for the first time. The spin attributed to the electron was 1/2. The unit for angular momentum is $\hbar$, where $\hbar$ is Planck's constant h divided by 2π.

Proton Spin Determined

Closely associated with the spin is a magnetic moment for the particle. The magnetic moment vector is oriented in the same direction as the spin vector for particles of positive charge, while it is opposite to the spin for particles of negative charge. Thus, in addition to the charge interaction between particles, there will also be an interaction between the magnetic moments, if the particles have spin. If the nucleus has spin, the magnetic moment associated with it will interact with the magnetic moment of the electron due to its orbital motion and spin angular momentum. Such an interaction would give rise to further fine structure of spectral lines, called *hyperfine structure*. Measurements of hyperfine structure will give information on the spin that the nucleus of an atom carries. Through such measurements, Dennison attributed spin 1/2 to the proton in 1927.

Dirac's Theory of the Electron

The relativistic equation for the electron was discovered by Dirac in 1928. In searching for an equation, he demanded that if the spatial derivatives appeared to first order, then time derivative also must appear only to first order to satisfy the demands of special relativity. These requirements necessitated the introduction of a wave function with four components. He found further that solutions of the equation exist for positive as well as negative energies. For nonrelativistic energies, he found that the wave function with four components reduces to one with only two nonzero components, providing a natural basis for the spin. He calculated the spin magnetic moment of the electron and the energy levels of the electron in the Coulomb field of a nucleus. He also developed the quantum field theory for the emission and absorption of radiation.

Events Leading to the Discovery of the Neutron

Getting back to the constituents of the nucleus, we have already given arguments why a model involving protons and electrons is not viable. Additional arguments against a proton-electron structure come from quantum mechanical considerations of an assembly of identical particles (1926), called the *spin-statistics theorem*. Particles of integral spin are described by wave functions which are symmetric under the interchange of any two particles (Bose-Einstein statistics), and those of half-odd integral spin are described by wave functions which are antisymmetric under the interchange of any two particles (Fermi-Dirac statistics). Consider the nucleus of nitrogen. It has atomic number $Z = 7$, and mass number $A = 14$. In the proton-electron model, it would have 14 protons and 7 electrons, a total of 21 spin one-half particles. This would make the nucleus of nitrogen have half-odd integral spin and, hence, obey Fermi-Dirac statistics. Experiments on the molecular spectrum of homonuclear N_2 molecule showed that the nitrogen nucleus must obey Bose-Einstein statistics; in other words, it must have integral spin. Thus again, the proton-electron model for the nucleus fails. The resolution of the situation came with the discovery of the neutron by Chadwick in 1931.

Chadwick's Discovery of the Neutron

Experiments, in 1930 and 1931, by Bothe and by the Curies, in which beryllium was bombarded by alpha particles from a natural radioactive source, revealed that a radiation was emitted having a high penetrating power. They measured the ionization produced by this radiation in an ionization chamber with a thin window. When they placed a hydrogen containing substance in front of the thin window, they found the ionization increased. It appeared that protons were ejected by this radiation from the hydrogen into the chamber. Their suggested explanation was that protons were being ejected by a Compton-like process in the hydrogenous material, and they estimated the energy of the gamma ray quanta around 50 MeV. Chadwick, in 1931, repeated the experiment and found that radiation ejects particles, not only from hydrogen, but also from a whole host of other materials. The protons from hydrogen appeared to have maximum velocities of one tenth the velocity of light. With other materials, the ionization released in the chamber appeared to be due to recoil atoms. He showed, by detailed analysis of the data including all the recoils, that the beryllium radiation could not be gamma rays. He could make sense of the data if he assumed that the beryllium radiation consisted of neutral particles of protonic mass. He coined the name *neutron* to describe this particle. This was the fourth elementary particle, after the electron, photon, and the proton, to be discovered. In 1932, a picture of the nucleus emerged in which protons and neutrons were its constituents. In the proton-neutron model, the nitrogen nucleus would consist of 7 protons and 7 neutrons, for a total of 14 spin-half particles, so that the spin of the nitrogen nucleus would be an integer and it would obey Bose-Einstein statistics. The conflict with the molecular spectra observations would be removed. The concept of an *isotope* where a chemical element characterized by its atomic number Z but different atomic masses specified by A could now be understood in terms of different neutron numbers N, with $Z + N = A$.

Photon Spin Determined (1931)

The direct determination of the photon spin was carried out in a little-known experiment by Raman and Bhagavantham using the then newly discovered Raman effect. By applying energy and angular momentum conservation in their observations of the scattering of photons by rotating molecules, they established that the photon spin is 1.

Nuclear Magnetic Moments Measured; Neutron Spin

In 1939, under the leadership of Rabi, a new method, called the *molecular-beam resonance method*, was developed to carry out precision magnetic moment measurements. In this method, the nuclear magnetic moment is obtained by measuring the Larmor precession frequency in a uniform magnetic field. Molecules, like HD and D_2 (D, deuterium, being the isotope of hydrogen with $A = 2$), are mostly in states having zero rotational angular momentum at normal temperatures. The magnetic moments of the proton and the deuteron (deuterium nucleus) were determined using this technique, and the values found were in very good agreement with those obtained using hyperfine structure measurements. A nonvanishing value for the deuteron magnetic moment means that the nuclear spin of the deuteron is nonzero. This implies that the neutron has a nonzero spin. If a spin $(1/2)$ was attributed to the neutron, just as for the proton, this measurement indicates that the spin of the neutron is aligned parallel to the proton spin in the deuteron. The results of the measurements were consistent with this assumption. The neutron spin was hereafter taken to be $1/2$ on the basis of these measurements.

Electron Spin from Dirac's Theory and Antiparticles

A remarkable achievement of Dirac's relativistic theory of the electron is that the spin of the electron comes out of the theory automatically. The equation had solutions of negative total energy. Dirac had the brilliant insight to interpret these solutions as antiparticles to the electron. This prediction of the antiparticle meant that there must exist in nature, particles of exactly the same mass as the electron, but opposite in charge. Such antiparticles are called *positrons*. He showed further that in an electron-positron collision, they would annihilate and produce gamma rays. Conversely, gamma rays could materialize into electron-positron pairs in the presence of matter. The extraordinary prediction of the existence of the positron by Dirac was experimentally verified by Anderson in 1932 while studying cosmic rays.

Discovery of Cosmic Radiation and the Positron

In a series of investigations which started in 1912, Hess showed that the earth was bombarded by extremely energetic particles coming from outer space. This was done by sending ionization chambers mounted on balloons and launched into high altitudes. A number of other workers, in other parts of the world, joined in investigating the nature of these

radiations. This field came to be known as studies of *cosmic rays*. Many
surprising results have turned up in these investigations, some of which
continue even to the present days. Anderson set up a cloud-chamber
at mountain top altitudes and took photographs of cosmic ray events
(1932). It was in one of these photographs that he obtained clear evidence of electron-positron production in cosmic rays.

Postulate of the Neutrino

Ever since the early days of discovery of β radioactivity, a number of
people were involved with further investigations of the properties of the
β particles. It was discovered that β particles of both signs (electrons
and positrons) are emitted in β decays. Early crude measurements of
β^- energy, by absorbing them in thin foils, showed them to be electrons.
The measurements of energy of β particles were continually improved
by the construction of magnetic spectrometers and other methods. At
the same time, progress was being made in more precise measurements
of atomic masses using mass spectrographs. It was established from a
number of such measurements that, although the nucleus undergoing
the β transformation was in a definite state and the product nucleus
was also in a definite state, the emitted β particle had a continuous
distribution of energies. Measurements showed that the energies of the
β particles continuously ranged from very low energies to a maximum
energy E_{max}, (the end-point energy of the β spectrum), where E_{max}
is equal to the energy difference between the parent nuclear and the
product nuclear states. These observations were very puzzling because
it seemed to imply lack of energy conservation in this process.

Not willing to abandon conservation of energy, Pauli, in 1930, came
up with the idea that possibly a neutral invisible particle is emitted
along with the β particle. He suggested that the two particles together
share the energy difference between the initial and the product nucleus
consistent with conservation of energy and momentum. If such a neutral
particle did indeed exist, its mass, as deduced from the energy distribution of the β particles at the end point, showed that it was consistent
with being zero within experimental errors. It was thus assumed that it
was a particle of zero rest mass. The name *neutrino* has been given to
this particle. Because the nuclear states are also characterized by definite values of angular momentum, the neutrino along with the β particle
must serve to conserve angular momentum. In terms of the angular momenta involved, measurements showed that, either both the initial and
product nuclei had integral angular momenta, or both had half-odd integral angular momenta, never otherwise. This is only possible if the
β particle and the neutrino together carry off integral units of angular

momenta. Because the electron has an intrinsic spin (1/2), the neutrino too must possess a half-odd integral spin, that is, it must obey Fermi-Dirac statistics, like the electron. For definiteness and simplicity, it was postulated to have a spin (1/2) just like the electron.

The neutrino eluded direct observation for a long time after Pauli's suggestion; persistent efforts, however, led to a direct observation of its properties by the use of some remarkable methods. Currently, some seventy years later, the question of whether the neutrinos have a mass is still a matter of vigorous experimental investigation involving many different techniques.

Field Theory of β Decay

Very soon after Pauli's suggestion, Fermi constructed a quantitative theory of β decay (1934). This was based on field theory. It was patterned after the theory of emission of radiation from an excited atom. He gave quantitative expressions for the mean lifetime for β decay, and also the expression for the continuous energy distribution of the β particles for comparison with experimental data. This theory served as a back bone for a lot of further experimental and theoretical work in many of the following years.

One feature of the theory of radiation from atoms is that parity is conserved in the electromagnetic interactions responsible for the emission of radiation. The theory of β decay invoked a new form of interaction, called *weak interaction*, information about which had to be gathered form studies on β decays. It was tacitly assumed that the weak interaction, like electromagnetic interactions, conserved parity. A big surprise was in store when it was found in 1957 that parity is not conserved in β decays.

Yukawa's Prediction of Mesons

Another significant development in theory occurred in 1935. Yukawa attempted to find a field theory of the nuclear force patterned after the theory Fermi had developed for β decay. In the theory of Fermi, the basic transformation taking a neutron into a proton involves the emission of the electron-neutrino pair with a certain assumed coupling. Using the virtual emission and absorption of electron-neutrino pair between the neutron and proton, Yukawa tried to generate the strong nuclear force between the proton and the neutron in the nucleus. He found the strength of the β decay coupling to be too weak to generate the strong nuclear force. He thus put forward another idea, namely, that the transformation of the neutron into a proton involves emission of another new particle with a new coupling. He generated the neutron-proton interaction via the virtual emission and absorption of this new particle between them. He could adjust the strength of the new coupling so that he could obtain the strength of the nuclear force of the right magnitude. To fit the short range nature of the nuclear force, he realized that this new particle cannot be massless, for it was known that exchange of massless photons gives rise to the long range Coulomb force. He introduced a mass for the new particle and fitted its value to obtain the right range of the nuclear force, about 10^{-13} cm. He needed about 300 electron masses to achieve this. Here was a new prediction on the existence of particles of mass intermediate between the electron and the proton. Such particles were later found in nature and are called *mesons*.

Nuclear Physics Developments (1930–1950); Isotopic Spin

In order to induce nuclear transmutations artificially, and enlarge on what was learned about nuclear transformations from natural radioactivity, it was necessary to accelerate particles to energies in the millions of electron volts range. The period between the late 1920's and early 1950's saw the construction of a number of accelerators. The Cockcroft and Walton generator, the Van de Graaff accelerator, the cyclotron, the betatron, the synchrotron, and the linear accelerator were all developed during this period and used in the exploration of nuclear transformations. Many radioactive isotopes were artificially produced for uses in biology, chemistry, medicine, etc. Systematic investigations of energy states of nuclei, determination of the quantum numbers relevant to these states, selection rules for transitions between states, etc. were the focus of attention. Enormous progress was made, including developments in using nuclear fission reactions for generation of energy. We do not go into this subject, and the many fascinating developments that took place, because that is not the primary focus of this book. However, before we leave this topic, we should mention one specific feature of the strong interactions revealed by the studies in nuclear physics, which has an impact on particle physics. It has been found that the specifically strong interaction between a pair of protons is the same as that between two neutrons and that between a neutron and a proton; it is a statement of the *charge independence of nuclear forces*. This experimental finding is very well expressed in terms of a new symmetry called *isotopic spin symmetry*, suggested by Heisenberg in 1932. The neutron and the proton are close in mass and both have spin $\frac{1}{2}$. In the limit when this mass difference is ignored, they can be considered as two substates of one particle called the *nucleon*. If the nucleon, in addition to its spin angular momentum (1/2), is endowed with a new property called *isotopic spin* with a value (1/2), the proton and the neutron can be considered as the nucleon with "up" and "down" isotopic spin projections along some axis in the isotopic spin space. The charge independence of nuclear forces becomes the statement that the nuclear forces are invariant under rotations in the isotopic spin space. The isotopic spin concept and its extensions have proved to be key concepts in classifying and understanding the behavior of many other newly discovered elementary particles.

Muon discovered

Since the discovery of cosmic rays, they were the only source of very high energy particles coming to the earth from outer space. Evidence for the existence of a number of new particles, the first of them being the positron, came from studies in cosmic rays. In 1937, Neddermeyer and Anderson were the first to point out the existence of particles of mass in the range between the electron and the proton masses. They observed these in their studies of cosmic rays with a counter controlled cloud chamber. Confirmation of these observations came from the work of Street and Stevenson, also in 1937. They estimated the mass of these particles to be in the range of about 200 hundred electron masses, and that the particles had a charge equal in magnitude to that of the electron. These particles were observed to come in both positive and negative charge states. The name *mesotron*, which later was shortened to *meson*, was given to such particles. Rossi, in 1939, was the first to report that these mesons were short-lived. He produced a first estimate of their mean lifetime. Williams and Roberts, in 1940, were the first to observe the decay of the meson into an electron and suggested a mean lifetime of 10^{-6} s for such decays. A more precise measurement of the mean lifetime was reported by Rossi and Nereson in 1942 as 2.15 ± 0.07 microseconds.

At first it was thought that these particles were the ones predicted by Yukawa in 1935. Closer examination of the interaction properties of this particle with matter showed that it was too weak to generate the strength required for the strong nuclear force. For a while, this caused confusion in the field. The situation was clarified by the suggestion from Marshak and Bethe that possibly two mesons are involved here, only one of which might give the nuclear force according to the Yukawa formulation. The situation was completely resolved with the discovery of the π meson somewhat later. The weakly interacting particle came to be known as the μ *meson* or *muon*.

Lamb Shift, $g-2$ measurements

Two crucial experiments were performed to put the Dirac theory of the electron to more stringent tests. One of the experiments, performed by Lamb, focused attention on one of the predictions of Dirac's equation applied to the hydrogen atom—the exact degeneracy of the $2S_{1/2}$ and the $2P_{1/2}$ levels. If the two levels are not degenerate, it must be possible to induce transitions between them and measure the energy difference. The $2S_{1/2}$ state is a metastable state and such atoms can be detected by their ability to eject electrons from a metal target. The $2S_{1/2}$ to $2P_{1/2}$ transition was induced with microwaves of the right frequency, and the decrease in the population of $2S_{1/2}$ state was measured from the decrease in the ejected electron current. It was found from these measurements that the $2S_{1/2}$ state was higher than the $2P_{1/2}$ state by an energy difference corresponding to a frequency of about 1051 MHz.

The other experiment, performed by Kusch and Foley, was to check the other prediction of Dirac theory, namely, that the gyromagnetic ratio g is 2. They set out to measure the difference of g from 2, using the atomic beam resonance technique, by measuring the frequencies associated with Zeeman splittings of energy levels in two different atomic states in a constant magnetic field. From these measurements their result for g was 2.00244 ± 0.00006, the difference being about $+0.00244$.

Field Theories—Quantization and QED

Soon after the formulation of quantum mechanics, the Schrödinger equation was applied successfully to solve a number of problems. It could not be used in problems where the particles became relativistic. Generalization of the Schrödinger equation was necessary to apply in the relativistic case. In answer to this quest, the Klein-Gordon equation was the first one to be written down. It represented a natural relativistic generalization of the Schrödinger equation. It describes particles of spin zero. In this equation, consistent with special relativity, space derivatives of second order and time derivative of second order appear on an equal footing. Because of this, the expression for the probability density, which in Schrödinger theory was positive definite, is not positive definite in the Klein-Gordon theory. This leads to difficulties with the physical interpretation of the theory. This started Dirac on a search for an equation which would lead to positive definite probability densities. It resulted in his discovery of the relativistic equation for the electron. Particles described by the Dirac equation were shown to have intrinsic spin of $\frac{1}{2}$.

Both the Klein-Gordon and Dirac equations, taken as equations for the wave function of a particle, still suffer from problems. They possess solutions for both positive and negative energies. While the solutions for positive energies can be naturally interpreted as the wave function of the particle, the solutions for negative energies do not have any simple physical interpretation. For the electrons, Dirac made use of the Pauli exclusion principle and assumed that all the negative energy states are completely filled and that only deviations from the completely filled state would have observable consequences. A hole created in the negative energy sea by exciting one of the particles to positive energy would appear as a particle of positive charge and mass equal to the mass of the electron. The hole manifests as the antiparticle to the electron. Such a reinterpretation is not available for the Klein-Gordon theory as there is no exclusion principle operating in this case.

The way to get around this problem is not to interpret these equations as the equations for the wave function of a single particle, but as the equations for field functions, which when quantized give the particles and the antiparticles of the field. For any free field, a relativistically invariant Lagrangian is chosen so that the Euler-Lagrange equations give the equations of motion of the field, for example, the Klein-Gordon or the Dirac equations. Procedures of Lagrangian mechanics were followed to construct the canonical conjugate to the field function (the canonical momentum), and then the Hamiltonian and other quantities. Quantization was carried out by introducing commutation relations between the field and its canonical conjugate. In carrying out these procedures, one carries out a Fourier mode expansion of the field function and its canonical conjugate. The Fourier expansion contains both positive and negative frequency components. The expansion coefficients which are ordinary numbers in a classical field theory become operators for quantization, called *annihilation*, and *creation* operators for particles and antiparticles. The creation operators operate on the vacuum state (defined as a state with no particles or antiparticles) and create, one, $2, \ldots, n$ particles or antiparticles. When the expression for the Hamiltonian was worked out, it was found to be possible to write its eigenvalues as a sum over all the modes of the product of the number of particles in that mode, multiplied by the energy eigenvalue of that mode, and another similar sum for the antiparticles. A further subtlety was encountered here. Fields which give rise to particles having half-odd integral spin, have to be quantized with anticommutation relations between creation and annihilation operators, while the fields which give rise to particles of integral spin have to be quantized with commutation relations between them. Otherwise one does not get a positive definite Hamiltonian or a positive definite occupation number. The quanta of the field were called *fermions* and *bosons*, respectively. An immediate consequence of the anticommutation relation in the fermion case is that the occupation number in a state can take on only the values 0 or 1, which is a statement of the Pauli exclusion principle.

The procedure outlined above for free fields, was extended to fields with interactions between them. However, the resulting field equations form a coupled set, and no general method has been found to solve the coupled set, except in some special cases. In the case when the interaction between fields is weak, the coupled set has been solved by developing a covariant perturbation theory. This has been done in the case of electrons interacting with the electromagnetic field. The resulting theory is called *quantum electrodynamics* (QED).

In the case of QED, the coupling between the fields can be expressed in terms of a dimensionless coupling constant, the fine structure constant, which has the value of approximately 1/137. Perturbation theory gives an expansion of the solution in terms involving powers of this number and hence might be expected to give a reasonably good answer with a finite number of terms. In higher orders of perturbation theory, the contributions depend on some integrals which are divergent. Covariant methods by which to separate the infinities and extract the finite parts in an unambiguous manner (called *renormalization*) exercised many a mind during the 1940's. Tomonaga, Schwinger, and Feynman developed these methods independently and applied them successfully to the calculation of the anomalous magnetic moment of the electron and many other processes. Bethe used the theory and calculated the Lamb shift in hydrogen and showed it to be remarkably close to the experimental value. The agreement between theory and experiment for many quantities is truly remarkable.

These developments have served as a model for developing field theories for other fundamental interactions. The Yukawa model for the interaction of nucleons and mesons was explored more thoroughly. As a model for nuclear interactions, the size of the dimensionless coupling constant needed was of the order of 15. Application of perturbation theory was out of the question in this case. Theorists turned their attention to finding suitable methods to deal with strong interactions in field theory. To this day, a half century later, despite better understanding of some of the problems and the development of a field theory of strong interactions called *quantum chromodynamics* (QCD), the calculation of bound state energy spectrum of strongly interacting particles is still largely unsolved.

Pion Discovered (1947)

Using the photographic nuclear emulsion method exposed to cosmic rays at high altitudes, Lattes, Occhialini, and Powell showed that cosmic rays contain charged particles of mass intermediate between those of the electron and the proton, subsequently called *mesons*. They showed that some of these particles interact with nuclei of the emulsion and disintegrate them, producing a residue of heavy particles. They also had other examples in which the initial meson slows down in its passage through the emulsion and decays into another meson. These two types of events were interpreted as one in which a π^- meson interacts with a nucleus and produces a disintegration, and the other in which the decay of a π into a muon plus a neutral particle occurs. Thus, it is shown that two mesons are present, one of which interacts strongly with nuclear particles. The particle that Yukawa proposed could be identified with the π meson also called *pion*. The flux of such particles in cosmic rays was not enough to measure the masses and other properties accurately. More precise measurements had to await the construction of higher energy accelerators in the laboratory to produce these particles in abundance and study their properties.

V Particles

Continuing studies of cosmic rays with counter controlled cloud chambers, Rochester and Butler in 1947, pointed to the existence of some new particles which left V-shaped forked tracks in the cloud chamber pictures. They were found to be rather rare. In about 5000 photographs of various events taken in 1500 hours of operation, only two photographs showed these V-shaped events. A few years earlier, Leprince-Ringuet had reported an event which suggested a new particle with mass about 990 times the mass of the electron. Further evidence for V particles in cloud chamber observations came from observation of 34 V events in a sample of 11,000 events, reported by Seriff and colleagues in 1950. The V-shaped events were interpreted as arising from the decay of a parent neutral particle into a pair of charged particles, which left V-shaped tracks in the chamber. From this sample, they made an estimate of the lifetime of the neutral particle and came up with a figure of 3×10^{-10} s. They also reported on the nature of the decay products.

Pions Produced in the Laboratory (1949)

A 184″ cyclotron, based on the principle of phase stability, was completed in the campus of the University of California at Berkeley in 1947.

It was a frequency modulated synchrocyclotron with the capability of producing beams of 180 MeV deuteron and 400 MeV alpha particles.

Using a 380 MeV alpha particle beam incident on a carbon target, a team which included Lattes produced π^+ mesons (positive pions) in the laboratory for the first time in 1949 and detected them by the nuclear emulsion method. The characteristic decays of π^+ mesons in the emulsion, resembling what had been seen by Lattes, Occhialini, and Powell in cosmic rays, were found. They also reported producing one positive meson for every four negative mesons in a target of carbon $(1/16'')$ thick.

In 1950, using a synchrotron with 330 MeV gamma-rays incident on various targets, hydrogen, beryllium, and carbon, Steinberger and colleagues reported finding that multiple gamma-rays are emitted. By studying the angular correlations of the produced gamma-rays, they showed that they come in pairs from the decay of a neutral meson. An estimate of the cross section for the production of these neutral mesons showed that it is similar to that for charged mesons. The cross sections in hydrogen and the cross section per nucleon in beryllium and carbon were found to be comparable. The meson they found was the neutral counterpart π^0 of the charged π mesons.

The neutral π meson was also found in cosmic rays through the study of the spectrum of gamma-rays in the atmosphere at a height of 70,000 feet using the nuclear emulsion technique. The spectrum seen was consistent with their being produced by the decay of neutral mesons. They estimated the mass of the neutral mesons to be $(290 \pm 20)m_e$ and their lifetime to be less than 5×10^{-14} s.

Pion Properties Determined (1951)

Soon after their production in the laboratory, the masses of the charged and neutral π mesons (hereafter also called pions) were determined. Attention turned to determining the intrinsic spin and parity of the pions. The spin of the π^+ was determined by using the principle of detailed balance in the reactions $\pi^+ + d \leftrightarrow p + p$. These experiments determined the spin of the π^+ to be zero. Experiments on the capture of π^- mesons in deuterium led to final states which had 2 neutrons in some 70% of the cases, and 2 neutrons plus a photon in about 30% of the cases. These measurements, along with some theoretical input, led to the determination of the intrinsic parity of the π^- to be odd. Combined with the information of spin zero for π^+ and the measured near equality of the masses of the π^- and π^+, there was a strong suggestion that they were different charge states of the same particle. It is reasonable to attribute negative parity to both of them. Since the neutral pion also was observed to have a mass close to those of the charged pions, all three charge states were grouped into a pseudoscalar isotopic spin 1 multiplet. Subsequent work over the years has amply verified these assignments to be correct.

Nature of V Particles Clarified: Λ^0 and K^0

Considerably more work accumulated much more data on the V particles and led to further understanding of these particles by 1951. The data on measurements of the momenta of the secondary particles together with ionization measurements enabled them to discriminate between protons and mesons. From the data it became clear that there were two kinds of V particles. One kind had decay products, a proton, and negative pion, and another kind decayed into oppositely charged pions. Measurements of the mass of the former type gave a value of 2203 ± 12 electron masses, and of the latter type around 796 ± 27 electron masses. What they observed was what we now call Λ^0 *hyperon* and K^0 *meson*.

Charged Hyperons

In 1953 nuclear emulsion studies of cosmic rays revealed the existence of new unstable charged particles with a mass larger than the proton. They looked like charged varieties of the Λ^0 hyperon. They were observed to decay into fast charged pions of either sign or into a slow proton. The observations were interpreted as two alternative decay modes of one particle, which decays to a neutron plus $\pi^\pm$ mesons or into a proton plus π^0 meson. Cloud chamber studies also found the decays to the proton plus π^0 final state. These observed particles are what we now call the $\Sigma^\pm$ *hyperon*.

V Particles, Hyperons, New Particles, Produced in the Laboratory

The frequency of occurrence of new particles in cosmic rays was such that it was not possible to obtain an accurate measure of their properties, such as mass, lifetime, spin, and parity. To study them better under controlled conditions, they would have to be produced in the laboratory. The energy reach of the Berkeley 184″ synchrocyclotron was not sufficient to produce these heavy particles. Simply increasing the size of accelerators in order to reach even higher energies did not seem like an economically efficient option. It would require so much more iron for the magnets and a corresponding increase in costs. Fortunately, the invention of the strong focusing principle for accelerators by Courant, Livingston, and Snyder in 1952 made the access to much higher energies possible at only modest increase in costs.

A succession of increasing energy accelerators was produced in the mid 1950's. These machines helped to establish in a quantitative way the mass spectra and other properties of elementary particles, some of which were already known from cosmic rays. The Cosmotron operated in 1953 at the Brookhaven National Laboratory and produced protons of 2.2 GeV energy. The Berkeley Bevatron, a proton synchrotron, accelerating protons to 6.2 GeV, and the Alternating Gradient Proton Synchrotron at the Brookhaven National Laboratory, accelerating protons to 30 GeV energy, were in full operation from the mid 1950's to the late 1960's and produced many results. The development of electron linear accelerators started in the mid 1930's and picked up pace in the 1950's. By 1955, the first 1 GeV electron linear accelerator was functioning at the Stanford Linear Accelerator Center. Before it reached the full energy in 1955, it was functioning at lower energies in the few years prior. The detectors for particles ranged from cloud chambers to diffusion chambers, bubble chambers, scintillation counters, nuclear emulsions, spark chambers, and Cherenkov counters.

The first particles to be studied with these newer machines were the V particles. The properties of the Λ^0 hyperon and of the K^0 were more precisely determined. The charged counterparts of the neutral K particle were also found as well as the charged Σ hyperons and their neutral counterpart. Two further hyperons, the Ξ^- and Ξ^0, were found at a mass higher than that of the Σ hyperons. The properties of the Δ resonances seen in pion nucleon scattering, some of which were found at lower energy cyclotrons, were further clarified and extended at the newer machines. The antiproton and the antineutron were discovered at the Bevatron machine around 1956. A whole host of new mesons were found in the 1960's. In the early to mid 1950's, a large number of mesons and baryons had been found and studied; the situation called for the classification and understanding of the observed particle spectra.

Associated Production of New Particles

From the observed decay lifetime of the Λ^0 hyperon into $p + \pi^-$, one can obtain a value for the strength of the $\Lambda p \pi$ coupling. If the same interaction is responsible for the production of the Λ^0 by $\pi^- p$ collisions, we can calculate the cross section for the production of Λ^0. The cross section for production turns out to be ten orders of magnitude smaller than what is observed. From this we must conclude that the decay interaction and the interaction responsible for production must be different; the production proceeds via strong interactions, while the decay occurs due to weak interactions. Similar conclusions are arrived at for the other hyperons also—strong production and weak decay.

To explain this paradox, Pais introduced the concept of *associated production*. According to this concept, these new particles can only be produced in pairs, while in decay, only one of the new particles appears in the reaction. Examples of strong reactions which produce these particles in pairs are: $\pi^- + p \to \Lambda^0 + K^0$; $\pi^- + p \to \Lambda^0 + K^+ + \pi^-$; $p + p \to \Lambda^0 + p + K^+$; $p + p \to \Sigma^+ + p + K^0$. No single K production without the associated Σ or Λ has been observed. There is a further problem with the Ξ hyperons: it is observed that $\Xi^- \to \Lambda^0 + \pi^-$ is a weak decay, which would be understandable if the Ξ were in the same family as the nucleon. If that were the case, $\Xi^- \to N + \pi^-$ (where N is a nucleon) should be a decay which proceeds via strong interactions. This decay mode for the Ξ has been looked for and has never been seen. Clearly something more was needed. Gell-Mann and Nishijima supplied the solution by introducing the quantum number called *strangeness*.

Gell-Mann, Nishijima Scheme

Gell-Mann, and independently Nishijima, in 1954–1955, tried to extend the concept of isotopic spin to the new particles. In doing so, another quantum number called *strangeness* had to be introduced. Using this, they gave a generalization of the expression relating the charge Q (in units of $|e|$) to the isotopic spin, which will include all the new particles and be in accord with all the observations. The formula they gave was

$$Q = I_3 + \frac{B + S}{2},$$

where I_3 is the projection of isotopic spin on the "3" axis, and B is the baryon number: 1 for all particles which ultimately decay into the nucleon and 0 for all the mesons. The number S, the strangeness, is 0 for the nucleon and the pion, and is different from 0 for all the new particles. The members of the various multiplets are:

Baryons (B=1)

$$S = -2, \ I = 1/2, \ \Xi^0(I_3 = 1/2), \ \Xi^0(I_3 = -1/2),$$

$$S = -1, \ I = 1, \ \Sigma^+(I_3 = 1), \ \Sigma^0(I_3 = 0), \ \Sigma^-(I_3 = -1),$$

$$S = -1, \ I = 0, \ \Lambda^0(I_3 = 0),$$

$$S = 0, \ I = 1/2, \ p(I_3 = 1/2), \ n(I_3 = -1/2).$$

Mesons (B=0)

$$S = +1, \ I = 1/2, \ K^+(I_3 = 1/2), \ K^0(I_3 = -1/2),$$

$$S = 0, \ I = 1, \ \pi^+(I_3 = +1), \ \pi^0(I_3 = 0), \ \pi^-(I_3 = -1),$$

$$S = -1, \ I = 1/2, \ \bar{K}^0(I_3 = 1/2), \ K^-(I_3 = -1/2).$$

In this classification scheme, the K mesons are not members of an isotopic spin 1 multiplet. They are put in two isotopic spin 1/2 doublets, where $\bar{K}^0, K^-$ are the antiparticles of K^0, K^+. There are two distinct neutral K particles, unlike the case of the π^0, which is its own antiparticle.

With these assignments, it has been verified that all processes, which occur through strong interactions, conserve strangeness, while in weak decay interactions strangeness is not conserved. It is observed that a strangeness change of 2 units occurs with much less probability than a strangeness change of 1 unit. Thus the charged Σ's and Λ's decay to the nucleon state with the weak interaction rate, while the decay of the Ξ's to the nucleon state has not been seen. This classification scheme, which came about in 1954–1955, was the precursor to the SU_3 symmetry scheme to be proposed later by Gell-Mann and Ne'eman in 1961.

Yang-Mills Field Theory (1954)

In 1954, Yang and Mills constructed a field theory in which global isotopic spin invariance was made into a local one. The demand for invariance under local isotopic spin transformations necessitates the introduction of gauge fields. These fields have come to be known as *Yang-Mills gauge fields*. At the time this theory was introduced, it was considered an interesting theoretical exercise, and beyond that it did not have any impact. Nearly two decades later it became a very important ingredient in unifying electromagnetism and weak interactions, and in describing strong interactions.

The Tau-Theta Puzzle

Another significant finding emerged from the more accurate determination of the properties of the K mesons. Some of these mesons were observed to decay into two pions (called the Θ meson), while others decay into three pions (called the τ meson). When the masses and the lifetimes were not accurately known, these two particles were considered different. After it was found that the masses and lifetimes were the same within experimental error, it became clear that it would have to be an extraordinary coincidence if two different particles were involved. With improved analysis using a special plot suggested by Dalitz, spin and parity analyses could be done on the decay data. The analyses indicated that the spins of these particles were zero, and the parity would have to be even for the Θ particle with the two pion decay mode and odd for the τ particle with the three pion decay mode. If Θ and τ are one and the same particle, say K, parity would have to be violated in the weak decay of the K. This opened up the question of parity conservation in weak interactions, in particular, those responsible for β decays.

Parity Violation Observed (1957)

Lee and Yang made a thorough analysis of parity violation and proposed an experiment to test whether parity was conserved in β decay. In 1957, Wu and collaborators performed the experiment and found that parity was indeed violated in nuclear β decay. A number of other experiments established that the slow decays of many elementary particles involved the same interaction as in nuclear β decay. Parity violation in the weak decays of many other baryons and mesons was established.

CP Conservation (1957)

Charge conjugation (C) and parity operation (P) are discrete symmetries of field theories. The operation C changes particles into antiparticles, and parity operation involves changing a right-handed coordinate system into a left-handed one by a reflection through the origin. Symmetry under the C operation implies that no change would occur if we replaced all the particles by their corresponding antiparticles. Symmetry under P operation implies that there is no preference of left-handed over right-handed coordinate systems in their description. In an analysis of many weak decays, it was found that along with loss of parity symmetry, charge conjugation symmetry is also not valid.

In the case where the weak decay has neutrino as one of its products, it is easy to see how this might come about. Assuming the neutrino has zero mass, its spin gets aligned in or opposite to its direction of motion. The alignment is determined by the eigenvalues of an operator called *helicity*; if it is left-handed (spin aligned antiparallel to the momentum), helicity has the eigenvalue -1, while if it is right-handed (spin aligned parallel to the momentum), the eigenvalue is $+1$. An experiment designed to measure the helicity of the neutrino was performed by Goldhaber, Grodzins, and Sunyar in 1958. It determined the helicity of the neutrino to be -1, that is, left-handed. There is no right-handed neutrino, which is a violation of reflection symmetry (parity). If we carry out a charge conjugation operation on the neutrino, we get an antineutrino, but its left-handed nature is unchanged. The charge conjugation operation gives us a left-handed antineutrino. If such a particle exists in nature, we would conclude that C symmetry holds. Lee, Oehme, and Yang, and independently Landau, pointed out, from theoretical considerations, the possibility that P and C may each be violated, but the combined symmetry operation (CP) may still be a good symmetry. In the case of the neutrinos, this would imply that only left-handed neutrinos and right-handed antineutrinos exist in nature. Pais and Treiman suggested that studying neutral K-meson decays would be a good test of CP conservation.

Neutral K Mesons and CP Violation

Gell-Mann and Pais made the following interesting observations on the neutral K mesons. The K^0 and $\bar{K}^0$ carry different values of strangeness at their production by strong interactions. They undergo decay by weak interactions, and both K^0 and $\bar{K}^0$ can undergo decay into two charged pions. Thus when weak interactions are included, these states mix. If CP is conserved, we can form linear combinations of K^0 and $\bar{K}^0$ states, one of which is even under CP, and the other is odd under CP. Let us call the CP even state K_1^0 and the CP odd state K_2^0. If CP conservation holds for the decay, K_1^0 will decay into two pions, while the other, K_2^0, will decay into three pions which are not in a symmetrical state. The two body final state is favored by phase space and hence K_1^0 has a shorter lifetime compared to K_2^0. The consequence of this is that, if one starts with a pure beam of K^0, the K_1^0 part of this beam will decay rapidly into two pions near where the K^0 was produced, while the decays of the K_2^0 part will occur quite a bit farther away. These components have actually been found experimentally, the short lifetime being about 10^{-10} s, and the long lifetime being about 10^{-8} s.

The situation is more subtle than described above. Subsequent work by Fitch and Cronin in 1964 found that CP is not exactly conserved; there is a small amount of violation. This means that the short- and long-lived components are not quite CP eigenstates but contain a mixture with a small amount of the opposite CP eigenstate. The short-lived object decays mainly into two pions while the long-lived object decays mainly, but not entirely, into three pions. The discovery that the long-lived object, which should only decay into three pions if CP is conserved, actually has a small two pion decay mode, too, shows that CP is not conserved.

SU_3 Symmetry

In the period between 1955 and 1965 a large number of new particles were produced with the Brookhaven AGS proton machine, the Berkeley Bevatron, and some other machines. This included the discovery of new baryon resonances and a number of new mesons. On the baryon side, states which seemed like excited states of the Λ, the Σ, and the Ξ hyperons were found. On the meson side, a number of mesons which seemed to have a spin 1, and also some further pseudoscalalar mesons, were found.

Theorists were looking for symmetry schemes which could lead to an understanding of the mass spectrum of these particles. Fermi and Yang, already in 1949, had introduced an important idea that the pion might be a composite, built as a bound state of a nucleon and an antinucleon. Taking this notion a step further, Sakata, in Japan, tried an extension of the isotopic spin scheme in which, in addition to the proton and the neutron, the strange particle Λ was introduced as a basic entity, all three forming a triplet. He introduced their antiparticles (antitriplet) as well. He envisaged a unitary symmetry (SU_3), an extension of the isotopic spin symmetry, in which the triplet of particles transformed into one another. He tried to build all the non-strange and strange mesons and baryons as bound states with the triplet (and the antitriplet) as constituents. The model met with moderate success; it was able to accommodate the pseudoscalar mesons as combinations of the triplet and the antitriplet, but failed rather badly when it came to the baryons, which had to be built out of two triplets and one antitriplet. For the baryons, he found many more states than were being seen experimentally.

Gell-Mann, and independently Ne'eman, in 1961, proposed that the octet representation of SU_3 be used to accommodate the mesons and the baryons. This avenue of approach met with considerably more success, for both the mesons and the baryons. Subsequently, in 1964, Gell-Mann, and independently Zweig, proposed the constituent quark model, according to which constituents were introduced which belonged to the fundamental triplet representation of SU_3, the baryons being built out of combinations of three triplets, while the mesons would be built out of triplet-antitriplet combinations. The constituent particles were called *quarks* by Gell-Mann, and *aces* by Zweig. The name *quark* has since come to be universally accepted by the community. Since three quarks make up a baryon, each quark has to carry a baryon number of 1/3 and be a fermion. (For simplicity, spin 1/2 was assumed for the quark.) Further, to get all the baryon charges right, the quarks had to be assigned fractional charges $Q|e|$ (in units of the fundamental charge $|e|$), where one quark [now called the *up (u)* quark] had to be assigned $Q = +2/3$, and the other quarks, [now called *down (d)* and *strange (s)*] had to be assigned $Q = -1/3$ each. The strange quark was also assigned strangeness of -1. One of the consequences of the SU_3 model is the prediction that a baryon with strangeness -3 (bound state of three s quarks) must exist. This particle, called the Ω^- was found in experiments in 1964 and provided a remarkable confirmation of the correctness of the SU_3 picture. An intriguing feature of the model is that, despite its enormous successes, quarks have never been found as free particles in nature in spite of many active searches for them. These failures to find quarks raised the question as to whether they were real constituents of hadrons, a question which was only answered somewhat later by studies in the deep-inelastic scattering of high energy electrons on nucleons.

Other Theoretical Developments (1950–1970)

The theory of QED developed by Feynman, Schwinger, and Tomonaga was used extensively to calculate higher-order corrections to various processes and confronted with experiments with amazing success. Renormalization procedures in this theory led to the formulation of the invariance of physical quantities under change of the renormalization scale. The renormalization group equations express such invariance requirements in a succinct manner. These and other developments showed that the perturbative solution of QED was enormously successful and could serve as a model for treating other interactions. The field theory of weak interactions was a case in hand. Schwinger proposed in 1957 that weak interactions may be mediated by the exchange of massive intermediate vector bosons. For processes involving small momentum transfers, the interaction is characterized by an effective dimensionful coupling constant (the Fermi coupling), which is much smaller in value than the fine structure constant involved in QED. Still the theory involving massive charged intermediate bosons proved to be non-renormalizable. The non-renormalizability was traced to the fact that the effective coupling constant in weak interactions is a dimensionful one, in contrast to the dimensionless coupling constant of QED. A search for a field theory which would unify electromagnetism and weak interactions was started.

The Yukawa model of the interaction of π mesons with nucleons described by a field theory with pseudoscalar interactions has a coupling constant which is dimensionless. It, just like QED, is renormalizable in a perturbative treatment. In its application to nucleon-nucleon interactions or pion-nucleon scattering, the value of the coupling constant needed to fit the data is very large, making perturbative treatment meaningless.

A non-perturbative treatment of the pion-nucleon scattering problem was attempted by Dyson and collaborators at Cornell University in 1952, using an approximation known as Tamm-Dancoff approximation. In this method the amplitude for the one meson - one nucleon state is approximated by letting it couple only to a very few neighboring states, so that an integral equation for the amplitude can be derived. To the lowest order, the kernel of the integral equation involves only the lowest power of the coupling constant. The integral equation was solved by numerical methods without expanding in powers of the coupling constant, and phase-shifts for the scattering of pions on nucleons were derived. It was an interesting feature of this method that, in the isotopic spin 3/2 and total angular momentum 3/2 of the pion-nucleon system, the phase shift for the state indicated a resonance behavior at an energy corresponding to exciting the Δ^{++} isobar. Unfortunately, the theory was not covariant. It was further found by Dyson that a consistent scheme for renormalization does not exist, thus leading to an abandonment of this approach.

Other approaches to treating strong interactions using local quantum field theory were under development in the early 1950's. In 1953, Gell-Mann, Goldberger, and Thirring derived dispersion relations for the forward scattering of pions on nucleons by imposing causality conditions on the commutators of field operators. In 1954, Yang and Mills investigated, as mentioned earlier, the consequences of demanding that isotopic spin invariance be a local gauge invariance. They showed that this necessitates the introduction of gauge fields in the problem, now known as *Yang-Mills fields*. The work on dispersion relations was developed further by Bogolyubov and collaborators in the Soviet Union in 1956. They derived dispersion relations in field theory for the pion-nucleon scattering amplitude in the general case. The formulation of an axiomatic field theory framework for the S-matrix was initiated in 1955. Mandelstamm proposed a representation for the scattering amplitude in 1958 and derived dispersion relation in two variables, the energy and the angle of scattering.

The S-matrix theory of strong interactions was actively developed by Chew and collaborators in this period, deriving a number of interesting results. A new method for performing the sum over partial waves to obtain the scattering amplitude in potential scattering was put forward by Regge in 1959. This involved the introduction of poles in the complex angular momentum plane, now called *Regge poles*. Chew and collaborators, in 1962, investigated the consequences of proposing that all baryons and mesons are Regge poles which move in the complex angular momentum plane as a function of the energy. The paths on which the poles move are called *Regge trajectories*. They are characterized by a set of internal quantum numbers, parity, baryon number, isotopic spin, strangeness, etc., and interpolate between physical values of spin J for mesons and $J - \frac{1}{2}$ for baryons. Properties of the Regge trajectories are further dictated by the requirements of analyticity and unitarity of the S-matrix. The Regge trajectories are straight lines when plotted on the (J, M^2) plane, where J is the spin of the particle and M its mass. They argued that the high energy behavior of the amplitude for some reaction is dictated by the presence of Regge poles in what is called a *crossed channel* reaction. Regge trajectories were useful in providing information on the asymptotic properties of scattering amplitudes. They conjectured that the experimentally observed fact of total cross sections (for a number of processes) approaching a constant limit implies the existence of a Regge pole with vacuum quantum numbers—the so-called *Pomeron trajectory*. Other Regge trajectories lay below the Pomeron trajectory. A prediction of the theory was that the forward diffraction peak must shrink with increase in energy. This was clearly observed in the experimental data.

The year 1961 saw the emergence of Goldstone's theorem: If a global symmetry of the Lagrangian for some system is spontaneously broken, then there must necessarily appear massless bosons, now called *Goldstone bosons*. This theorem has since played a very important role in particle physics. Another notable work in that year was the recognition by Salam and Ward that demanding local gauge invariance would be a good way to construct quantum field theories of interacting fields. Glashow, in the same year, suggested using the gauge group $SU_2 \times U_1$ for the interaction of leptons, which would require a neutral weak gauge boson in addition to the charged weak gauge bosons and the photon. In 1964, an example of a field theory with spontaneous breaking of gauge symmetry, giving rise, not to a massless Goldstone boson, but to massive vector bosons, was constructed by a number of people: Higgs, Brout, and Englert, and Guralnik, Hagen, and Kibble. This mechanism has since been called the *Higgs mechanism* for the generation of masses of vector bosons. It was also the year in which Salam and Ward proposed a Lagrangian for synthesizing electromagnetism and weak interactions, and produced some crude estimate for the mass of the quanta which mediate weak interaction. Weinberg, and independently Salam, in 1967, put forward a field theory, based on the gauge group $SU_2 \times U_1$, for the unification of electromagnetic and weak interactions. This theory used the mechanism for the spontaneous breaking of gauge symmetry to generate masses for the weak gauge bosons, charged as well as neutral. The charged weak gauge boson was predicted to have a mass around 80 GeV, and the neutral weak gauge boson a mass about 90 GeV, while the photon was left massless. It also predicted that the neutral weak gauge boson should have an interaction comparable to that of the charged weak gauge bosons. It was hoped that such a theory with spontaneous symmetry breaking would be renormalizable.

CP violation in the neutral K-meson system was observed by Christenson et al. in 1964. Wolfenstein put forward the super-weak theory of CP violation in neutral K mesons. The need for an additional quantum number (now called *color*) carried by quarks was pointed out by Bogolyubov et al., in 1965, in an attempt to resolve conflict with Fermi statistics for the ground state baryons. In 1966, in analogy to the electromagnetic interactions being mediated by vector photons between charged particles, Nambu proposed that strong interactions between quarks may be mediated by massless vector fields whose quanta are now called *gluons*. This may be said to be the beginnings of *quantum chromodynamics (QCD)*. Also in 1966, Han and Nambu proposed the three triplet quark model of hadrons, each triplet being distinguished by a new quantum number, which we now refer to as *color*. The emergence of a connection between CP violation and the baryon asymmetry of the universe occurred in 1967. It was also in 1967 that, a generalization of the Higgs mechanism of mass generation for the Yang-Mills type of gauge field theories was given by Higgs and Kibble; Faddeev and Popov in the Soviet Union solved some of the difficulties in the formulation of Feynman rules for Yang-Mills type gauge field theories by introducing a special method, now referred to as the *Faddeev-Popov method*.

In the late 1960's, the S-matrix theory of strong interactions led to a new model for hadrons. Hadrons were pictured as different modes of vibration of relativistic string. This period may be said to be the origin of string theories. From that period until the early 1980's, various theoretical aspects of the picture were clarified; from the point of view of application to hadrons, however, not many results can be seen. In 1984 some of the major hurdles in the theory were overcome and string theory started blooming again. It is at present one of the hottest topics in theoretical and mathematical physics.

Other Developments in Experimental Techniques (1950–1990)

Synchrotrons based on the phase stability principle were constructed. The construction of particle accelerators based on the strong focusing principle were also undertaken. The Cosmotron at Brookhaven National Laboratory, the Bevatron at Berkeley, the AGS proton synchrotron at Brookhaven National Laboratory, are some of the examples of accelerators in the U.S. which contributed a lot to the experimental data on particle physics of this period. The first GeV electron linear accelerator was completed at Stanford Linear Accelerator Center (SLAC) in 1955 and provided, in its earlier stages of development, data on the electromagnetic structure of nuclei, and later on, through deep inelastic scattering studies, provided information on the structure of the nucleon itself.

The European physicists, who had remarkable achievements to their credit during the first half of the twentieth century, had their work badly interrupted due to the second world war. Many leading personages migrated to the U.S. The post-war rebuilding of science in Europe started around 1950. Despite the hoary traditions of many of the institutions, no single European country by itself could spearhead this renaissance. The creation of a European laboratory, by a consortium of European nations, in the form of "Conseil Européen pour la Recherche Nucléaire", now commonly referred to as CERN, occurred in this period. One of the aims of CERN was to provide high energy accelerators for physicists in Europe to do front-line research in high energy physics. Since its inception, CERN has played a leading role in developing colliding beam machines and associated detectors for particle physics experiments. These efforts have contributed a great deal toward our understanding of the ultimate structure of matter.

Hand in hand with the development of accelerators in the U.S. and Europe, detector developments made significant progress. Solid as well as liquid scintillation counters were developed. Use of photomultiplier tubes to view the scintillations went together with the developments of scintillation counters. Semi-conductor detectors were developed. Spark chambers which could track particles were invented. Proportional counters were developed and were useful in neutron detection. Bubble chambers were invented in 1953, and dominated the scene for over two decades in obtaining evidence for the existence of new particles and new phenomena. Flash tube chambers came on the scene in 1955. Cherenkov detectors, based on the radiation emitted by charged particles in matter traveling at a velocity greater than the velocity of light in the medium, were developed and used successfully in the discovery of the antiproton in 1956. Invention of the multiwire proportional chamber in 1968 represented a major step in accurately detecting and measuring particle properties. They could be used as particle track detectors and had the further capability of measuring energy loss. With small spacing of the sense wires, they were good in experiments where the data rates were high. Drift chambers were soon to follow on the heels of the development of multiwire proportional chambers.

A realistic proposal to explore high energy processes by colliding one beam of particles against another was put forward for the first time in 1956. The advantage of such a method lies in the fact that all the energy in the beams is available for new particle production, although the rate of occurrence of processes is not as high as in beams colliding with fixed targets. Considerable developments have occurred since the idea was initially put forward; ingenious methods, called *beam cooling methods*, to increase the intensities in the colliding beams have been invented. Colliding beam accelerators are simply called *colliders* and the intensity in these machines is characterized by a quantity called *luminosity* of the collider. Colliders of electrons on positrons have been in operation since the early machines started operation at Frascati, Italy in 1961. Since then a number of e^-e^+ colliders at steadily increasing energies and luminosities have been constructed and have produced clean results in different parts of the world. In this connection, mention must also be made of the establishment in Hamburg, Germany of the DESY (Deutches Elektronen-Synchrotron) laboratory, which has developed electron-positron colliders and also the electron-proton collider named HERA (Hadron-Electron Ring Accelerator).

Electrons and positrons have been found to be structureless points up to the highest energies explored. Colliders involving them are preferred to those involving hadrons as one does not need to be concerned with the effects of the structure of the beam particles on the new phenomena being explored. The first of these was the Intersecting Storage Ring Collider (ISR) for protons operating at CERN in 1971, involving two 31 GeV proton beams. Since then, other proton-proton (and proton-antiproton) colliders or electron-proton colliders have been proposed and put into operation in Europe and the USA, starting in the 1980's and continuing well on into the present century.

Detectors having cylindrical geometry to surround the collision points of the beams in colliders have been developed. Cylindrical proportional and drift chambers are deployed in such experiments as central detectors. Provision of high magnetic fields for the measurement of momentum of particles is an important element in the design of detectors. The trend has been toward assembling multipurpose detectors which are combinations of a number of component detectors, each component acquiring data on specific properties of the particles originating at the collision point, all at one time. Studies at LEP in CERN, and at SLC in Stanford, have been done with a number of multipurpose detectors. The data obtained from these experiments have significantly advanced our understanding of elementary particles and their interactions.

Direct Observation of Neutrinos (1956)

Going back to the mid 1950's, Reines and Cowan, in 1956, succeeded in detecting free antineutrinos for the first time. The source of antineutrinos was a nuclear fission reactor. These were incident on a large target containing a liquid scintillator rich in hydrogen and induced the reaction $\bar{\nu}_e + p \rightarrow e^+ + n$. The occurrence of this reaction was confirmed by the detection of the gamma pulse from the annihilation of the positron, followed by a delayed gamma pulse from the capture of the neutron on the proton in the target, the delay time being the (known) slowing down time of the neutron prior to its capture.

Neutrinos of Different Flavor (1957)

In 1957, Nishijima pointed out the need for a new property characterizing massless neutrinos. The decay of the muon into an electron with no associated neutrinos (for example, $\mu \rightarrow e + \gamma$) has been searched for and not found. This transformation would be forbidden if the muon and the electron carried different lepton numbers, and these numbers were required to be separately conserved. Since muon-decay to electron plus two neutrinos is observed, the two neutrinos cannot be identical; one neutrino must carry off electron lepton number and the other must carry off muon lepton number, such that each type of lepton number can be conserved in the decay. The electron and its neutrino form a family and the muon and its neutrino form a second family. These families are said to carry *electron flavor* and *muon flavor*, respectively.

Experimental Discovery of Neutrinos of Different Flavor (1963)

The existence of a muon neutrino distinct from the electron neutrino was experimentally established by Lederman, Schwartz, and Steinberger in 1963, using neutrinos from pion and kaon decays. These neutrinos produced only muons through interaction with nuclei of a target, and no electrons were produced.

Quark-Lepton Symmetry and Charm Quark Proposal (1964)

Two lepton families, the electron and its associated neutrino and the muon and its associated neutrino, were established by 1964. On the quarks side, however, only three quarks were known: the u, d, and s quarks, of which the first two were considered as members of one family with isotopic spin 1/2. Bjorken and Glashow, on the basis of lepton-quark symmetry argued, that the s quark belonged to a second quark family, the other member of the family yet to be found. They called the missing member of the second family, the *charm* quark (c). This quark was not found until much later.

Bjorken Scaling and Its Experimental Discovery (1969)

Experimental studies of the deep inelastic scattering of electrons on nucleons were in progress at the Stanford Linear Accelerator Center. The data were analyzed in terms of two structure functions associated with the nucleon. These structure functions are in general functions of two variables: (1) q^2, the square of the four-momentum transferred, and (2) ν, the energy transferred to the nucleon by the incident electron. Both these variables take on large values in the domain of deep inelastic scattering. Bjorken, through arguments on the behavior of the commutators of two currents at almost equal time at infinite momentum, came to the conclusion that the structure functions will depend only on the ratio (ν/q^2) in the limit that $\nu \to \infty$ and $q^2 \to \infty$. This is referred to as *Bjorken scaling*. Precisely such a scaling was found by Friedman, Kendall, and Taylor from their experiments on deep inelastic electron-nucleon scattering at SLAC in 1969.

Parton Model (1969)

Feynman put forward a model of the structure of the proton which could explain the observed Bjorken scaling. He considered the proton to be made up of partons, each parton carrying a fraction x of the momentum of the proton. At extremely high incident electron energies and momentum transfers, the time duration of the interaction of the electron with a parton is so much shorter than the time duration of interaction of partons among themselves that the partons can be considered as structureless and free. The scattering cross section of the electron on the proton can be obtained by summing the cross sections for electron-parton scattering and integrating over all the parton momenta with a parton momentum distribution function in the proton. From the kinematics of the electron-parton collision, it can be easily shown that $x = Q^2/(2M\nu)$, where $Q^2 = -q^2$, and M is the mass of the proton. The parton distribution functions, which are functions of x, completely determine the cross section for the deep inelastic electron-proton scattering. Feynman's x is clearly the same as the Bjorken scaling variable except for some constant factors. Since the proton is known to contain quarks, the partons may be identified with quarks. The deep inelastic scattering process is determined by the quark distribution functions. For given Q^2, small x corresponds to large energy transfers ν, and the quark distribution function for small x gives a high resolution view of the proton. Quarks which have $x \simeq 1/3$ correspond to the quarks (valence quarks) of the constituent quark model of the proton and provide a low resolution view of the proton.

Renormalization of Yang-Mills Field Theory (1971)

In 1971 three remarkable pieces of work were completed by 't Hooft, and by 't Hooft and Veltman, which have had enormous impact on the further development of the field of particle physics. In one paper, 't Hooft gave a rigorous proof of the fact that Yang-Mills field theories are renormalizable. In another paper, 't Hooft proved that Yang-Mills field theories with spontaneously broken gauge symmetry are also renormalizable. In a third paper by 't Hooft and Veltman, a new method, *dimensional regularization*, was given for the regularization of gauge field theories. Earlier, in 1967, Weinberg had already proposed a Lagrangian for electroweak synthesis based on $SU_2 \times U_1$ gauge group and used the mechanism of spontaneous breaking of gauge symmetry to generate masses for the (weak) gauge bosons. With 't Hooft's proof on the renormalizability of Yang-Mills theories with spontaneously broken gauge symmetry, it followed that the field theory containing electroweak synthesis was renormalizable. Thus as far as the calculation of higher order processes were concerned, the electroweak field theory was on a par with QED, and higher order processes could be calculated, just as in QED, without ambiguities. Of course, in the electroweak theory, many more parameters appear (for example, masses of the fermions and gauge bosons), for which experimental input is necessary, than is the case for QED.

Experiments Find Weak Neutral Current Effects (1973)

One of the consequences of electroweak unification was the prediction of the neutral counterpart Z^0 of the charged weak gauge bosons $W^{\pm}$ with comparable couplings to fermions. This would mean that there must exist neutral current weak processes which occur at rates comparable to those of the charged current weak processes. In particular, if the charged current reaction, $\nu_{\mu} + \text{nucleus} \rightarrow \mu^- + X$ occurs, in which a muon and hadrons are produced in the final state, there must also occur the neutral current process $\nu_{\mu} + \text{nucleus} \rightarrow \nu_{\mu} + X$. The neutrino will not be seen, and the signature for a neutral current process will be the appearance of hadrons alone in the final state. Exactly such events were seen in the Gargamelle bubble chamber exposed to the neutrino beam at CERN. Events corresponding to the elastic scattering $\bar{\nu}_\mu e^- \to \bar{\nu}_\mu e^-$ were also observed. These experiments showed that neutral current effects were indeed being observed. To get a quantitative measure of the effects, more sensitive experiments were planned. Experimenters at SLAC measured the parity-violating asymmetry in the scattering of polarized electrons off a deuteron target. This involves measuring the difference between the deep-inelastic scattering cross sections for right- and left-handed electrons on deuterons $e_{R,L}d \to eX$. A good measurement of this asymmetry yielded an accurate value for the weak mixing angle.

Yang-Mills Theories and Asymptotic Freedom (1973)

Investigations by Gross and Wilczek, and by Politzer, on Yang-Mills gauge field theories revealed the existence of a very interesting property of these theories. The interaction between particles mediated by the gauge fields vanishes as the distance between the particles tends to zero (or the square of the four-momentum transfer between the particles tends to infinity). Since the particles behave as free particles in the asymptotically high energy region, this behavior came to be called *Asymptotic Freedom*. This feature provides a natural explanation for the parton model of hadrons and hence for Bjorken scaling.

QCD Formulated (1973)

Fritzsch, Gell-Mann, and Leutwyler formulated a Yang-Mills gauge field theory in which local invariance under color transformations was demanded. This necessitates the introduction of color gauge fields. In a theory with three quark colors, there is an octet of massless colored gluons. Gross and Wilczek showed that the exchange of the colored gluons between colored quarks gives rise to an interaction which will have the property of asymptotic freedom. If the gauge symmetry is not broken, they pointed out that the theory has severe infrared singularities which will prevent the occurrence of non-singlet color states. Thus it was proposed that observed hadrons are color singlets. Colored objects would be infinitely massive and will not be found in nature. This theory encompasses all the observed facts, Bjorken scaling, parton model, and quarks as partons with color. This theory, called quantum chromodynamics (QCD) is considered as the fundamental theory of strong interactions.

Standard Model Formulated (1973–1974)

Around this period, the ideas put forward earlier by Weinberg and Salam, and by Glashow, on electroweak unification gradually led to the formulation of the so-called *standard model*. The model Lagrangian is based on the gauge group $SU_2 \times U_1$. The left-handed fermions form (weak) SU_2 doublets, while the right-handed fermions are in SU_2 singlets. Of the four original gauge fields, three acquire masses by the spontaneous breaking of gauge symmetry, via the Higgs mechanism, and become the massive W^+, W^-, and Z^0 bosons. One of the original gauge bosons, which is left massless, is identified with the photon. The fermions of the theory are the leptons and quarks. The theory would have triangle anomalies unless the fermions in the theory had appropriate hypercharges such that all the anomalies cancel among themselves. The hypercharge assignments are such that cancellation of anomalies does indeed take place. Thus, the electroweak theory is renormalizable. The fact that the quarks carry color is irrelevant for the electroweak sector; in summing over the colors, one gets only a numerical factor for the number of colors.

The quarks also have strong interactions. For dealing with this part, use was made of the Lagrangian for QCD which was available from the work of Fritzsch, Gell-Mann, and Leutwyler; and Gross and Wilczek; and Politzer. Since QCD is also renormalizable, adding it to the electroweak theory produced a renormalizable gauge field theory which is capable of dealing with electroweak and strong interactions. This is the standard model. Calculations of strong interaction corrections to electroweak theory, called *QCD corrections*, are feasible in the realm of high energies where the strong interaction effects are small, due to the asymptotic freedom of QCD, and perturbation theory can be used. The time was ripe for mounting experiments to test the predictions of the standard model.

Discovery of Charm Quark; Hidden Charm (1974)

Ting's group at the Brookhaven (AGS) proton synchrotron studying the reaction $p + \mathrm{Be} \to e^+ e^- + X$ and Richter's group at SLAC studying $e^+ e^- \to$ hadrons, $e^+ e^-$, $\mu^+ \mu^-$ simultaneously reported the finding of a very sharp peak in the produced $e^+ e^-$ spectrum corresponding to a mass of 3.1 GeV. This particle was given the name J/ψ. Detailed studies of this particle have revealed that it is a vector meson. In terms of the quark picture, it was identified as a 1S bound state of charm quark and its antiparticle ($c\bar{c}$ bound state) with no net charm (hence called *hidden charm*). Based on this interpretation, an estimate of the mass of the charm quark around 1 GeV–1.6 GeV could be made. Following the initial discovery of the $c\bar{c}$ bound state, excited states, 2S, 3S, etc., of the system were also discovered. The bound system has been called *Charmonium*. The charm quark together with the strange quark completed the second generation of quarks after the (u,d) of the first generation.

Charm Hadrons Found (1975–1977)

According to the constituent quark model, charm hadrons, which are bound states of charm quark with u, d, or s quarks (or antiquarks), must exist. Σ_c^{++}, a (uuc) combination, and Λ_c^+, a (udc) combination, were both found to exist in 1975. The D mesons—$D^+,(c\bar{d})$; $D^0,(c\bar{u})$; $D^-,(\bar{c}d)$—were all found in 1976. These mesons are the analogs of the K mesons of the first generation of quarks. Evidence of mesons, which may be called *strange-charm mesons*, were also found soon after this time. D_s^+ ($c\bar{s}$ combination) and D_s^- ($\bar{c}s$ combination) were found in 1977. Evidence for a charm antibaryon $\bar{\Lambda}_c^-$, $(\bar{u}\bar{d}\bar{c})$ was also found in 1976.

Tau Lepton Found (1975)

At SLAC SPEAR e^+e^- ring, Perl and associates found events in e^+e^- annihilation, where the final products were $e^\pm\mu^\mp$+missing energy, with no other associated charged particles or photons. Most of these events were found at a center of mass energy of about 4 GeV. The missing energy and the missing momentum in these events indicated that at least two additional particles were produced. Perl proposed that these events can be explained if a pair of heavy leptons were produced with the lepton mass in the range 1.6 GeV to 2 GeV, one lepton decaying to electron and two neutrinos and the other decaying into muon and two neutrinos. The new heavy lepton was given the name of τ lepton and is a member of a third family of leptons. It, and its associated neutrino ν_τ, carry a tau lepton number and form the third generation family.

Discovery the of Bottom/Beauty Quark; Hidden Bottom/Beauty (1977)

The study of the reaction in which 400 GeV protons were incident on a nucleus leading to a final state in which a pair of muons of opposite sign along with some hadronic debris was produced revealed the existence of a strong and narrow peak in the signal corresponding to a mass of the $\mu^+\mu^-$ system of 9.5 GeV. This strong enhancement at a dimuon mass of 9.5 GeV was attributed to a new particle called Upsilon (1S), [Υ(1S)].

The process observed was interpreted as $p + $ nucleus $\rightarrow \Upsilon(1S) + X$, $\Upsilon(1S) \rightarrow \mu^+ \mu^-$. It was proposed that this new particle was the bound state of yet another quark, called the *bottom* or *beauty* quark b, and its antiparticle $\bar{b}$. The $\Upsilon(1S)$ is a $b\bar{b}$ bound state analogous to J/ψ which is a $c\bar{c}$ bound state, and just as charm is hidden in J/ψ, bottom (or beauty) is hidden in Υ meson. Very shortly after this, the existence of $\Upsilon(1S)$ was confirmed by the PLUTO collaboration in $e^+ e^-$ DORIS collider ring at DESY. From a measurement of the electron decay width of Υ, a charge assignment of $(-1/3)|e|$ for the bottom (beauty) quark was found to be preferred. Thus, the bottom member of the third family of quarks was found. Further work discovered other features of this hadron. An estimate of the mass of the b quark is found from the interpretation of $\Upsilon(1S)$ as a bound state of $b\bar{b}$ and is between 4.1 GeV and 4.4 GeV. Higher excited states, (2S), (3S), etc., have been found here as for the charmonium system and are called the *bottomonium* in this case.

The standard model requires that there must exist the "top" member of this family with charge $+2/3|e|$ to pair off with the bottom quark so that anomalies in electroweak theory can cancel. To complete this new family, the "top" member remained to be found if the standard model is to be proven right.

Efforts at Grand Unification

The electroweak model unifies electromagnetism and weak interactions by starting from a larger symmetry group than U_1. The question naturally arose as to whether there were any larger symmetry groups in which all three interactions the strong, the electromagentic, and the weak—could be unified. The question was answered in the affirmative by a number of people. In 1974, Georgi and Glashow put forward SU_5 as the gauge group. At some large energy scale denoted by M, they assumed that the symmetry is SU_5. At this scale there is only one coupling; the couplings, g_3 of SU_3, g_2 of SU_2, and g_1 of U_1, are all equal. As the energy is lowered, these couplings change (or run) according to the renormalization group equations, which makes them look different. Estimation of the unification mass scale puts it at about 10^{16} GeV. Other efforts were by Pati and Salam (1973) and others who worked on SO(10). The general problem of embedding the Standard Model group in a larger grand unifying group was studied by Gell-Mann, Ramond, and Slansky in 1978.

One consequence of these grand unification models is the prediction that the proton will not be a stable particle and will decay, violating baryon number. The lifetime for proton decay could be calculated in these unification models and was of the order of 10^{29} to 10^{33} years, depending upon which grand unified symmetry group was considered. These ideas spawned a number of experiments in different parts of the world looking for proton decay. Because of the expected extreme rarity of proton decay events, detectors looking for such events are very large and are usually constructed deep underground to cut down the effects of cosmic rays.

Supersymmetry

A new kind of symmetry operation, which generalizes the space-time symmetries of quantum field theories and transforms bosons into fermions and vice versa, was discovered in the early 1970's. Until that time, all the symmetry transformations left bosons as bosons and fermions as fermions.

One of the reasons for the interest in supersymmetry stems from the attempt to unify all forces in nature including gravity, which naturally occurs at the scale of the Planck mass 10^{19} GeV. The scale of electroweak physics, on the other hand, is set by the vacuum expectation value of the Higgs field which is about 250 GeV. The vast ratio in these scales, some seventeen orders of magnitude, was considered a "hierarchy" problem.

In the electroweak theory, it was found that the Higgs particle gets renormalization contributions to its mass which are such that the mass ultimately reaches the unification mass value. In other words, the low mass scale of Higgs mechanism is not stable with respect to radiative corrections. One natural way to keep the low mass scale stable is to introduce supersymmetry. The radiative corrections get contributions from both bosonic and fermionic intermediate loops of particles and these would cancel if there were exact supersymmetry.

But supersymmetry is not exact. Otherwise, nature would have exhibited degeneracy between bosonic particles and fermionic particles; for example, there are no spin zero particles having the mass of the electron, the muon, the proton, etc. Hence, it must be a broken symmetry, and the question of interest is the scale at which supersymmetry breaking occurs. This scale must be related to the electroweak scale of 250 GeV so that the large hierarchy in mass scales from the W's and Z's to the Planck mass scale can be understood. Supersymmetric theories with these low energy characteristics have been constructed which, besides providing stability to the low mass scale, predict the existence of supersymmetric partners to all the known particles at mass values which can be looked for at currently existing accelerators or at future proposed facilities. So far searches for the supersymmetric particles have turned up nothing. If, in future searches, evidence for supersymmetry is found, it could point the way for the unification of all the fundamental forces including gravity.

Weak Neutral Currents (1978)

The earlier indication of the existence of weak neutral currents was confirmed in a very beautiful and imaginative experiment performed at SLAC. A signal for the existence of weak neutral currents is that parity violation will occur. One can calculate how much parity-violating asymmetry should be expected in the inelastic scattering of longitudinally polarized electrons from hydrogen and deuterium. The experiment measured exactly this quantity in deuterium and found a value in excellent agreement with theoretical expectations. This established the existence of weak neutral current of the right magnitude and characteristics beyond any doubt.

Another place where neutral currents will have an effect is in atomic transitions. The electron circulating around the atomic nucleus has, in addition to the electromagnetic interaction with the nuclear charge, a weak interaction due to the virtual exchange of the neutral weak boson. This latter interaction is parity violating and the atoms in a medium should exhibit small optical activity. If plane polarized light is passed through the medium, the optically active atoms will rotate the plane of polarization of the light. The expected rotation angle is very small because the neutral current effect is so small. The experiments are difficult to perform and the early experiments were inconclusive. With improvements in techniques, optical rotation of the plane of polarization of the expected amount was observed. This again confirmed the existence of weak neutral currents.

Evidence for Gluons (1979)

The experimental analysis of e^+e^- annihilation into hadrons reveals the hadrons as a jet of particles, back to back, coming off from the annihilation vertex. The interpretation given is that the annihilation produces a quark-antiquark pair, which, as they separate, subsequently materialize into color neutral hadrons by picking off quarks or antiquarks from the vacuum. It is these color neutral hadrons that are seen, in the experiments, as thin jets of hadrons. The jets go in opposite directions because, in the e^+e^- collider, the annihilation occurs with the center of mass of the annihilating pair at rest. If QCD is to describe quark interactions through the virtual exchange of colored gluons, one should be able to see the gluons coming from the annihilation vertex also. Thus in addition to back to back jets of hadrons, one should see three jets of hadrons produced with zero total momentum for the jets. Three jet events were indeed observed at the DESY-PETRA electron-positron collider in 1979. The data were completely consistent with a picture in which quark-antiquark pairs are accompanied by hard non-collinear gluons.

Gluon Spin Determined (1980)

In e^+e^- annihilations into three jet events, the angular correlations between the axes of the three jets have behaviors which depend upon the spin of the gluon. The expected angular correlations assuming spin 0 or spin 1 for the gluon can be worked out assuming QCD. Experimental data were found to favor the scenario in which the gluon was attributed spin 1, and disfavored spin 0. Thus gluon, like the photon, carries spin 1.

Hadrons with b Quarks Found (1981)

Based on the fact that mesons and baryons containing the charm quark were found a few years earlier, it was reasonable to expect that mesons and baryons containing b-quarks should also be found at higher energies. Indeed, mesons containing b-quarks, were found in 1981 at the Cornell e^+e^- storage ring with the CLEO detector. In a center of mass energy range of 10.4 GeV to 10.6 GeV, corresponding to the $\Upsilon(4S)$ state, the experimenters observed a good enhancement of single electrons produced from the annihilation vertex. They interpreted these electrons as coming from the following sequence. First, $\Upsilon(4S)$ is produced from the annihilation which, if it is above the threshold for $b\bar{b}$ production, produces b and $\bar{b}$. These pick up quarks (or antiquarks) from the vacuum and become new mesons, B and $\bar{B}$. Each B meson, decays into a lower mass state X plus an electron and a neutrino: $B \rightarrow e + \nu + X$.

If the interpretation given above is correct, the B meson should also decay according to: $B \rightarrow \mu + \nu + X$. Thus, one should observe single muon signals also, just like the single electron signals. If the $\bar{B}$ meson also decays into a muon, one can expect to see a signal of two muons, one from the decay of the B and the other from the decay of the $\bar{B}$. All such signals were seen confirming the correctness of the interpretation and thus the existence of the B meson.

In the same year, evidence for a baryon containing the b-quark was found at the CERN ISR pp collider. This was a heavy baryon whose mass was measured to be 5.4 GeV, electrically neutral, and decayed into a proton, D^0, and π^-. It was found to be produced in association with another hadron which decayed semi-leptonically into a positron. The interpretation which fit the observation involved an associated production of "bottom (beauty)" states in pp interactions. The observed baryon fitted into the quark composition (udb), which could be called a Λ_b^0.

Discovery of $W^\pm$ and Z^0 (1983)

The discovery of the W and Z bosons was a long time in coming although there were many indirect indications for their existence. They were found in the $Sp\bar{p}S$ collider at CERN by UA1 and UA2 collaborations. They looked for events with large transverse energy electrons and events with large missing transverse energy. These events pointed to the existence of a particle of mass about 80 GeV with a two-body decay. The interpretation that fit the data best was as follows: $p\bar{p} \to W^\pm X$, $W^\pm \to e^\pm \nu_e$. The mass of about 80 GeV is very nearly the same as predicted for the weak vector boson mass in the theory of electroweak unification. It is thus clear that these experiments found the charged weak bosons.

In the same year, the UA1 collaboration reported on observation of electron-positron pairs, which appear to have originated from the decay of a particle of mass about 95 GeV. Observation of an event with $\mu^+\mu^-$ pair also pointed to its origin from the decay of a particle of mass about 95 GeV. These observations are consistent with the process $p\bar{p} \to Z^0 + X$, $Z^0 \to e^+e^-$, $\mu^+\mu^-$. UA2 collaboration also found events leading to electron-positron pairs which could have originated from the decay of a particle into e^+e^- or into $e^+e^-\gamma$. The mass deduced from a study of four of these events was about 92 GeV. The mass values for the observed particle suggest that it is the neutral weak gauge boson of the electroweak model.

With these discoveries, the standard electroweak model was placed on a firm footing. By 1986, the properties of the weak gauge bosons were more precisely determined, and no deviations were found from the predictions of the standard model. However, high precision quantitative tests of the standard model could not be done as the experimental data were limited in their precision. Such tests came from the Large Electron Positron (LEP) collider at CERN, and from SLC at SLAC, which started operating a few years later.

High Energy e^+e^- Experiments at LEP and SLC

High precision tests of the standard electroweak model had to await the construction and operation of the Large Electron Positron (LEP) collider at CERN. The Stanford Linear Collider (SLC), which was completed in 1989 and had polarization capabilities in the beams, was a unique device with which to carry out precision tests of the standard model.

The LEP collider is in the form of a circular ring of 27 km circumference inside which a vacuum pipe contains counter-rotating bunches of high energy electrons and positrons. These bunches are arranged to collide and annihilate at four collision points. In the first phase, LEP1 (1989–1995), the energy in each beam was enough to produce the Z^0 and sweep through the peak. The high luminosity of the beams enabled the production of millions of Z^0's, and a detailed quantitative study of their properties could be undertaken. At each of the four collision points, highly sophisticated multipurpose detectors surrounded them to catch all the products formed in the annihilations. These detectors have the capability of making a precise determination of the identity of the particles produced, their momenta, and other properties. The four detectors are called *ALEPH, DELPHI, L3*, and *OPAL*. They are huge, arranged in the form of cylinders surrounding the beam pipe, typically weighing several thousands of tons, and occupying a volume of roughly $(12 \text{ m})^3$. Typically, experiments mounted at each detector are run by collaborations of several hundred physicists and engineers from many universities and national laboratories spread all over the world. Details regarding these detectors will be found listed in a later section.

The SLC is an outcome of extensive upgrade of the two-mile long linear accelerator that was functioning previously at SLAC. The upgrades involved raising the energy of the electrons and positrons to 50 GeV, facilities to reduce the beam size to small dimensions, sets of magnets for transporting the separate electron and positron beams in opposite directions from the linear accelerator, and bringing them to a collision point. The electron and positron beams were in pulses of thirty per second with each pulse containing about 10^{10} particles. An elaborate system was put in place to focus the colliding bunches to an incredible size no larger than four microns by four microns. SLC started functioning in 1989. In 1992 a new source for producing polarized beams of very high intensity was added so that polarized beams of electrons and positrons could be produced and made to collide to produce Z^0's.

The detector used initially with SLC was an upgrade of the Mark II detector, used earlier at the SPEAR and PEP rings. In 1991 the upgraded Mark II was replaced by a complete detector system specially constructed for use with SLC, the SLC Large Detector (SLD).

The four CERN LEP experiments ALEPH, DELPHI, L3, and OPAL, and the SLC at SLAC with the SLD detector have all produced data of great precision. Quantitative high-precision tests of the standard electroweak model have been carried out, and extensive searches for hints of new physics have been made. The most significant results may be summarized as follows:

- All experiments have measured the line shape of the Z^0 resonance. This gives an accurate measure of the mass of Z^0 and its width. Since the various ways in which this particle can decay contribute to its width, the number of lepton families that exist in nature can be determined from an accurate measurement of the width of the Z^0. All experiments have concluded that the number of such lepton families is just *three*.

- All LEP experiments have verified a property called *universality*; this is a property which says that the particles in the three families all behave in the same way with respect to the weak interactions. In other words, the only differences between the electron, muon, and tau are their masses; their weak interactions are universal. One of the big questions, for which there is no answer, is why these particles behave in such identical fashion despite the fact that they have such widely different masses. This property of universality is also shared by the three quark families.

- In the standard electroweak model, the mechanism for generating masses for the gauge bosons as well as for the fermions is the Higgs mechanism. The Higgs scalar particle is predicted to exist, and the search for it has been one of the chief occupations of all the experiments. Unfortunately, the mass of the Higgs particle is not known, even approximately, so searches have to be made in all regions of masses. Despite intensive searches, none of the LEP experiments see a signal for the Higgs particle in the range of masses accessible to them. The experiments have put a lower limit on the mass of the Higgs particle. This limit, which is about 100 GeV with data from LEP1, has been pushed up to about 114 GeV with data from LEP2.

- With LEP2 upgrade of the energies of the electron and positron beams, production of W-boson pairs has been possible. Accurate measurements of the masses of the W's and their various decay channels have been completed.

- Searches for supersymmetric particles in the minimal supersymmetric extension of the standard model have been carried out. No supersymmetric particles have yet been found.

- SLD at SLC had the additional feature that it could collect data on Z^0 with polarized beams. Due to the polarization, there is expected to be a left-right asymmetry at the position of the Z^0

resonance. The left-right asymmetry, A_{LR}, is defined as the difference $\sigma_L - \sigma_R$ divided by the sum $\sigma_L + \sigma_R$, where $\sigma_L(\sigma_R)$ is the cross section for an incident left-(right-)handed incident electron. This asymmetry has been measured very precisely at SLD, with the uncertainties due to systematic effects mostly canceling. The asymmetry is extremely sensitive to the electroweak mixing angle parameter of the standard model, $\sin^2 \theta_W$. Precise measurement of the left-right asymmetry has enabled them to determine the world's best value for this parameter. The result derived for this parameter has significant implications for the value of the Higgs mass.

• At the time of this writing, the ALEPH experiment has reported seeing a very slight excess of events which could be interpreted as a possible signal for a Higgs particle at a mass close to 115 GeV. DELPHI experiment has also reported finding something similar. LEP was scheduled to be shut down in the end of September 2000 to make way for the LHC. Due to the developments in regard to the Higgs particle, the authorities at CERN have acceded to a request from the ALEPH collaboration to extend the running of LEP at the highest energy until at least early November, so that all four experiments could accumulate more data. No clear signal for the Higgs particle at this mass value has been seen.

Discovery of the Top Quark (1995)

After many years of searching for the top quark (which is a member of the third quark family and a partner with b-quark), it was finally discovered in 1995 by the CDF and D0 collaborations working at the Tevatron in Fermilab. The top was produced in proton-antiproton collisions at a center of mass energy of 1.8 TeV, at an incredibly high mass value of 175 GeV! Since the initial discovery, there have been further confirmations of the existence of this quark, and measurements of its properties indicate that it is indeed the t-quark of the standard model, the partner to the b-quark. Without the t-quark, the standard model would be in trouble, because anomaly cancellation would not be complete and the standard model would not be renormalizable. Finding it, and measuring its mass, has pegged one of the important parameters of the standard model. This knowledge is of great help in further searches for the Higgs particle.

More on Neutrinos

In 1955, at about the same time Cowan and Reines were attempting to detect antineutrinos from Savannah and Hanford nuclear reactors, R. Davies pioneered a radiochemical method for detecting neutrinos.

He based his experiment on an idea Pontecorvo had put forward in 1948 in an unpublished report from Chalk River, Canada. In it Pontecorvo had suggested detecting, by radiochemical extraction, ^{37}Ar isotope produced by the reaction ^{37}Cl$(\bar{\nu}, e^-)$ ^{37}Ar. It was already known that ^{37}Ar decays by electron capture, with a 34-day half-life, to ^{37}Cl. Thus the reverse reaction, ^{37}Cl(ν, e^-) ^{37}Ar, must occur. Davies argued that, if the neutrinos emitted in ^{37}Ar electron capture are identical to the antineutrinos coming from reactors, then the reaction ^{37}Cl$(\bar{\nu}, e^-)$ ^{37}Ar should also readily occur. He decided to look for it by irradiating drums, one containing 200 liters and another 3900 liters of carbon tetrachloride, with antineutrinos coming from outside the reactor shield. He removed any argon that would have been produced by flushing the system with helium gas, and counted them. From these measurements he could only place an upper limit of 2×10^{-42} cm^2/atom for the cross section of antineutrinos coming from the reactor. This upper limit was too big compared to theoretical expectations and no conclusions could be drawn regarding the identity of neutrinos and antineutrinos from this measurement. Cosmic rays produced a background ^{37}Ar activity in the tanks (through ^{37}Cl(p, n) ^{37}Ar), which prevented Davies from exploring lower values of the cross section.

Still, he continued further to see if he could make the method sensitive enough to detect neutrinos from other sources, in particular, neutrinos coming from the sun. He performed an experiment with the 3900-liter tank buried under 19 feet of earth. With this overlay, cosmic ray nucleonic component should be reduced by a factor of one thousand and he could be sensitive to the much lower cross sections of interest. The neutrinos from the pp reaction chain in the sun have a maximum energy of 0.420 MeV, lower than the threshold of 0.816 MeV for the ^{37}Cl(ν, e^-) ^{37}Ar reaction. Thus, this method could not detect the pp neutrinos but could see the higher energy neutrinos coming from the carbon-nitrogen cycle. An estimate of the flux of neutrinos from pp in the sun was 6×10^{10} neutrinos/(cm^2 s). From his initial measurements, Davies could only place an upper limit for this flux at 10^{14} neutrinos/(cm^2 s).

Two things became clear from these studies. First, the radiochemical method for detecting neutrinos is feasible; its sensitivity, however, has to be increased, and cosmic ray and other background effects would have to be reduced. Second, a better understanding of the energy generation processes in the sun and better estimates of the flux of solar neutrinos from the different nuclear reaction chains were necessary.

Improvements on both these fronts have been carried out in the last four and half decades such that some new and significant results are emerging on neutrinos.

- On the theoretical front, a much better model of the sun's energy generation and transport has emerged from a lot of work Bahcall and associates have done over a number of years. The fluxes of neutrinos form the main nuclear reaction sequences, and some of the side reactions that occur have all been calculated carefully. These fluxes are referred to as the fluxes from the Standard Solar Model. The bulk of the neutrino flux is due to the pp reaction chain, which cuts off at 0.42 MeV. Among other smaller contributors to the flux, the one of particular interest from the point of view of Davies' experiment is the boron neutrinos from $^8_5$B. These neutrinos are spread in energy from about 1 MeV to about 14 MeV, with a maximum in the flux occurring at about 5 MeV. The flux of boron neutrinos is about five orders of magnitude smaller than the flux from the pp reaction chain.

- On the experimental front, Davies set about improving the sensitivity of the chlorine-argon radiochemical method and reducing backgrounds by working deep underground in the Homestake mine. Over many years of operation, Davis collected data on the flux of the boron neutrinos and found it smaller than that expected from the standard solar model by a factor between 2 and 3.

- The difference between the number of neutrinos observed from the sun and what is expected on the basis of the standard solar model has come to be called the *solar neutrino problem*.

The reduced flux seen with respect to the boron neutrino flux is also seen in the study of the dominant neutrino flux from the pp chain. Radiochemical detectors, using the reaction ^{71}Ga$(\nu, e^-)^{71}$Ge and counting the number of ^{71}Ge, are sensitive to the neutrino flux from the pp chain. The experimental collaborations SAGE and GALLEX have done the experiments. Their results also show numbers which are smaller than that expected from the standard solar model. Two other neutrino experiments Kamiokande and SuperKamiokande, involving large water detectors in the Kamioka mine in Japan, are sensitive to the same range of energy of neutrinos as those in the chlorine experiment of Davies. In these detectors, the measurement of the neutrino flux is done by an entirely different method. They observe the flashes of Cherenkov light produced by the electrons scattered elastically by the incident neutrinos in water, with an arrangement of a large number of photo multipliiers looking into the water tank. These measurements also give a result which is about half the expected value.

Explanations of these observations are based on two possibilities. First, the standard solar model may not be accurate enough to give the theoretical fluxes. Second, the reduced fluxes seen in the experiments may be due to a phenomenon called *neutrino flavor oscillations*. The standard solar model has been examined very carefully and there seems to be a consensus of opinion that its results for the neutrino fluxes can be trusted. Explanation in terms of neutrino flavor oscillations is being increasingly favored.

Neutrino flavor oscillation phenomenon involves the following picture. The basis of weak interactions involves neutrinos with definite lepton numbers which label the different flavors of leptons, electron-like, muon-like, etc. If the neutrinos have a nonvanishing rest mass, the basis of the mass eigenstates will be mixtures of flavor eigenstates (and vice versa), and the two bases are related by unitary transformations. In such a situation, a neutrino born with a certain flavor from a weak interaction decay, say electron-like, will, after propagating through a certain distance in vacuum or in matter, have a finite probability for changing its flavor to, say, muon-like or tau-like flavor. If the detector is one which detects electron neutrinos only, it will measure a reduced flux of electron neutrinos. All the detectors above are electron neutrino detectors. It is possible they are measuring reduced electron neutrino fluxes due to the occurrence of flavor oscillations.

A new heavy water detector for neutrinos has come into operation in 1999 at the Sudbury Neutrino Observatory, in Sudbury, Canada, called *SNO*. This detector has the capability to detect not only electron neutrinos but also neutrinos of other flavors. The data from SNO are expected to resolve whether the solar neutrino deficit is due to flavor oscillations or to deficiencies in the standard solar model.

Efforts are being made to check the idea of flavor oscillations with neutrino beams produced from high energy accelerators. There are a number of such projects which are either under way or being planned for the near future.

Since the existence of rest mass of the neutrino is related to the phenomenon of neutrino flavor oscillations, efforts are also being made to measure the electron neutrino mass directly by high precision studies of the beta decay of tritium near the end point of the electron energy spectrum.

Future Outlook

In summary, through the dedicated work of a very large number of workers in the field in the last three decades, the standard model of elementary particles and their interactions has evolved. With the discovery of the top quark in the last five years and the precision measurements carried out at LEP and SLC, the standard model has been put on a secure foundation. Yet there are many questions remaining to be answered. At the center of these questions: Where is the Higgs particle? Only when it is found and its properties measured, can we hope to understand the generation of fermion masses.

There are still other unanswered questions: Are there still more generations? Where is Supersymmetry? Does string theory have anything to do with elementary particles? Are quarks and leptons truly pointlike, or are they composite? Doubtless, there is lot of work to be done in the years ahead.

The other sector which is exciting is the neutrino sector. An unambiguous observation of neutrino oscillations will be a very exciting event establishing the existence of nonvanishing neutrino masses, which has far-reaching implications.

CHAPTER 2

Historical Overview of Accelerators and Colliding Beam Machines

---

Accelerators

Devices for producing high energy particles are the particle *accelerators*. Studies of elementary particles start with the production of these particles by accelerating protons or electrons to high energies and bombarding a target with them. This target may be a stationary target or another focused beam, a *fixed target setup* or a *collider setup*, respectively. Apart from the energy to which the particles are accelerated, another factor of importance is the intensity of the beam particles produced by the accelerator, or the luminosity in the collider.

Early particle accelerators depended upon production of high voltages applied to electrodes in an evacuated tube. These go under the name of *Cockcroft-Walton* generators [1] and *Van de Graaff* [2]; machines. These early devices are limited to achieving energies only in the few MeV range because of high voltage breakdowns, discharges, etc., and are suitable for explorations in nuclear physics. They are not of much interest for studies in elementary particle physics, which typically require energies in the hundreds of MeV to GeV ranges.

To achieve the much higher energies of interest to particle physics, particle accelerators have been constructed either as circular accelerators using a combination of magnetic and radio frequency (rf) electric fields such as *cyclotrons*, *synchrocyclotrons*, and *synchrotrons*, or as linear machines using only high frequency electric fields, called *linear accelerators* or *linacs*. In these machines acceleration to high energies is achieved by repeatedly applying a radio frequency electric field in the right phase, at the right time to the particles, so that each application increases the energy of the particles by a small amount. Other particle accelerators like betatrons and synchrotrons achieve a steady acceleration of particles by the electric field induced by a varying magnetic flux.

Cockcroft-Walton Generators

This is one of the earliest accelerators built using electrostatic principles. It involves constructing circuits for voltage multiplication using transformers and capacitors. The transformer supplies alternating current of certain voltage to a capacitor-rectifier circuit in such a way that the AC current travels up a line of capacitors and is distributed to rectifiers and returned to the ground through another line of capacitors. The DC current flows through the rectifiers in series. The rectified voltage is a multiple of the input voltage to the transformer, the multiplying factor determined by the number of capacitors in the line. Although these accelerators are now primarily of historical interest, the particles accelerated by these devices are injected into other machines for accelerating them to very high energies. Such a use goes under the name of an *injector*.

Van de Graaff Accelerator

In this device a sphere is charged to a high potential by a moving belt on which charge is continually sprayed and transported. The charge is removed at the sphere through a system of brushes and the potential of the sphere is raised. The accelerating tube has the sphere at one end of it and is the high voltage electrode. It is capable of accelerating positive ions to several MeV of energy and currents of order $100\mu A$ are achievable. The whole system is enclosed in a sealed tank containing a special gas mixture to prevent electrical breakdowns.

To produce higher energies, tandem Van de Graaffs have been constructed in which one starts by accelerating negative ions toward the sphere, stripping them of their electrons at the sphere, and then accelerating the positive ions. This way one can double the energy of the positive ions. Van de Graaff accelerators have played an important role in investigations into properties of nuclei. They are not of much use in elementary particle physics.

Cyclotron

Invented by Lawrence [3], this accelerator employs a combination of magnetic field and oscillating electric field to accelerate charged particles. The particles to be accelerated travel in bunches and a magnetic field is used to bend the path of the particles into circular arcs. At certain times when they reach certain parts of the path, the phase of the electric field is such that the particles are accelerated there. In other parts of the path, that is at other times, the phase of the electric field is of no concern because the particles are shielded from the field. If the frequency of the electric field is chosen so that it is in the accelerating phase for the particles, they get repeated pushes at the accelerating parts of the path. Even a modest amplitude of the electric field is sufficient for achieving high energies.

The angular velocity of the rotation of the particle of charge q and mass m is equal to qB/m and is independent of the velocity or the radius of the orbit, as long as the particle is nonrelativistic. It is called the *cyclotron frequency*. If the frequency of the alternating electric field is chosen equal to the cyclotron frequency, the condition for repeated acceleration at the gaps is met. For magnetic fields in the range of 1 to 2 Tesla, and for accelerating protons, this frequency works out to be in the radio frequency range—hence the term *radio frequency (rf) electric field* is used to refer to it.

The workings of the cyclotron can be understood as follows. A constant magnetic field is created in a region occupied by structures called the *dees*. The dees are hollow metallic semicircular cylinders shaped in the form of the letter D (and a reversed D) such as is obtained by cutting a cylindrical pill box across a diameter. The two D shapes are separated by a gap along the straight section of the cut pill box. The dees are connected to an rf power supply. Positive ions, say protons, are injected into the center of the dees. Under the constant magnetic field, they are bent into a semicircular path inside the dee until they come to the straight edge of the dee. The rf electric field is in such a phase as to accelerate the ions across the gap between the dees. When they are moving inside the dees, they are shielded from the electric field. The ions acquire incremental energy only across the gap and in the next dee travel along a path of larger radius and reach the other straight edge of the dee. If, by this time, the phase of the electric field is again such as to accelerate the ions, they acquire some more energy. They go into a semi-circular path of larger radius and the process of acceleration repeats across the gap between the dees and the particles acquire more and more energy. The condition for the repeated increases in energy will be achieved if the rf frequency is equal to the cyclotron frequency of the ion. When the radius of the orbit in the dee becomes nearly equal to the radius of the dee, the beam is extracted for external use by deflecting the ions with a negatively charged plate. In the late 1930's, protons of energy up to 25 MeV were accelerated in a cyclotron. Increases beyond this energy could not be achieved with a fixed rf frequency machine because the cyclotron frequency decreases due to the relativistic increase of the proton mass.

Synchrocyclotron

There is a limit to the energy which can be achieved in a cyclotron. When the energy increases, the particle's mass increases and the cyclotron frequency qB/m is reduced. To continue accelerating the particles across the gaps beyond this energy, the rf frequency must also be reduced to keep pace with the changed cyclotron frequency. Machines have been built which incorporate these features. They are called *synchrocyclotrons*. These machines can accelerate particles to much higher energies as long as one scales up the size of the dees. Synchrocyclotrons have been used to accelerate protons in the 100 to 1000 MeV range. Examples of such synchrocyclotrons are the 184-inch Berkeley machine (maximum energy 720 MeV), the CERN synchrocyclotron (maximum energy 600 MeV), and the Dubna synchrocyclotron (maximum energy 680 MeV). However, there is a limit to further increases in energy using this method because the magnet becomes prohibitively expensive.

Betatron

The betatron was invented by Kerst [4]. It is a machine for accelerating electrons. The electrons are injected into a doughnut shaped evacuated ring and are accelerated by the induced electric field due to a changing magnetic flux within the doughnut ring. The magnetic field B_0 needed to keep the particle in a circle of radius R satisfies the relativistic equation $eB_0R = p$, where e is the charge of the particle and p the relativistic momentum. When the flux ϕ linked with the orbit changes with time, the particle will feel the induced electric field and will change its momentum. It can be shown that to keep the electron in the orbit for a large number of turns, it is necessary to satisfy the condition $\phi = 2\pi R^2 B$, where ϕ is the flux enclosed by the orbit, B the magnetic field at the orbit, and R the radius of the orbit. If this condition is satisfied, the electron orbit does not shrink or expand. The electron in this orbit can be accelerated by increasing ϕ and B together. The electrons gain energy as they move in the circle of fixed radius. Typically, the energy increase in each turn is about a few hundred electron volts. Thus to increase its energy substantially, a large number of rotations is required, and for all that time, the particles must stay focused in a bunch.

Betatron Oscillations

The magnetic field at the position of the orbit of the electrons in a betatron can be resolved into two components, one along the axis of symmetry of the machine (perpendicular to the plane of the doughnut) defined as z, the other in the radial direction r. If the particles are to be in stable motion in the orbit, any displacements in either the z or the r directions must bring restoring forces into play so that the equilibrium orbit is attained again. Detailed analyses show that, if the z component of the magnetic field at the equilibrium orbit position is given by $B_z = B_0(\frac{r}{R})^{-n}$, with n positive, the index n must satisfy the condition $0 < n < 1$, in order for the forces to be of the restoring type. Then the beam executes *betatron oscillations* in the z and the r directions with frequencies given by $\omega_0\sqrt{n}$ and $\omega_0\sqrt{(1-n)}$, respectively, where $\omega_0 = \frac{eB_0}{m}$. The condition for a stable beam implies that the magnetic field must decrease from the center of the machine to the outer edge. Clearly, the amplitude of these oscillations must be smaller than the dimensions of the vacuum chamber holding the beam. Otherwise the beam will be lost to the walls.

Synchrotron: Principle of Phase Stability

E. McMillan proposed that a new kind of accelerator called the *synchrotron* be built based on the principle of *phase stability* [5]. To understand this principle, consider a particle in a cyclotron just about to pass the gap between the dees. Suppose it has the right velocity to cross the gap when the electric field between the dees is going through zero. If there are no energy losses, the particle will go on moving in this orbit indefinitely at constant speed. Such a particle may be said to be in a synchronous orbit. Suppose we have another particle which arrives at the gap somewhat earlier than the particle in the synchronous orbit. Then it will see a nonzero electric field and will be accelerated and gain energy. Because of its increased relativistic mass, its angular velocity will decrease. It will take longer for it to arrive at the next gap and it will be a little later in phase than previously. This will keep on occurring until such a particle will cross the gap at zero electric field. But, this particle still has higher energy than that required for it to cross the gap at zero field. Going into further rotations, this particle will tend to cross the gap when the electric field is in a decelerating phase. This will reduce its energy and bring it back to its synchronous value. Thus the situation is such that the disturbed orbits will oscillate both in phase and energy about the synchronous values which are constant. Such oscillations are referred to as *synchrotron oscillations*. To increase the energy, it is necessary to increase the synchronous value. This can be done by either decreasing the frequency of the electric field or increasing the strength of the magnetic field.

A device which makes use of the decrease in the frequency of the electric field (while keeping the magnetic field constant) is the synchrocyclotron. A device which works using the other possibility, namely increasing the magnetic field, accelerates the particles like in a betatron, and it is called a *synchrotron*. In this latter case, the machine will be shaped in the form of a doughnut ring with one or more gaps in the ring where the alternating electric field can be applied. A magnetic field necessary to maintain a particular orbit radius is applied. For accelerating electrons, one increases the magnetic field while holding the frequency of the electric field constant. For nonrelativistic protons on the other hand, it is necessary to increase the freqency also as the magnetic field increases.

Alternating Gradient Strong Focusing Machines

To reach very high energies with a synchrotron, one can increase either the radius of the machine or the strength of the magnetic field. From a practical point of view, however, there is a limit to how high the magnetic field can be raised. Since the particles circulate in a doughnut shaped evacuated vessel and go around many revolutions, it is necessary to hold the particles in their motion, to great precision, to the middle portion of the doughnut shaped region. To cut costs, this doughnut shaped region should be as small as possible. These requirements translate into keeping the vertical and horizontal oscillations (betatron oscillations) of the beam in the doughnut small. It is found that the amplitude of the vertical oscillation varies inversely with the frequency of the vertical oscillation. Thus this frequency must be high. The frequency of the vertical oscillation is shown to be $\omega_0\sqrt{n}$ and can be increased by increasing n for a fixed magnetic field. But there is a limit to how high n can go since stability in the radial motion demands that $n < 1$. This clearly leads to an impasse. A way around this impasse was the invention of the principle of *strong focusing* by Christofilos in an unpublished paper [6] and, independently found by Courant, Livingston, and Snyder [7]. They showed that if the magnet is built of alternate sections of large positive and negative values of n, such a combination is similar to combinations of converging and diverging lenses of equal strength and can be shown to be converging on the whole. They showed that the oscillation amplitude can be diminshed substantially. Using quadrupole magnets, phase stable synchronous acceleration can still be achieved with large reductions in radial oscillation amplitudes compared with a machine with uniform n values. The first strong focusing proton synchrotron using the alternating gradient principle, the AGS (Alternating Gradient Synchrotron), was built at Brookhaven National Laboratory and accelerated protons to 30 GeV energy. The route to much higher energies, in the range 10 to 100 TeV, seems feasible using the alternating gradient principles.

Some Fixed Target High Energy Accelerators

Notable among the many synchrotrons that have been operated around the world for fixed target experiments are the 3 GeV protron synchrotron called the Cosmotron at the Brookhaven National Laboratory (BNL) (1952), the 1 GeV machine at Birmingham (1953), the 12 GeV electron synchrotron at Cornell University (1955), the 30 GeV AGS machine at BNL, the 1000 GeV superconducting synchrotron at the Fermi National Accelerator Laboratory (FNAL), and the 500 GeV SPS machine at CERN. These machines have proved to be tremendously important in making rapid progress in elementary particle physics research in the last half century.

Synchrotron Radiation

Synchrotrons which accelerate electrons suffer from an important limitation. The charged particles in the circular machine emit electromagnetic radiation called synchrotron radiation. It is found that the energy radiated by a particle in every revolution is given by $E_{\mathrm{rad}} = \frac{4\pi}{3} \frac{\alpha \beta^2 \gamma^4}{\rho}$, where ρ = radius of curvature of the orbit, $\alpha = \frac{1}{137}$, β = particle velocity in units of the velocity of light (one in our units), and $\gamma = (1 - \beta^2)^{-1/2}$. Since the relativistic energy of the particle is given by $E = \gamma m_0$, where m_0 is the rest mass energy of the particle, the value of γ is much higher for electrons than for protons of the same momentum. The loss by synchrotron radiation is very much more significant for electron synchrotrons than for proton synchrotrons because it varies as the fourth power of γ. With the above formula, one finds typically, for a 10 GeV electron circulating in a circle of radius 1 km, the energy loss by synchrotron radiation is about 1 MeV per revolution, while for 20 GeV electrons, it rises to 16 MeV per revolution. Compensation for this huge energy loss becomes an important feature in an electron machine. Hence, the use of linear accelerators to accelerate electrons to high energies.

Linear Accelerator

Electrons by virtue of their low mass become relativistic already at energies of the order of an MeV. Circular machines such as the cyclotron, the betatron, or the synchrotron are not very suitable to accelerate them to energies higher than a few hundred MeV. This is because of the strong radiation emitted by charged particles under accelerated motion, the *synchrotron radiation*. The energy loss by synchrotron radiation varies as the fourth power of the energy of the particles and inversely with radius of the orbit. Thus, at some stage in a circular machine, the amount of energy the particles lose by synchrotron radiation becomes greater than the energy they gain from the rf source. Clearly the method to cut the synchrotron radiation losses is to avoid using circular machines and accelerate the particles in linear machines instead.

The linear accelerator can be used to accelerate electrons, protons, or even heavier ions. It also uses multiple pushes given to the particles in the beam to accelerate them to high energies. Modern linear accelerators make use of the electromagnetic field established inside a hollow tube of conducting material, called a *wave guide*. Standing waves are formed inside the cavity of the wave guide. The cavities act as resonators and are referred to as *rf cavities*. Because standing waves can be considered as a superposition of two waves travelling in opposite directions, particles which move with the same velocity as one of the travelling waves will be accelerated. For high energy electrons, whose velocity is almost that of light (c), the electromagnetic wave in the cavity must move with the phase velocity. Achievement of this condition is made possible by the insertion of suitable partitions inside the cavity and exciting the cavities at the right frequency. Considerations of phase stability in linear accelerators proceed much as those in circular accelerators. It is achieved during that part of the cycle of the rf when the potential increases rather than when it decreases.

The first proton linear accelerator was built in Berkeley in 1946 and the first electron linear accelerator was successfully put into operation in Stanford around 1955. Linear accelerators, which accelerate electrons and positrons with high intensity to 50 GeV of energy, have been constructed and operated at the Stanford Linear Accelerator Center.

Colliding Beams Accelerator

In fixed target experiments, all the energy of the accelerated particle is not available for exploring the high energy frontier because some of the energy is associated with the motion of the center of mass and is not available for the production of new particles. Achievement of highest center of mass energies is possible if one could perform experiments in a frame in which the center of mass of the colliding particles is at rest. This is possible in a collider setup. Most of the colliders in operation are synchrotrons with counterrotating bunches of particles in circular rings.

The laboratory reference frame is defined as one in which a particle of energy E is incident on a target particle of mass M at rest, while the center of mass reference frame is one in which the center of mass of the (incident particle-target particle) system is at rest. These reference frames are related by Lorentz transformations along the beam direction in the laboratory. The square of the total four-momentum is denoted by the symbol s and is a relativistic invariant. Using this, one can find relations between the components in these two frames. In the center of mass frame, since the total three-momentum is zero, the square of the total four-momentum is simply the square of the total energy $W = E_1 + E_2$ of the particles labeled by the labels 1 and 2, respectively. Thus, in the center of mass frame, $s = W^2$. All the center of mass energy is available for production of new particles. Written in terms of laboratory frame variables, since a particle of energy E and three-momentum $\vec{p}$ is incident on a stationary target of mass M, $s = W^2$ is also equal to $(E + M)^2 - \vec{p}^2 = 2(EM + M^2)$. Thus for $E >> M$, W grows at best as $\sqrt{E}$. To improve on this, one could look into doing experiments in the laboratory in which the energy associated with center of mass motion is reduced to zero.

A first proposal for attaining very high energies in collisions by means of intersecting beams of particles was made by Kerst et al. [8]. The *colliding beams* accelerator is a realization of this idea. In the colliding beam setup, one has two beams of relativistic particles directed one against the other arranged such that the total three-momentum of the two particle system is zero in the collision region. In this case, $s = W^2 = (E_1 + E_2)^2$, where E_1 and E_2 are the energies of the particles in the two beams. If $E_1 = E_2 = E$, we have $W^2 = 4E^2$, or $W = 2E$. In other words, in the colliding beam frame of reference, W grows like E and all the energy in the beam is available for reactions. The only practical question is one of obtaining sufficient intensity in each beam so that there will be a reasonable number of events in which reactions occur.

Luminosity in Colliding Beams

In colliding beams setup, the collision rate is of prime concern. Colliders that have been built so far involve protons on protons or on antiprotons, electrons on positrons, and electrons on protons. The rate R of reaction in a collider is given by $R = \sigma L$, where σ is the cross section for the interaction of the particles in the beam, and L is called the *luminosity*. It is given in units of $\text{cm}^{-2} \text{ s}^{-1}$ and one would clearly like to have for it as high a value as possible. For two oppositely directed relativistic beams of particles traveling in bunches in the beams, it is given by $L = fn \frac{N_1 N_2}{\sigma_x \sigma_y}$, where f is the revolution frequency, n the number of bunches of particles in each beam, and N_1 and N_2 the number of particles in each bunch. The product $\sigma_x \sigma_y$ is the cross sectional area of the beams with a length σ_x in the horizontal (x) and length σ_y in the vertical (y) directions. L can be increased if the cross sectional area of the beams is decreased. There are special methods to achieve this, but there is a limit to how much this can be increased because of space charge effects of the beams. Typical values of L for electron-positron colliders are around 10^{31} to 10^{32} $\text{cm}^{-2} \text{ s}^{-1}$. These values are much smaller than what is available in a fixed target setup, which, for a proton synchrotron beam of few times 10^{12} particles per second impinging on a liquid hydrogen target about 1 m long, is about 10^{37} $\text{cm}^{-2} \text{ s}^{-1}$.

Proton-Proton and Proton-Antiproton Colliders

Kerst et al. [8] showed that it is possible to have sufficiently intense beams in machines such as the proton synchrotrons for the event rates to be nonnegligible. The first such collider machine was the ISR (Intersecting Storage Ring), constructed at CERN. It contained two rings of magnets (with vacuum chambers inside) adjacent to each other and stored proton beams of 30 GeV energy circulating in opposite directions. There were eight locations in the ring where the proton beams were brought to intersect. These were the locations where collisions occurred and the products of the collision were studied with suitable detectors set around the collision regions. This machine, with W=60 GeV in the center of mass frame, is equivalent to a fixed target proton synchrotron beam of energy about 2,000 GeV. A colliding beam accelerator involving protons and antiprotons, the S$p\bar{p}$S was the successor to the ISR at CERN. The first W particles and Z^0's were produced with it. The TEVATRON at FNAL started running in the collider mode in the year 2000 (p$\bar{p}$, 2×1 TeV). The Large Hadron Collider (LHC) is under construction at CERN and is expected to start working in the year 2005 (pp, 2×7 TeV). Another example of a colliding beam accelerator is HERA which has studied collisions of 30 GeV electrons with, initially, 820 GeV and, later, 920 GeV protons in storage rings.

e^+e^- Collider Rings

Colliding beam accelerators involving storage rings of electrons and positrons of steadily increasing energies have been constructed and used in
the study of particle creation in electron-positron annihilations. Examples of e^+e^- circular colliders which have had a large impact on high
energy physics are: AdA at Frascati (2×250 MeV; 1961), Princeton-
Stanford machine (2×500 MeV; 1961), VEPP-1 at Novosibirsk (2×0.7
GeV; 1963), ACO at Orsay and ADONE at Frascati (2×1 GeV), SPEAR
at SLAC (1972) and DORIS at DESY (Hamburg; 1974) (both 2×4 GeV),
PEP at SLAC (2×15 GeV), PETRA at DESY (Hamburg) (2×19 GeV),
CESR at Cornell (2×9 GeV), TRISTAN in Japan (2×33 GeV), and
BEPC at Beijing, China (2×2.2 GeV). At present the highest energy

Figure 2.1: An aerial view of the CERN site with the LEP circular ring
shown. (Courtesy CERN)

e^+e^- collider ring is the Large Electron Positron Collider (LEP) at CERN and has reached beam energies of about 2×108 GeV.

CERN is the European laboratory in which a consortium of European countries have invested for doing research in particle physics. Over the years it has maintained very active experimental and theoretical programs of investigations in particle physics. With LEP, it has been possible to produce Z^0's copiously, and four large international collaborations have made a precision study of their properties in detail. In

Figure 2.2: The SLD detector was developed and built at the Stanford Linear Accelerator Center, a high energy physics research facility operated on behalf of the U. S. Department of Energy by Stanford University. (Courtesy SLAC)

Figure 2.1 on page 61, a picture of the CERN site located at the France-Switzerland border together with the LEP ring is shown. At the highest energies of operation (208 GeV in center of mass), it has also produced pairs of W particles, enabling a precision study of these particles.

e^+e^- Linear Collider

An exception to the circular storage ring machines is the linear electron-positron collider at SLAC, called the *Stanford Linear Collider (SLC)* and completed in 1989. SLAC is a high energy physics research facility operated on behalf of the U. S. Department of Energy by Stanford University. There is a collection of experimental facilities at SLAC which are shown in Figure 2.2 on page 62. Shown in this last figure are: 3 km long linear accelerator, accelerating electrons and positrons to 50 GeV energy; End Station A for fixed target experiments; SPEAR storage ring, now used as a synchrotron radiation source; PEP, the 30 GeV colliding storage ring, now upgraded to PEPII to serve as a B-factory, for colliding

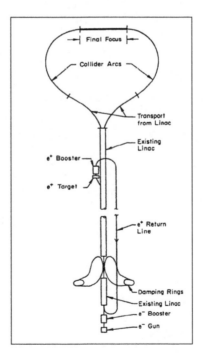

Figure 2.3: A schematic diagram showing the SLAC linear e^-e^+ collider.

9 GeV electrons with 3.1 GeV positrons; and finally, SLC, the 100 GeV electron-positron collider. Associated with this complex is also a facility for research into the design and construction of future accelerators.

The SLC has successfully operated for a number of years. Together with the SLD detector, it has produced information on Z^0's supplementing and complementing the information available from LEP at CERN. The operation of SLC, a schematic diagram of which is shown in Figure 2.3 on page 63, represents an important milestone in the further development of linear accelerators. The experiences gained by operating SLC will be of great help in the route to attainment of even higher energies by colliding beams from linear accelerators. Plans are currently being discussed for a 1 TeV electron-positron linear collider, called *Next Linear Collider (NLC)*; it is possible we may have one of these facilities operating toward the end of the first decade of the 21st century. Feasibility of colliding muon beams of very high energies is also being considered.

CHAPTER 3

Historical Overview of Detector Developments

Development of Detectors for Particle Physics

The remarkable developments in elementary particle physics have been possible, on the one hand, with the development of higher and higher energy accelerators and methods for increasing the intensities of the particle beams and, on the other hand, with the development of elaborate, multi-purpose, complex detectors, capable of carrying out measurements of many parameters, such as energy, momentum, position, time, mass of the particles with as much precision as possible. The handling of the large volumes of data produced by these detectors has only been possible because of the simultaneous developments in computing power. The construction and operation of these complex detectors in actual experiments is constantly evolving and has developed into a fine art, and many volumes have been written on the subject. Here is given a brief historical overview of the development of the detectors. For more details, refer, among many books, to *Experimental Techniques in High Energy Physics*, by T. Ferbel [9], and *Particle Detectors*, by C. Grupen [10].

The detection of particles is based on an understanding of a number of physical phenomena that occur in the interaction of the particles with the detector medium, and the relative importance for the energies of the particles concerned. The physical effects on which the detection of the particles is based depend on the type of the particle, and whether it is electrically charged or neutral. For charged particles, brief descriptions of the theory of ionization energy loss and the radiation energy loss in media will be covered below.

For charged particles passing through matter, the most important basic phenomena are the energy loss through ionizing collisions with the particles of the medium (called *ionization loss*), and the loss of energy by bremsstrahlung (called *radiation loss*). Ionization loss is important for heavy, moderately relativistic charged particles, while radiation loss is the dominant energy loss mechanism for highly relativistic charged particles. Electrons and positrons become relativistic already at energies of a few MeV, and at high energies their energy loss is almost all due to radiation.

For the photon, detection is based on its electromagnetic interactions with charged particles. Photons of low energy, less than 0.1 MeV, lose energy mostly by photoelectric absorption in atoms. Between 0.1 MeV and 1 MeV, there is competition between photoelectric absorption and Compton scattering. Those of moderate energy, roughly between 1 MeV and 6 MeV, lose energy by Compton scattering off the electrons in atoms. Those of high energy, roughly higher than 6 MeV, lose energy by producing electron-positron pairs in the vicinity of atomic centers. The secondary particles produced by photons are charged particles, which can be detected by methods used for the detection of charged particles.

For detecting neutrons, different methods have to be employed, depending on the energy. In all these methods, secondary charged particles are produced by the neutron interactions, which are then detected by the methods used to detect charged particles. Neutrons in the energy range of tens of MeV to about a GeV produce recoil protons through elastic (n, p) scattering. Neutrons of energy higher than 1 GeV produce showers of hadrons through inelastic interactions with nuclei of the medium, and the hadronic showers are detected.

Neutrinos are notoriously difficult to detect because of their extremely weak interactions with matter. For their detection, one depends upon charged current reactions of the form $\nu_x + n(or\ p) \rightarrow x^{\mp} + p(or\ n)$, where the subscript x on the left side denotes the neutrino flavor, and $x^{\mp}$ on the right side denotes the charged lepton associated with the flavor x. (Similar reactions hold for antineutrinos, also.) Inelastic neutrino reactions on nuclei can also be used because these produce charged secondaries. The charged particles produced are detected by the methods for the detection of charged particles. Detection of neutrinos requires huge targets and high neutrino fluxes to have any significant number of interactions. Another technique, called the *missing energy* method, is used to determine that a neutrino has been produced. This method was used in the discovery of the W-bosons produced at the $Sp\bar{p}S$ collider at CERN. The detector surrounding the production location had full coverage in solid angle so that the energy and momenta of all the produced particles except the neutrino were measured. With this full coverage, the fact that a neutrino is emitted can be deduced from the missing energy and the direction of the missing momentum needed to balance energy and momentum.

Before we go into the details of the detectors, we describe briefly the physical processes initiated by charged particles and photons in media. A good understanding of these basics is very important for the successful construction and operation of the detectors.

Ionization Energy Loss and the Bethe-Bloch Equation

The fundamental expression for ionization loss was derived a long time ago, and goes under the name of *Bethe-Bloch equation* [11,12]. The mean rate of energy loss per cm, for a particle of momentum p and mass m in material of density ρ, is represented by $-\frac{dE}{dx}$ and given by,

$$-\frac{dE}{dx} = z^2 C \frac{Z}{A} \frac{\rho}{2\beta^2} \left[\ln \frac{2m_e \beta^2 \gamma^2 E_{max}}{I^2(Z)} - 2\beta^2 \right],$$

where z is charge, in units of electron charge, of the incident particle; Z and A are the atomic number and atomic weight of the atoms of the medium respectively; $\beta = p/\sqrt{p^2 + m^2}$ is the velocity of the incident particle in units of the velocity of light (taken as 1 here) and $\gamma^2 = 1/(1-\beta^2)$; m_e is the rest mass energy of the electron ($c = 1$); E_{max} is the maximum kinetic energy that can be transferred to a free electron in one collision; and $I(Z)$ is the average ionization energy of an atom of atomic number Z. From energy and momentum conservation, the expression for E_{max}, when the incident particle with mass m and momentum p collides with an electron of mass m_e, is $E_{max} = 2m_e \beta^2 \gamma^2 / [1 + (m_e/m)^2 + 2m_e\gamma/m]$. As to other symbols in the expression, $C = 4\pi N r_e^2 m_e$, where N is the Avogadro's number and $r_e = \alpha/m_e$, is the classical electron radius, where $\alpha = \frac{1}{137}$ ($c = 1$). For $I(Z)$, Bloch [12] suggested the simple expression $I(Z) = I_H Z$, where I_H is the ionization energy of the hydrogen atom; many efforts have been put in by various individuals to get better values for this quantity since Bloch's original suggestion. Because this quantity occurs only in the logarithm, this approximation does not cause a serious error. A more serious correction arises from the fact that in a medium other than a gas, the particles of the medium cannot be considered as isolated for all but the closest collisions. The correction is called *density effect* and was first quantitatively evaluated by Fermi [13]. The correction can be expressed as a modification of the square bracket in the above expression for average energy loss by subtracting a term δ from it. Since Fermi's work, the density correction has been refined by many workers; we draw attention here to the work of Sternheimer [14]. The above expression for the average energy loss by ionization, corrected for the density effect, is found to be valid for most charged particles (other than the electron) over a wide range of energies, from a few MeV to several GeV. *Stopping power* is another name given to this function.

We should note that the stopping power is a function only of β of the incident particle, varying roughly as $1/\beta^2$ for low β and increasing as the logarithm of γ for high β. The minimum of the stopping power function (rather a broad minimum) occurs between these $\beta\gamma$ values corresponding to a few times (p/m). The range R for a particle which suffers loss of energy through ionization only, is obtained by performing the integral $R = \int_0^E dE/[dE/dx]$. It roughly represents the distance the particle travels before it is stopped. The concept of range is rather limited in its application. Fluctuations in energy loss affect the range leading to considerable variations in value. These fluctuations give rise to what are called *straggling* effects.

In the above discussion, we did not include electrons as incident particle. The region of energies over which electrons lose energy by ionization is rather limited, the upper limit being tens of MeV. Above this energy, radiative loss is the dominant energy loss. Even in the energy range in which energy loss by ionization is to be considered, the above expression has to be modified for two effects. First, the maximum kinetic energy given by the electron to an electron in the medium is only $(1/2)m_e(\gamma-1)$, and second, the effects of the identity of the two particles has to be taken into account. We do not give the expressions here, but refer to the article on stopping powers of electrons and positrons [15].

Effects of Multiple Scattering of Charged Particles

When charged particles pass through matter, they are scattered by the electric fields of the nuclei and electrons present in the matter. The path of the particle, instead of being perfectly straight, is a zig-zag one, characterized by a large number of small angular deflections. These angular deflections represent the effect of multiple scattering the charged particle suffers. Molière studied the distribution of scattering angles in Coulomb scattering. He found that for small scattering angles, the distribution is Gaussian, with mean scattering angle zero, and a width which depends on the charge z, the momentum p of the particle, and the properties of the medium. The width is characterized by the root mean square of the projected scattering angle Θ^p_{rms}, given by

$$\Theta^p_{rms} = \frac{13.6 \text{ MeV}}{\beta p} z \sqrt{\frac{x}{X_0}} [1 + 0.038 \ln x/X_0],$$

where β is the velocity of the particle in units of the velocity of light (recall $c = 1$), x is the distance traversed in the medium, and X_0 is the radiation length in the medium. Here, $X_0 = [A/(4\alpha\rho N Z^2 r_e^2 \ln 183 Z^{-1/3})]$, where Z and A are the atomic number and atomic weight of the atoms of the medium, respectively, α is the fine structure constant, r_e the classical electron radius, ρ the density, and N the Avogadro number.

It is clear that multiple scattering effects put a maximum limit on the momentum of a particle measured by deflecting it in a magnetic field. The higher the momentum of the particle, the smaller its deflection. Due to multiple scattering effects, the deflection angles less than Θ^p_{rms} do not make sense. Corresponding to this root mean square angle, there is a maximum momentum that can be determined by using the bending in a magnetic field. To minimize multiple scattering effects, it pays to have as little material as possible.

Energy Loss by Radiation: Bremsstrahlung

Highly relativistic charged particles lose energy predominantly through radiation emission. The charged particles deflected by the Coulomb field of nuclei emit some of their kinetic energy as photons. This process is called *bremsstrahlung*. The cross section for the bremsstrahlung process was first calculated by Bethe and Heitler, hence called the *Bethe-Heitler cross section* [16]. The energy loss by bremsstrahlung by a relativistic particle of mass m, charge ze, and energy E is given by,

$$-\frac{dE}{dx} = 4\alpha\rho\frac{N}{A}Z^2 z^2 (\frac{m_e}{m})^2 r_e^2 E \ \ln(183 Z^{-1/3}),$$

where α is the fine structure constant, Z and A are the atomic number and atomic weight respectively of the particles of the medium, and r_e is the classical electron radius α/m_e. In the above, only the effect of the nuclear Coulomb field has been taken into account. There are also Z atomic electrons. Interactions with these also give rise to bremsstrahlung. A simple way to correct for these effects is to replace the Z^2 factor in the above with $Z^2 + Z = Z(Z+1)$. Below we will assume that this has been done. (More sophisticated ways of taking into account screening effects of the nuclear Coulomb field and of the atomic electrons are available, but we do not enter into them here.)

We note that, unlike ionization energy loss, radiation energy loss varies linearly with the energy of the particle and inversely as the square of the mass of the particle. Hence for heavy particles, radiation energy loss is negligible unless the particle is extremely relativistic. For light particles such as the electron, radiation energy loss becomes very important at energies of the order of tens of MeV.

We may rewrite the expression for the radiation loss as

$$-\frac{dE}{dx} = \frac{E}{(m^2/m_e^2)X_0} \text{ with } X_0 = \frac{A}{4\alpha\rho N Z(Z+1)r_e^2 \ln 183 Z^{-1/3}}.$$

The quantity X_0 is the *radiation length*. The radiation length is usually given for electrons; for other particles of mass m, the effective radiation length has a further factor m^2/m_e^2. Integrating the equation for the radiation energy loss, we have $E(x) = E_0 \exp\{-x/[(m^2/m_e^2)X_0]\}$; that is, the energy of the particle, with initial energy E_0, after passing through a length x in the medium, is exponentially reduced to $1/e$ of its initial value in a distance equal to the effective radiation length for the particle of mass m. If the medium in which the particle travels has a mixture of different elements, then X_0 is obtained from, $X_0^{-1} = \sum_i f_i/X_{0,i}$, where f_i is the fractional abundance of element i and $X_{0,i}$ is the radiation length for element i. Sometimes, the thickness traversed in the medium is measured in g cm^{-2}, obtained by multiplying the thickness in cm by the density ρ of the medium.

At highly relativistic energies, a process related to bremsstrahlung is the direct production of electron-positron pairs. This can be thought of as the process in which the bremsstrahlung photon is virtual and converts into the electron-positron pair. This has to be added to the real radiation loss to get the total loss of energy.

There is an energy called the *critical energy* E_c, at which the energy loss due to ionization equals the loss due to radiation. The value of E_c for electrons is approximately given by $E_c \simeq (550/Z) MeV$, valid for $Z \geq 13$. For muons, this critical energy is scaled by $(m_\mu/m_e)^2$. In Cu ($Z = 29$) for muons, this is about 810 GeV.

This concludes the discussion of the physical processes involved with charged particles.

Physical Processes with Photons

We now proceed to a discussion of the physical processes which affect photons in the medium. The photon, unlike charged particles, either suffers an absorption in an atom of the medium or suffers a large angle scattering usually accompanied by a change of energy for the photon. To the former category belong processes called *photoelectric effect* and *pair production*, while Compton effect belongs to the latter category. In photoelectric effect, the photon is absorbed by an atom of the medium and an electron is ejected from the atom. In pair production, which is only possible for photons of energy greater than twice the rest mass energy of the electron (1.1 MeV), the photon disappears and produces an electron-positron pair in the neighborhood of the nucleus of an atom of the medium. Compton effect involves the scattering of photons on the (nearly) free electrons in the medium.

Each of these physical effects is characterized by a cross section for the process. The calculations of the cross sections for each of these fundamental physical processes have been carried out using quantum mechanics. These cross sections have characteristic behaviors depending on the property of the atoms of the medium and on the energy of the photon. Once these are known, one can calculate how a beam of photons is reduced in intensity as the beam propagates through the medium.

The intensity of a photon beam $I(x)$, at a location x in the medium, is related to its initial intensity I_0, through the relation $I(x) = I_0 e^{-\mu x}$, where μ is called the *mass absorption coefficient*. If x is measured in g cm^{-2}, then μ, having the dimension of cm^2 g^{-1}, is related to the cross sections by $\mu = \frac{N}{A} \sum_i \sigma_i$, where N is Avogadro's number, A is the atomic weight, and σ_i is the cross section for a process labeled i. (If, however, x is to be in *cm*, μ must be multiplied by the density of the medium and will be given in units of cm^{-1}.) It turns out that μ is a strong function of the photon energy. For low energies of photons, 0.01 MeV to 1 MeV, photoelectric effect is the dominant effect. For medium energies, in the range of 1 MeV to about 6 MeV, Compton effect is the dominant process. For energies higher than about 6 MeV, the contribution from pair production overtakes that due to Compton effect and becomes the dominant one.

Atomic Photoelectric Absorption

The cross section for the absorption of a photon of energy E by a K-shell electron in an atom is large and contributes about 80% of the total cross section. Clearly, the ejection of the electron from the atom is possible only if E is greater than the binding energy of the K-electron. For energies not too far above the K-shell binding energy, the photoelectric cross section σ_γ^K has the form

$$\sigma_\gamma^K = \frac{4\sqrt{2}}{\epsilon^{7/2}} \alpha^4 Z^5 \sigma_\gamma^{el},$$

where $\epsilon = (E/m_e)$, and σ_γ^{el} is the Thomson elastic scattering cross section for photons, equal to $(8\pi/3)r_e^2$. The Thomson cross section has the numerical value 6.65×10^{-25} cm^2. For $\epsilon \gg 1$, the energy dependence is a much less steep function and is more like $(1/\epsilon)$. The high power dependence on Z is a characteristic feature of atomic photoelectric absorption.

As a result of the vacancy created by the ejection of the K-electron, an electron from a higher shell may fall into it, emitting a real X-ray photon with an energy equal to the difference in energies of the two shells. It is also possible that this energy is emitted, not as a real photon, but as a virtual photon, and may be absorbed by an electron in the same atom, and this electron might leave the atom with an energy equal to the virtual X-ray energy minus the binding energy of that electron. Such emitted electrons, whose energies are much smaller than that of the ejected K-electron, are called *Auger electrons*.

Scattering of Photons by Quasi-Free Electrons

In media, the contribution to the scattering of gamma rays of energies between 1 MeV and a few MeV comes from the electrons in the atoms. These electrons can be treated essentially as free because their binding energies are small compared with the photon energies. Photon scattering by free electrons was observed and analyzed by Compton and goes under the name of *Compton effect* (see section under "Compton Effect"). The expression for the total cross section for the scattering of the photon by a free electron is called *Klein-Nishina formula* [17]. It is

$$\sigma_{KN} = 2\pi r_e^2 \left[\frac{1+\epsilon}{\epsilon^2} \left\{ \frac{2(1+\epsilon)}{1+2\epsilon} - \frac{1}{\epsilon} \ln\left(1 + 2\epsilon\right) \right\} \right. \\ \left. + \frac{1}{2\epsilon} \ln\left(1 + 2\epsilon\right) - \frac{1+3\epsilon}{(1+2\epsilon)^2} \right].$$

Here, r_e is the classical electron radius, and $\epsilon = E_\gamma/m_e$ is the ratio of the energy of the gamma ray E_γ in units of the electron mass energy m_e (with $c = 1$). This is the cross section for one electron; for gamma ray scattering from an atom, we must multiply the above expression by Z, as there are Z electrons per atom, and they contribute incoherently to the scattering $\sigma_{KN}^{atom} = Z\sigma_{KN}$.

For high gamma ray energies, ϵ is large, and this cross section drops like $\ln \epsilon/\epsilon$. Energy and momentum conservation in the photon-electron collision leads to an expression for the energy of the scattered photon E'_γ in terms of the energy of the incident photon E_γ and the angle θ_γ, by which the gamma ray is scattered in the laboratory:

$$\frac{E'_\gamma}{E_\gamma} = \frac{1}{1 + \epsilon(1 - \cos\theta_\gamma)}.$$

The back-scattered gamma ray ($\theta_\gamma = \pi$) has a minimum energy given by $E'_\gamma = E_\gamma/(1 + 2\epsilon)$. In this circumstance, the scattered electron receives the maximum energy. The angle of scattering of the electron with respect to the initial direction of the photon in the laboratory has a maximum value $\pi/2$.

Production of Electron-Positron Pairs by Gamma Rays

The threshold energy of the gamma ray at which electron-positron pair production is possible in the Coulomb field of a nucleus is $2m_e$ (taking $c = 1$). For photon energies in the interval $\frac{1}{\alpha Z^{1/3}} > \epsilon \gg 1$, the nuclear Coulomb field is not shielded by the atomic electrons, and the expression for the pair production cross section per atom is

$$\sigma^\gamma_{e^+e^-} = 4\alpha r_e^2 Z^2 \left(\frac{7}{9} \ln 2\epsilon - \frac{109}{54} \right).$$

When there is complete shielding, $\epsilon \gg \frac{1}{\alpha Z^{1/3}}$, the cross section becomes

$$\sigma^\gamma_{e^+e^-} = 4\alpha r_e^2 Z^2 \left(\frac{7}{9} \ln 183 Z^{-1/3} - \frac{1}{54} \right).$$

It is found that for very high photon energies, the pair production cross section becomes independent of the photon energy and reaches a value which can be written as

$$\sigma^\gamma_{e^+e^-} \approx \frac{7}{9} \frac{A}{N} \frac{1}{X_0},$$

where X_0 is the radiation length in g cm^{-2}, A the atomic weight, and N Avogadro's number.

Just as in the bremsstrahlung process where the contribution from atomic electrons gave a modification of Z^2 to $Z(Z+1)$, in pair production also, the contribution from the atomic electrons can be taken into account by the same procedure. In all the above pair production cross section formulae, the multiplying factor Z^2 should really be $Z(Z+1)$.

The total photon absorption cross section is the sum of the cross sections for the photoelectric, Compton, and pair production processes. Other photon processes, such as nuclear reactions induced by photons, photon scattering by nuclei, etc., have rather small cross sections and can be completely ignored in connection with processes of the detection of photons.

This ends the discussion of physical processes initiated by photons and which are relevant with respect to detectors of photons.

Energy Loss by Strong Interactions

If the charged particles we considered above are hadrons, then in addition to the electromagnetic interactions contributing to their energy loss in a medium, their *strong interactions* may also contribute to the energy loss. The total cross section for the strong interaction of a hadron with the particles of the medium is made up of elastic and inelastic parts. The inelastic parts are a reflection of the high number of secondary particles that can be produced in a collision at high energies.

We can define an average absorption coefficient μ_{abs}, so that the absorption of hadrons in passing a path length x through matter is described by $I = I_0 \exp(-\mu_{abs}x)$, where I_0 is the initial number of hadrons and I the number surviving after passing through a thickness x of matter. The coefficient μ_{abs} is related to the inelastic cross section by $\mu_{abs} = \frac{N}{A}\sigma_{inel}$, where N is Avogadro's number, A is the atomic weight, and σ_{inel} is the inelastic cross section. The dimension of μ_{abs} is cm^2 g^{-1}, if thickness is measured in g cm^{-2}. If we multiply μ_{abs} by the density ρ of the material, we get it in units of cm^{-1}, in which case the thickness traversed is measured in cm. The inverse of μ_{abs}, λ_{abs} (of dimension [cm]), represents a mean absorption distance in which the number of particles is on the average reduced to $1/e$ of the initial value in passing through the medium.

We can also define a total absorption coefficient μ_{tot}, which has contributions from both the elastic and inelastic processes. The σ_{inel} will be replaced by $\sigma_{tot} = \sigma_{el} + \sigma_{inel}$. Clearly, $\mu_{tot} > \mu_{abs}$, as $\sigma_{tot} > \sigma_{inel}$.

The magnitude of the strong interaction total cross sections is usually expressed in barns. The unit is 1 barn $= 10^{-24}$ cm^2. At high energies, the total cross sections grow typically from 0.04 barns in water ($Z = 1, A = 1$) to about 3.04 barns in uranium ($Z = 92, A = 238$). Correspondingly, λ_{tot} varies from about 43 g cm^{-2} in water to 117 g cm^{-2} in uranium. For materials with $Z \geq 6$, λ_{tot} is typically larger than the radiation length X_0.

This ends the discussion of all the physical processes which are particularly relevant for detector development. We describe below the various detectors (by no means an exhaustive list) which have played a large role in the discoveries of new particles and high energy phenomena.

Zinc-Sulphide Screen

The earliest detectors depended on direct optical observations. Examples are the detection of X-rays through the observation of the blackening of photographic films and the detection of α particles with screens coated with zinc-sulphide. Faint flashes of light, called *scintillations*, are produced when the α particles impinge on the screen. The scintillations were viewed with the eye and counted, obviously a laborious and tiresome task. Rutherford and his collaborators used this method to study the scattering of alpha particles by thin foils of various materials, including gold. The famous Rutherford's scattering law was established using this simple equipment.

Cloud Chamber

Another early detector was the Wilson cloud chamber. C. T. R. Wilson made use of the fact that when a charged particle goes through a chamber containing a supersaturated vapor, the vapor has a tendency to condense on the ions created along the track of the particle and form droplets on them. If the chamber is illuminated, the droplets become visible and can be photographed with stereo cameras to get a three-dimensional view of the particle's track. By placing the cloud chamber in a uniform magnetic field, charged particle tracks curve in the magnetic field depending on the sign of the charge they carry. The momentum of the charged particle can then be determined by making a measurement of the curvature of the track and using the laws of charged particle motion in the magnetic field. Counting the number of droplets along the track can be used to give a measure of the particle's energy. Knowledge of the energy and the momentum of the particle allows one to calculate its mass. The positron and the muon were two of the particles discovered in cosmic rays studies using the cloud chamber. These were profound discoveries; the discovery of the positron established Dirac's idea of antiparticles, and the discovery of the muon established that there indeed were particles in the mass range between that of the electron and that of the proton as suggested by Yukawa.

Bubble Chamber

Another detector similar to the cloud chamber is the bubble chamber [18]. In this chamber, the medium is a liquefied gas; hydrogen, deuterium, neon, and organic liquids have been used. The liquefied gas is contained under pressure close to its boiling point in a chamber with extremely smooth inner walls and fitted with pistons. If one expands the chamber with the liquid in such a state, the pressure is lowered, and the liquid goes into a superheated liquid phase. When it is in this phase, if a charged particle enters it, bubbles are formed tracking the ionization left by the particle. The bubbles grow in size and the growth can be stopped at any time by stopping the expansion of the chamber. The size of the resulting bubbles determines the spatial resolution of the

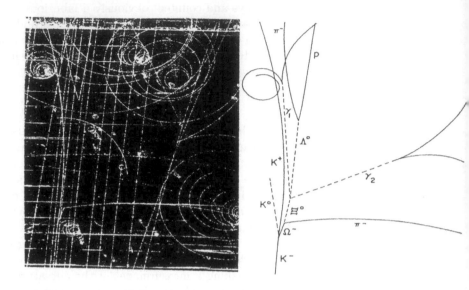

Figure 3.1: A bubble chamber picture of Ω^- decay. (Figure from V. Barnes et al., *Physical Review Letters* 12, 204, 1964; reproduced with permission from N. Samios and the American Physical Society © 1964.)

bubble chamber. High resolution bubble chambers with resolutions of a few μm have been operated. The bubbles along the tracks of charged particles are photographed with several cameras to help in the three-dimensional reconstruction of the track. More recently, development of holographic readout systems enables three-dimensional reconstruction of events with high spatial resolution. The bubble density along the tracks is proportional to the ionization energy loss along the tracks.

The bubble chamber has been used at accelerators by adjusting the timing of its entry into the superheated phase with the time of entry of the particle beam. At moderately high energies, the interactions can produce events with a number of secondary particles, all of which can be recorded with high spatial resolution. Bubble chambers have usually been operated in magnetic fields. An example of a bubble chamber picture by which the Ω^- was discovered is shown in Figure 3.1 above.

The magnetic field enables a determination of the momentum of the particles by measuring the radius of curvature of the track and a determination of the sign of the charge they carry. Many hadronic resonances were discovered using such chambers at Brookhaven National Laboratory, Lawrence Berkeley Laboratories, CERN, and other laboratories, in the mid 1960's. These chambers are also useful in the study of rare complex events arising, for example, in neutrino interactions.

There are some limitations in the use of bubble chambers for very high energy experiments, which must be recognized. The repetition rate of this detector being low, it will not cope with the high event rates expected in many high energy experiments. Triggering the chamber by depending on the time of entry of the particle into the chamber is not possible because the lifetime of the ions is not long enough; the superheated phase has to be prepared prior to the arrival of the beam into the chamber. The basic shape of the chamber does not lend itself for use around high energy storage ring experiments where the entire solid angle around the collision region must be covered. Also, there is not enough mass in the detector to completely contain all the high energy particles and make measurements of the total energy deposited. Identification of high energy particles, based on measurements of specific ionization loss alone, may not work, because there may be no good separation between them. For this, one also needs the momentum of the particle. A good determination of the particle momentum depends on how well the radius of curvature of the path in a magnetic field can be determined, which, in turn, depends on how long a path length we have to work with. For very high momenta, the required path length may exceed the dimensions of the chamber. Despite these limitations, small bubble chambers have been put to use, because of their high spatial resolution, as vertex detectors, in experiments with external targets. These have enabled measurements of small lifetimes of particles as low as 10^{-14} s.

Spark Chamber

A commonly used track detector is the spark chamber. It consists of a number of parallel metal plates mounted in a region filled with a mixture of gases, helium and neon. The plates are so connected that every alternate plate is connected to a high voltage source, while the other plates are connected to ground. The high voltage source to every second electrode is triggered by a coincidence between two scintillation counters, one placed above the set of plates, and the other below them. When a charged particle goes through the system of plates, it ionizes the gas mixture in the space between the plates. The high voltage is so chosen that sufficient gas multiplication occurs along the track of the particle to get a spark discharge between the plates. The spark discharges are clearly visible and can be photographed. More than one camera must be used if we want the data for construction of the track in three-dimensional space. (There are also electronic ways of recording data from spark chambers. In this case the plates must be replaced by a set of wires, and one must locate which wires are fired.) Before the next event can be recorded, a clearing field must be applied to clear away the ions from the previous discharge. This causes a dead time which can be of the order of several milliseconds.

Streamer Chambers

A streamer chamber is a rectangular box with two planar electrodes, the space between which is filled with a suitable gas mixture. A sharp high voltage pulse of short rise time and small duration is applied to the electrodes after the passage of a charged particle, approximately parallel to the electrodes. Every ionization electron released along the track of the charged particle initiates an avalanche in the intense uniform electric field between the electrodes. The electric field being only of short duration (several nanoseconds), the avalanche formation terminates when the field decays. The avalanche multiplication factor can reach as high as 10^8. Many of the atoms in the avalanche are raised to excited states and emit radiation. The result is that one sees, from a side view, luminous streamers in the direction of the electric field all along the particle's trajectory. On the other hand, if we could view them through the electrode, the streamers would look like dots, affording the possibility of increased spatial resolution. To make this view possible, the electrode is made of a fine grid of wires rather than plates. Streamer chambers are capable of giving pictures of particle tracks of superior quality. There are special methods to improve the resolution obtainable with streamer chambers by reducing the diffusion of particles in the interval between the passage of the charged particle and the onset of the electric field, but we do not go into these details. More details may be found in reference [19].

The successful operation of the streamer chamber depends upon the generation of the electric field with the special characteristics mentioned above. Such fields are generated with a Marx generator. It consists of a bank of capacitors which are charged in parallel, each to a voltage V_0. Then these are arranged, by triggering spark gaps, to be connected in series so that the voltage across the bank of capacitors is nV_0, where n is the number of capacitors in the bank. Through a special transmission line, the high voltage signal across the capacitor bank is transmitted to the electrodes of the streamer chamber without any losses.

For fast repetitive operation of the streamer chamber, the chamber must be cleared of all the electrons from the previous event. (The positive ions are not a problem because of their low mobility.) Because a large number of electrons is present in the streamers, clearing them with clearing fields will take too long, and the dead time of the chamber will be too long. To solve this problem, an electronegative gas, such as SF_6 or SO_2, is added. These have enormous appetite for electrons. The electrons attach themselves to these molecules in very short times. Such gases are called *quenchers* and allow recycling times of the order of a few hundred milliseconds.

Scintillation Counters

These counters represent considerable development over the simple zinc-sulphide scintillation screen of Rutherford's alpha particle scattering days. The principle of operation of a scintillation counter may be outlined as follows. The energy loss suffered by a particle impinging on a scintillator substance triggers the emission of light. This light is delivered, by using suitably constructed light guides, to a device such as a photomultiplier, which, together with suitable electronics, records the light as an electrical signal and, hence, the particle that caused the emission of light.

Scintillating materials may be inorganic crystals, such as NaI, CsI, or LiI, doped with some materials to produce activation centers. They may also be some organic compounds in the form of solids or liquids. In inorganic scintillators, the conversion of the energy of the incident particle into light is due to the energy levels in the crystal lattice. In organic scintillators, the process is different. Organic scintillators are usually mixtures of three compounds. Two of these components, which are active, are dissolved in an organic liquid or mixed with some organic material to form a plastic. One of the active components is such that, the energy loss due to a particle incident on it triggers the emission of fluorescent radiation in the ultraviolet. The ultraviolet light is absorbed in a short distance in this mixture. To get the light out, the other active component is added. Its function is to be a *wave length shifter*; that is, it absorbs the ultraviolet light and emits it again at a longer wave length in all directions. This second active compound shifts the wave length so that it overlaps the peak of the sensitivity of the photomultiplier device. It is thus clear that liquid or solid scintillators can be built in any shape to suit the experimental need.

The lifetimes of excited levels in the scintillating material essentially determine the decay time for the light from the scintillator. Inorganic scintillators have decay times of the order of microseconds, while for organic scintillators, they are much shorter, typically of the order of nanoseconds. The organic scintillators are useful as triggering and time measurement devices. Scintillators, both inorganic and organic, are used as components in high energy physics experiments.

Cherenkov Detectors

When a particle travels through a medium such that its velocity v is greater than the velocity of light in the medium c/n, where n is the refractive index of the medium, radiation is emitted with certain special characteristics. This was predicted by Cherenkov [20] a long time ago. The physical mechanism responsible for the emission of radiation is due to the rapidly varying (time dependent) dipole polarization of the medium induced by the fast moving particle. When the velocity of the particle is greater than the velocity of light in the medium, the induced dipoles tend to line up along the direction of the particle motion and produce a directed coherent radiation field. When the velocity of the particle is less than the velocity of the light in the medium, the induced dipoles are randomly oriented, and there is no radiation. The emitted radiation is confined to a cone around the track of the particle with the cone angle θ being given by $\cos\theta = 1/(n\beta)$, where β is the velocity of the particle in units of the velocity of light. For real values of θ, since $|\cos\theta| \leq 1$, $n\beta \geq 1$. There is a threshold value of $\beta = \beta_{thr} = 1/n$, at which $\cos\theta = 1$. Thus, at the threshold value of β, the Cherenkov radiation is emitted exactly in the forward direction $\theta = 0$. Further, because the medium is dispersive, n is a function of the frequency of the radiation, ω, and only those frequencies will be emitted for which $n(\omega) > 1$. Corresponding to $\beta_{thr} = 1/n$, the threshold energy for a particle of mass m, to emit Cherenkov radiation, is given by $E_{thr} = \gamma_{thr} m$, with $\gamma_{thr} = 1/\sqrt{1 - \beta_{thr}^2}$.

The refractive indexes for water, plexiglass, and glass, are 1.33, 1.48, and 1.46–1.75, respectively. Correspondingly, the γ_{thr} for these media are, 1.52, 1.36, 1.22–1.37 respectively. These do not correspond to very high values of the energy. Large γ_{thr} values can be obtained for n close to 1. This is the case for gases. Helium, CO_2, and pentane at STP have $(n-1)$ values, 3.3×10^{-5}, 4.3×10^{-4}, and 1.7×10^{-3}, respectively, with corresponding γ_{thr} values about 123, 34, and 17, respectively. A material called *silica-aerogel* has been developed, which gives $n - 1$ in the range from 0.025 to 0.075, and is useful to bridge the gap between the gases and the transparent liquids and solids.

The expression for the number of photons dN radiated as Cherenkov radiation in a path length dx of the medium is

$$\frac{dN}{dx} = \alpha z^2 \int_{\beta n > 1} d\omega \left(1 - \frac{1}{\beta^2 n^2}\right),$$

where α is the fine structure constant $(1/137)$, and z is the number of units of charge on the particle. From this the number of photons emitted in a path length L, in the frequency range between ω_2 and ω_1, is

$$N = z^2 \alpha L \int_{\omega_2}^{\omega_1} \sin^2 \theta \, d\omega.$$

We can evaluate this integral approximately by assuming the integrand is essentially constant as a function of ω. Then, the number of photons emitted becomes

$$N \approx z^2 \alpha L \sin^2 \theta (\omega_1 - \omega_2).$$

If we consider the visible range with wavelengths from $\lambda_1 = 400$ nm to $\lambda_2 = 700$ nm, evaluating the above for a singly charged particle $z = 1$, we get $\frac{N}{L} \approx 490 \sin^2 \theta$ cm^{-1}. The amount of energy loss by Cherenkov radiation does not add a significant amount to the total energy loss suffered by the charged particle.

There are two ways that Cherenkov detectors have been put to use. One of these is as a threshold detector for mass separation. The other is for velocity determination and is called a *differential Cherenkov detector*.

First, we consider the threshold detector. Suppose there are two particles of masses m_1 and m_2 (and $m_2 > m_1$) and let us suppose that the refractive index is such that the particle with mass m_2 does not produce Cherenkov radiation; that is, β_2 is slightly less than $1/n$. (Let us take $\beta_2 \simeq (1/n)$.) Only the particle of mass m_1 produces radiation. The number produced is proportional to

$$\sin^2 \theta = [1 - 1/(\beta_1^2 n^2)] \simeq [1 - \beta_2^2 / \beta_1^2].$$

This simplifies to

$$\sin^2 \theta = \frac{1}{\beta_1^2} \left[\frac{1}{\gamma_2^2} - \frac{1}{\gamma_1^2}\right].$$

If the two particles have the same momentum, the expression in the square bracket in the equation just above is $(m_2^2 - m_1^2)/p^2$, so that

$$\frac{N}{L} \approx 490 \frac{m_2^2 - m_1^2}{p^2} \text{ cm}^{-1}.$$

If the quantum efficiency of the photomultiplier for the photons is ϵ, the number of photo electrons N_e is

$$N_e \simeq 490 \epsilon L \frac{m_2^2 - m_1^2}{p^2} \text{ cm}^{-1}.$$

To get N_e photoelectrons, we need a length L, where

$$L = \frac{N_e p^2}{490\epsilon(m_2^2 - m_1^2)} \text{ cm.}$$

We see that the length L needed to separate particles of momentum p goes up as p^2. Putting in numbers, for separation of kaons ($m_K = 494$ MeV) and protons ($m_p = 938$ MeV), at $p = 10$ GeV with $\epsilon = 0.2$, to get $N_e = 10$, we need a minimum length in the medium of $L = 16.05$ cm. All this calculation assumes that we can find a material with the right refractive index such that the proton of 10 GeV is just below threshold for the emission of Cherenkov photons. It may be difficult to achieve this in practice. What is done in practice is to use two threshold detectors. Suppose one uses silica-aerogel and pentane. The kaon will give a Cherenkov signal in both, while the proton will give a signal only in the silicon-aerogel. Comparing these rates can tell us the relative numbers of kaons and protons.

The differential Cherenkov detector accepts only particles in a certain velocity interval. Suppose we have a transparent medium of refractive index n in which all particles with velocity above a minimum velocity β_{min} give Cherenkov radiation. Let the Cherenkov photons go from the dense medium into a light guide containing air, which guides the photons into the photomultiplier. The Cherenkov photons from the minimum velocity particles are emitted in the forward direction ($\theta = 0$) and they pass through the dense medium-air interface. As the velocity of the particles increases, the emission is confined to an angle θ given by $\cos^{-1} 1/n\beta$. At a value of $\beta = \beta_{max}$, the angle θ becomes equal to the critical angle for total internal reflection at the interface. These photons will not be seen by the photomultiplier. This is one way to construct a Cherenkov detector which accepts particles with β in the range $\beta_{min} < \beta < \beta_{max}$. Another way we could do this would be to have an optical system which focuses the conical emission of Cherenkov light and at the focus have a diaphragm which lets only the light confined to a small angular range into the photomultiplier. If we change the radius of the diaphragm, we are in effect looking at particles of velocity in different intervals.

Differential Cherenkov detectors with correction for chromatic aberrations have been developed having resolution in velocity $(\Delta\beta/\beta) \approx 10^{-7}$ and are called *DISC detectors*. With these detectors, pions and kaons can be separated at several hundred GeV.

At storage ring colliders, it is not possible to use differential Cherenkov detectors. Storage ring detectors have to be able to detect particles coming in all 4π solid angle directions. Here one uses what are called *RICH* (Ring Imaging Cherenkov) *detectors* (refer to Figure 3.2). Around the

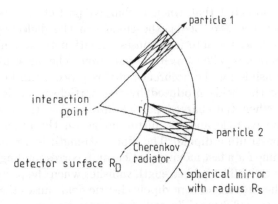

Figure 3.2: A schematic diagram showing the use of Ring Imaging Cherenkov detector at a collider. (Reproduced with permission of Cambridge University Press, Figure 6.14 from *Particle Detectors* by C. Grupen. Permission from C. Fabjan is also gratefully acknowledged.)

interaction point, a spherical mirror of radius R_{sph} is provided whose center of curvature coincides with the interaction vertex. Another concentric spherical surface of radius $R_{det} < R_{sph}$ serves as the detector surface. The space between these two spherical surfaces is filled with a Cherenkov radiator. As soon as a particle enters the radiator medium from the inner sphere, it emits Cherenkov photons within a cone of angle θ around the direction of the particle.

By reflection from the outer spherical mirror, a ring is formed on the detector surface whose radius r can be measured. Since the focal length f of the spherical mirror is $R_{sph}/2$, $r = f\theta = R_{sph}\theta/2$, this gives $\cos\theta = \cos 2r/R_s$. For Cherenkov radiation, $\cos\theta = 1/(n\beta)$, so we can immediately know the particle velocity, $\beta = \frac{1}{\cos(2r/R_s)}$. Thus a measurement of the radius of the Cherenkov ring on the detector gives the particle velocity. The Cherenkov ring detector must not only detect photons but also measure their coordinates in order to determine r. For this purpose, a *multiwire proportional chamber* (MWPC—see below for its description), with some photon sensitive mixture added to the gas, is installed in the chamber. In order to get a good working detector, a number of other technical problems have also to be solved but we do not go into these details here.

Transition Radiation Detector

It has been observed that when a charged particle passes through a boundary between two media, the change in the dielectric properties of the medium at the interface causes radiation to be emitted. Such emitted radiation is called *transition radiation*. The physical mechanism that is responsible for the occurence of this process can be understood by analyzing the fields produced by the particle when it crosses the boundary. When the charged particle approaches the boundary, the charge of the particle and its mirror image on the other side of the boundary constitute a dipole. The dipole strength is a rapidly varying function of time for a fast moving particle and, correspondingly, the fields associated with it. The field strength vanishes when the particle is at the interface. The time dependent dipole electric field causes the emission of electromagnetic radiation. The radiated energy emitted can be increased by having the charged particle travel through many interfaces created, for example, by sets of foils with intervening air spaces.

The characteristics of the transition radiation emitted have been studied. It was found that the radiant energy emitted increases as the energy of the particle increases. Since the total energy E of a particle can be expressed as $E = \gamma m$, where m is the rest mass energy of the particle (with $\gamma = 1/\sqrt{(1 - \beta^2)}$, where β is the velocity of the particle in units of the velocity of light), for extremely relativistic particles, γ is very large, and the larger the γ, the larger the radiated energy. Because of this property of the transition radiation, it is highly attractive to use it to measure total energies of particles. Further, a frequency analysis of the transition radiation shows that the photon energies are in the X-ray region of the spectrum. The fact that the radiated energy is proportional to γ arises mainly from the increase in the average X-ray photon energy. For an electron of few GeV energy, the average energy of transition radiation photons in a typical radiator is about 20 keV. The angle θ between the direction of emission of the photons and that of the charged particle is found to have an inverse dependence on the γ of the particle. In a radiator made up of a periodic stack of foils separated by air gaps, it is found that there is no transition radiation for $\gamma < 1,000$. When $\gamma > 1,000$, constructive interference effects occur between radiation coming from the various interfaces, and one sees strong radiation. The threshold for a periodic structure radiator is $\gamma = 1,000$.

Transition radiation detectors are made with a set of foils of materials of low Z followed by a MWPC X-ray detector. The foils must have low Z, because the X-ray photons produced suffer strong photoelectric absorption, which varies as Z^5. Low Z foils allow the X-rays to escape out of the foils. Once out, they can be detected using an X-ray detector.

An MWPC filled with xenon or krypton can be used for this purpose. The large Z values for xenon and krypton ensure good absorption for X-rays in the MWPC.

The threshold of $\gamma = 1,000$ in periodic radiators allows one to separate particles at high energies which are otherwise not separable. Consider separating 15 GeV electrons from 15 GeV pions. Both are extremely relativistic. The electrons have $\gamma = 30,000$, while the pions have $\gamma = 111$; through a periodic radiator, the electrons will produce strong transition radiation while the pions will not. Similarly, pion-kaon separation is possible for energies of 140 GeV. At these enregies $\gamma_\pi > 1,000$, while $\gamma_K \simeq 280$, and pions will contribute to transition radiation, while kaons will not.

Nuclear Emulsion Method

The photographic film method has been transformed, with the development of the nuclear emulsion method, for recording tracks of charged particles. A thin glass plate is coated with an emulsion consisting of specially fine grained silver halide crystals (size, $\approx 0.1~\mu m - 0.2~\mu m$) mixed in a gelatin. The thickness of the emulsion coating can be anywhere from a few tens of microns to two thousand microns. The ionization released in the passage of the charged particle in the emulsion leads to a reduction of the silver halide compound to metallic silver along the track of the particle. On photographic development and fixing of the emulsion plate, these metallic silver particles are left while the unexposed silver halide molecules are dissolved away. The end result is that the track of the charged particle is made visible by the presence of metallic silver left along the track. Two requirements on the detector we would want are good spatial resolution and good sensitivity. To get a good spatial resolution with this method, one must start with especially small silver halide grains. On the other hand, if the emulsion is to be sensitive enough to record particles with minimum ionization, the grain size cannot be too small. These competing requirements force a compromise on the grain size. Special nuclear emulsion plates meeting the requirements have been made by the photographic industry and have been used in cosmic ray experiments and in some accelerator experiments. They are very good as vertex detectors having a spatial resolution of 2μ m. Measurements on particle tracks in nuclear emulsions used to involve a lot of scanning labor. With CCD cameras and suitable pattern recognition software, the work with nuclear emulsions can be automated on the computer.

Ionization Chamber

This is a device for the measurement of ionization energy loss of a charged particle or the energy loss suffered by a photon in a medium. If the particle or the photon is totally contained and does not leave the chamber, the chamber measures the total energy of the particle or the photon.

Ionization chambers can be constructed in planar or in cylindrical geometry. In the planar case, the chamber has two planar electrodes which are mounted parallel to the top and bottom walls of a rectangular leak-proof box. A uniform electric field is set up between the two electrodes by connecting them to a voltage supply. The container is filled with a gas mixture, or a noble-gas liquid, or even a solid. This is the medium which is ionized by the passage of a charged particle (or radiation) through it, and the electrons and the positive ions drift in the electric field and reach the appropriate electrodes. The electric field is not strong enough to cause any secondary ionization from the primary ions and electrons. When the primary charges move to the plates, they induce a charge on the capacitor plates, which is converted to a voltage signal and measured. The electrons drift much faster than the positive ions; in typical cases, the collection times for electrons are of the order of microseconds, while for the ions they are in the millisecond range. By suitable choice of the time constant of the circuit, the electron signal amplitude can be made independent of where the electron originated inside the chamber.

In cylindrical geometry, the leak-proof box is cylindrical. One electrode, the anode, is a wire on the axis of the cylinder, and the other electrode, the cathode, is the wall of the cylinder. The space between the anode and the cathode contains the gas mixture, or a noble-gas liquid. The two electrodes are connected to a voltage supply, and a field is created in the space between them. Unlike the planar case, the field here is not uniform, varying as $(1/r)$, where r is the radial distance from the axis of the cylinder. The passage of a charged particle through the region between the electrodes causes ionization, and the electrons drift to the anode and the ions to the cathode. If the radius of the anode wire is not too small, the electric field is not sufficient to create additional ionization from the primary particles. When the drifting particles move toward the electrodes, they induce charges on the electrodes which can be converted to a voltage signal and measured. The signal is mainly due to the electron drifting to the anode and can be shown to depend only logarithmically on the point of origin of the electron. Cylindrical ionization chambers, filled with suitable mixture of gases, have been made, whose pulse duration is of the order of a few tens of nanoseconds.

If a gas mixture or noble-gas liquid is used as the working medium in the ionization chamber, it must not contain any electronegative impurities in it. Otherwise, the chamber does not function properly.

Silicon Microstrip Detectors

If an ionization chamber uses a solid as the working medium instead of a gas, we get a solid state detector. Because the density of a solid relative to a gas is high, a solid state detector can be used to detect particles of much higher energy than in a gaseous ionization chamber.

A solid state detector is obtained by creating a region of intrinsic conductivity between a p-type and an n-type conducting layers in the semiconductor. Such a structure can be made by taking boron-doped silicon (which is p-conducting) and drifting lithium into it. In this way, structures with very thin p and n regions separated by a relatively large intrinsic region can be produced. A typical structure might have thin p-type and n-type regions separated by several hundred micro meters. Strips may be put on the p-type region separated from one another by 20 μm, with a negative potential applied to each. When a charged particle passes through this structure, it creates electron-hole pairs in it. The holes migrate to the strips and induce a pulse, which can be read with suitable electronics. The distribution of charge on the readout strips allows a spatial resolution of the order of tens of micrometers. Silicon microstrip detectors are used as vertex detectors in e^+e^- colliders in close vicinity of the interaction point. If the decay vertex is clearly distinguished from the interaction point, this information can be used to calculate the lifetime of the unstable hadron which decayed. Lifetimes in the pico second range are accessible with silicon microstrip detectors.

If, instead of long strips, the strips are further subdivided into a matrix of pads and each pad is isolated from another, such a setup can be used to analyze complex events. Each cathode pad is read individually, and a two-dimensional picture can be obtained. This type of silicon detector is called a *CCD silicon vertex detector*.

Proportional Counters

If, in the cylindrical ionization chamber, the anode wire is made of very small diameter (or the anode voltage is increased), the field strength in the vicinity of the anode becomes high enough so that secondary ionization takes place. Every primary electron leads to an avalanche of 10^5 to 10^6 secondary electrons. This factor is called the *gas amplification factor* α. If the field strengths near the anode do not get too high, this amplification factor α is a constant, and the output is proportional to the number N of primary electrons. The chamber operated in this regime of electric fields is called a *proportional counter*. The output signal is proportional to the primary ionization deposited by the incoming particle.

The avalanche formation takes place very close to the anode wire, in a region of the order of a few mean free paths of the electrons, a few μm. If one calculates the amplitude of the signal contributions from the electrons and ions, one finds that the ions drifting slowly away from the anode produce a much larger signal than the electrons drifting to the anode. For example, for an anode radius of 30 μm, cathode radius 1 cm, Argon gas filling the chamber, and anode voltage of a few hundred volts, the rise time of the electron pulse is in the nanosecond range, while that of the ions is in the millisecond range. This fact has to be kept in mind in using a proportional counter in any experiment.

To observe the electron signal alone, one has to carry out a differentiation of the signal with a suitable resistance-capacitor combination circuit. Without this special circuit, the chamber cannot be used in places where high rates are involved. An additional factor that must be borne in mind is that the fluctuations in the primary ionization and its amplification by the avalanche process have an adverse effect on the energy resolution achievable with this detector.

Geiger Counter

In the proportional chamber, if the anode voltage is made higher than in the proportional regime mentioned before, a copious production of photons takes place during the formation of the avalanche. This leads to higher probability for producing even more electrons through the photoelectric effect. Photoelectric effect occurs also at points farther away than the location of the primary avalanche. The number of photoelectrons produced per electron in the initial avalanche increases very rapidly due to the contributions of secondary and further avalanches. In such a situation, the signal is not proportional to the primary ionization but depends only on the voltage applied. This mode of operation of the chamber leads to the Geiger counter. In the Geiger regime, the amplitude of the signal is due to some 10^8 to 10^{10} electrons produced per initial electron.

Once a charged particle has passed through the counter, the produced electrons quickly reach the anode, but the ions being much heavier take a long time to go to the cathode. On reaching the cathode, they will cause further ejection of electrons from the cathode, which will initiate the discharge again. To prevent this from happening, one has to choose the resistance R and the RC time constant of the circuit such that the instantaneous anode voltage is reduced below the Geiger threshold until all the positive ions have reached the anode. This time is usually of the order of milliseconds and contributes to the dead time of the counter. Another method that is adopted is called self-quenching. It has been found that vapors of methane, ethane, isobutane, or alcohols, or halogens such as Br_2 added to the gas in the counter act as good quenchers of the discharge. These additions absorb ultraviolet photons so that they do not reach the cathode and liberate further electrons, and the discharge stays located near the anode wire. The dead time of the counter limits the ability of the counter to handle processes occurring at high rates unless special measures are taken.

Multiwire Proportional Chamber (MWPC)

This device is a further development from the proportional counter. It has the capability of being used as a detector of tracks of particles and also measuring their energy loss. It can be used in experimental arrangements where the data rates are expected to be high. MWPC are extensively used in experiments in particle physics.

Instead of just one anode wire as in the proportional counter, the device has a series of anode wires stretched in a plane, between the two cathode planes. Typically, the anode wires are gold-plated tungsten wires of diameters in the range 10 μm to 30 μm, separated from one another by 2 mm and from the cathode planes by about 10 mm. The tension in the wires has to be such that they are mechanically stable against the electrostatic forces between the anode wires.

It has been found that such an arrangement functions as a series of independent proportional counters [21]. The avalanche generation in the multiwire proportional chamber proceeds much as in the proportional counter. Each anode wire has avalanche charges in its immediate vicinity proportional to the initial ionization, and the signal comes from the positive ions drifting to the cathode. The influence of the negative charges near one anode wire on a neighboring anode wire is negligible. If one views the negative pulses induced on the anode wire with some high time-resolution devices, such as fast oscilloscopes or fast analog-to-digital converters (called *flash ADC's*), one gets a series of pulses induced by the different avalanches originating with different initial electrons sequentially drifting into the near vicinity of the anode wire. These pulses have been observed to have a sharp rise time of 0.1 nanosecond and decay in about 30 nanoseconds. With a slower electronic device one will see only the time-integrated pulse.

The spatial resolution for an MWPC with continous cathode planes is about 600 μm. Even then, the position perpendicular to the wire

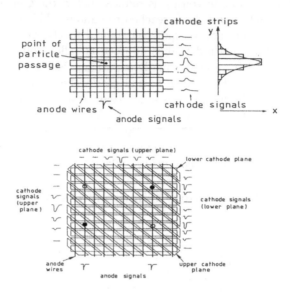

Figure 3.3: A schematic diagram showing how the tracks of two particles are correctly constructed from the information gathered by segmented cathode planes. (Reproduced with the permission of Cambridge University Press from Figures 4.28 and 4.29 in *Particle Detectors* by C. Grupen.)

is known but not the position along the wire. This situation can be improved if the cathode plane is segmented into a number of parallel strips oriented in a direction perpendicular to the direction of the anode wires. (The signal from each strip has to be read electronically, increasing the costs associated with the electronics.) The amplitudes of the pulses in the strips vary as a function of the distance of the strip from the avalanche. The center of gravity of these pulse heights gives a good measure of the avalanche position. In this manner, depending on the number of segments in the cathode, with judicious choice of the setups, spatial resolutions of the order of 50 μm along the anode wires can be obtained.

It turns out that, with only one cathode plane segmented, the information from the anode wires and the cathode strips does not lead to unique specifications of the coordinates from which to construct the tracks; at least two cathode planes are needed. To see this, let us consider just two particles going through the chamber simultaneously (see

Figure 3.3). Two anode wires, say a_1 and a_2, will give signals. The cathode strips also give signals. Suppose it is determined that strips c_1 and c_2 contain the centroids of the pulse amplitudes. Now there is an ambiguity in determining the coordinates of the particles because we could associate c_1 or c_2 with each of a_1 and a_2, giving four possibilities (two filled dots and two open dots in the figure). Only two of these is the correct answer, but which? To resolve this ambiguity, suppose the other cathode is also segmented (referred to as the upper cathode plane in the figure), its strips oriented at 45° angle with respect to the strips of the first cathode (called the lower cathode plane in the figure), its signals also recorded, and centroids determined. The additional information from the second segmented cathode shows that the filled dot locations are picked out rather than the open dots in the figure. Thus, the ambiguity is resolved. Of course, this additional information adds to the cost of electronics.

If the cathode strips are further segmented, so that we have a sequence of pads in each of the cathode planes, and signals from each pad are read and centroids determined, one can construct without ambiguity a number of simultaneous particle tracks passing through the chamber. This additional knowledge comes at further cost associated with the required electronics.

Microstrip Gas Detector

Microstrip gas detectors are MWPC highly reduced in physical dimensions. The reduction in dimensions is possible, because the wires of the MWPC are replaced by small strips which are evaporated on a thin, ceramic, quartz, or plastic substrate. The cathode is also in the form of strips evaporated on to the substrate, the anode and cathode strips forming an alternating structure on the substrate. The anode strips are about $5\mu m$ in size, the cathode strips are about 60 μm, and the anode to anode separation is about 200 μm. The substrate with the anode and cathode structures on it is mounted in a leak-tight box of height a few millimeters and filled with a suitable gas mixture providing the medium for ionization. The cathode strips can be further segmented to allow two-dimensional readout.

Such microstrip gas detectors offer a number of advantages. The dead time is very short, because the cathodes are so close to the anodes, and the positive ions of the avalanche have to drift only a very small distance. They have high spatial resolution. They are excellent devices for use as track detectors in high rate environments.

Planar Drift Chamber

The drift chamber is a further evolution of the MWPC in which the number of anode wires in the chamber can be reduced considerably (without deterioration of spatial resolution) and hence the costs. The time interval between the passage of a charged particle through the MWPC and the creation of a pulse at the anode wire depends on the distance of the passing particle from the anode. This time interval is found to be about 20 ns per millimeter. The design of the drift chamber is based on this principle. A cell of a drift chamber is constructed so that the electrons from the initial ionization first drift in a low field region, of the order of 1,000 V/cm, created by introducing potential wires, and later enter the high field avalanche region around the anode. For many gas mixtures, it is found that the drift velocity is practically independent of the field strength, so that there is a linear relation between distances and drift times. (If needed, one can introduce additional potential wires to make the field constant in the drift region.) Thus, one needs far fewer anode wires than in an MWPC, and/or their spacings can be increased without reduction in the spatial resolution of the chamber. In this chamber, in addition to recording the outputs as in the MWPC, the drift time of the charges is also measured.

Cylindrical Wire Chambers

Cylindrical wire chambers have been developed for use with storage ring colliders. They are capable of providing high spatial resolution for tracks, and have excellent solid angle coverage around the collision region. They are called central detectors because they are located immediately surrounding the beam pipe, and there may be other detectors outside these detectors.

The beam pipes are on the axis of the cylinder, taken as the z axis. Around this axis, between two end-plates, the anode and potential wires are stretched, so as to form concentric cylinders around the axis. One cylinder in cross section is shown in Figure 3.4.

The potential wires are larger in diameter than the anode wires. Thus layers of concentric cylinders of proportional chambers, or drift chambers, are formed. The cells for the drift chambers are hexagonal or trapezoidal in shape. The chamber is immersed in an axial magnetic field, and electric fields in the drift cells are perpendicular to the axial magnetic field (taken as the (r, ϕ) plane with r=radial distance, ϕ=azimuthal angle). The magnetic field is introduced for momentum measurement. Because there are electric fields $(\vec{E})$ in a direction perpendicular to the magnetic field $(\vec{B})$, one has to take into account $\vec{E} \times \vec{B}$

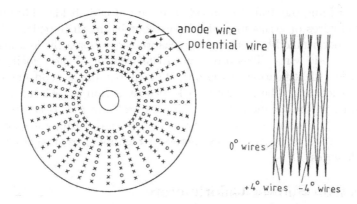

Figure 3.4: A schematic diagram showing the cross section of a cylindrical wire chamber with anode and potential wires indicated. Also indicated on the right are stereo wires for removing ambiguities. (Reproduced with the permission of Cambridge University Press from Figures 4.41 and 4.45 in *Particle Detectors* by C. Grupen.)

effects in track reconstruction. The signals from this type of chamber, from the anode and potential wires, are similar to those from the planar proportional and drift chambers. The position in the r, ϕ plane is obtained from measurements in the drift cells, but the z coordinate along the wire is not determined. To determine this position unambiguously, some of the anode wires are stretched at a small angle with respect to the wires parallel to the axis ("stereo" wires). These play the role that the segmented cathode pads play in resolving ambiguities in the MWPC with respect to the coordinate measurements obtained from the outputs of the electrodes.

Cylindrical Jet Chambers

This is part of a central detector used with storage ring colliders. They are specially designed so as to have a good capability for identification of particles by measuring energy loss extremely well.

The basic construction is like that of the cylindrical drift chamber (see description of "Cylindrical Wire Chambers" above). In the azimuthal direction (ϕ direction), this detector is segmented into pie shaped regions which provide the regions for the drift. There is an axial magnetic field for momentum measurement. In the drift cells there are also electric fields (in a direction perpendicular to the magnetic field) for providing the drift. To provide constant electric fields additional potential wires or strips may also be present in the drift region. Energy loss is determined by performing measurements of specific ionization on many anode wires. This provides a measurement of the velocity of the particle. Identification of the particle requires also a good measure of its momentum. The accurate track reconstruction is achieved with the bending in the magnetic field, taking care of any $\vec{E} \times \vec{B}$ effects on the tracks. To resolve ambiguities in the z position determinations, one can use "stereo" wires, or another method which involves staggering of the anode wires.

Electron Shower Calorimeters

In many processes, high energy electrons are ejected from the interaction region and it is necessary to measure their energies accurately. The electromagnetic shower calorimeters are devices which measure the energies of electrons and photons above several hundred MeV. The physical processes on which the detection is based are the processes of bremsstrahlung emission by the electrons and of creation of electron-positron pairs by the photons. These processes occur in the medium of the detector; a single starting electron develops to become a cascade shower through these processes. The detector is large enough in size that all the energy is deposited in the detector. It is called a total absorption detector. If the detector is such that it samples the amount of energy deposited at various locations along the direction of the development of the shower, it is called a sampling calorimeter. These sampling detectors may be liquid argon chambers or scintillation counters. In a liquid argon chamber, the signal is in the form of a charge pulse, while in a scintillator, the signal is in the form of a light pulse. These signals are recorded and analyzed by using appropriate electronics and photomultiplier systems, respectively. To get an idea of the size of the detector, for complete absorption of electrons, say of 10 GeV energy, one requires a size which is about twenty radiation lengths long. It is found that the relative energy resolution of such detectors improves as the energy increases, as $1/\sqrt{E}$. If the detector and readout systems are segmented, such calorimeters can provide information with good spatial resolution also. Such electromagnetic shower calorimeters are an integral part of any large multipurpose detector used at high energy accelerators.

Hadron Shower Calorimeters

As with electrons, hadrons create a cascade shower of hadrons through their inelastic collisions with nuclei of the detector material. Materials, such as uranium, tungsten, iron, or copper in which the interaction lengths are short, are good for use as a sampling hadron calorimeter. Typically, high energy hadrons, with energies greater than 1 GeV, produce hadronic cascades. They are detected as in the electromagnetic shower counter, through the charge pulse or the scintillation light pulse that is produced in the medium. Here, as in the electromagnetic shower detector, the relative energy resolution improves as the energy increases. However, the energy resolution of a hadronic shower detector is not as good as that of the electromagnetic shower detector. This is because a substantial fraction of the energy, such as that required to break up nuclei against their binding energies, is not measurable in the detector. Further any muons produced usually escape from the medium without being detected and carry away energy.

One feature of the shower detectors which can be useful for identification of the particle as an electron or hadron is based on the different lateral and longitudinal shapes that form as the showers develop. Further, muons can be distinguished from the other particles, as they typically penetrate through large amounts of material.

Time Projection Chamber (TPC)

The Time Projection Chamber (TPC) is an elegant method for recording tracks of particles and works in planar or in cylindrical geometries. The original idea is due to Nygren [22] The working of the cylindrical TPC can be schematically explained as follows (see Figure 3.5).

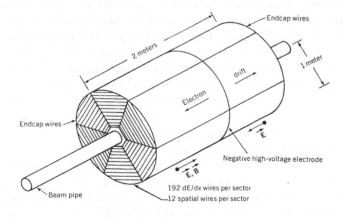

Figure 3.5: A schematic diagram of a TPC working in an $e^- e^+$ collider. (Reproduced with permission from the American Institute of Physics and D. Nygren, Figure 4 in *Physics Today*, October 1978.)

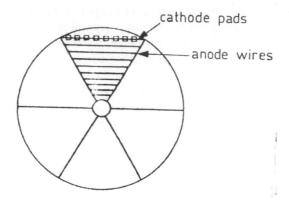

Figure 3.6: The diagram shows the segmentation of cathode pads for reading out the signals from the TPC. (Reproduced with the permission of Cambridge University Press from Figure 4.57 in *Particle Detectors* by C. Grupen.)

A cylindrical chamber is filled to a pressure of 10 atm with a gas mixture of Argon and methane (90:10) and divided into two parts by an electrode at its center. The axis of the cylinder is taken as the z axis, and the end-plates of the cylinder represent the (r, ϕ) planes. An electric field is maintained between the central electrode and the end-plates. This allows the charged particles produced by ionization to drift toward the end-plates. There is also an axial magnetic field present. It serves to limit the diffusion of particles perpendicular to the field; the electrons, in particular, spiral around the magnetic field in tight circles. The end-plates are divided into six pie shaped structures, each containing multiwire proportional chambers. The anode wires in each sector are stretched, parallel to one another, in the ϕ direction at increasing values of r, starting from close to the axis of the cylinder and working outward. Just behind the anode wires, there are sets of segmented cathode pads (see Figure 3.6).

The signals are read from the anode wires and the cathode pads. From these, the r, ϕ coordinates are obtained. The arrival times of the primary electrons are also measured and recorded. From these, the z coordinates are obtained. Thus, the TPC gives the r, ϕ, z coordinates for every collection of the primary electrons, which the charged particle produces by ionization along its track. The signal on the anode wire also gives information about the specific ionization energy loss of the particle. Together with the timing information, these serve to identify

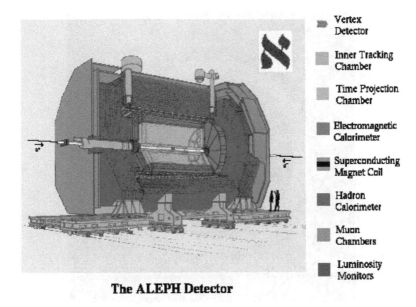

➡	Vertex Detector
▨	Inner Tracking Chamber
▨	Time Projection Chamber
▨	Electromagnetic Calorimeter
▰	Superconducting Magnet Coil
▨	Hadron Calorimeter
▨	Muon Chambers
▰	Luminosity Monitors

The ALEPH Detector

Figure 3.7: A cutaway view of the ALEPH detector; the size of the detector can be gauged by the people shown in the lower right hand corner of the picture. (Courtesy ALEPH collaboration and CERN.)

the particle. One very important useful feature of the TPC is that the detector contains so little material that effects of multiple scattering on the charged particles is minimized.

The TPC [23] has been used in the study of μ to e conversion at TRIUMF. It is also incorporated into the ALEPH detector at LEP in the study of e^+e^- annihilations.

Detectors at LEP

Our understanding of elementary particles has been advanced greatly by the precision measurements that have been carried out with four large multipurpose detectors, ALEPH, DELPHI, L3, and OPAL, working with the Large Electron Positron Collider (LEP) at CERN, and with the SLD detector working with the SLC at SLAC. The data obtained are in good agreement with the predictions of the standard electroweak model including higher order corrections. We end this section on detectors with a brief description of each of them.

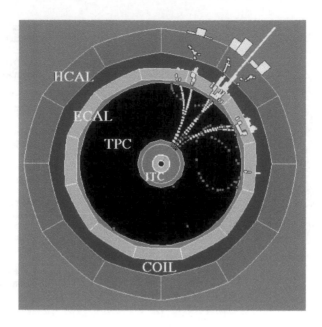

Figure 3.8: A view of the cross section across the detector showing its
different components. (Courtesy ALEPH collaboration and CERN.)

ALEPH Experiment

ALEPH collaboration, a large group of many physicists and engineers
from around the world, constructed the ALEPH detector for use at LEP
(see Figure 3.7). The detector is arranged in the form of concentric
cylinders around the beam pipe.

The interactions of electrons and positrons occur in the middle of the
detector (refer to Figure 3.8). A superconducting coil, 6.4 m long and
5.3 m in diameter, generates a magnetic field of 1.5 T for momentum
determinations. The return yoke for the magnetic field is in the form of
a twelve-sided iron cylinder with plates at its two ends. There are holes
in the end-plates to accommodate the beam pipe and the quadrupole
focusing magnets for the beams. (The beam itself is inside a beryllium
pipe of diameter 16 cm and has a vacuum of 10^{-15} atm.) The iron in the
return yoke cylinder has a thickness of 1.2 m and is instrumented as a
hadron calorimeter (HCAL), by being divided into many layers through
which streamer tubes are inserted. Outside the yoke, there are two
further layers of streamer tubes to give the position and angle of muons
that have gone through the iron. Going in from the superconducting coil,

there is the electron-photon shower calorimeter (ECAL) in the form of alternating layers of lead and proportional tubes. It has high angular resolution and good electron identification capability.

As we go further inward, we find the central detector of charged particles, in the form of a time projection chamber (TPC). It is 4.4 m long and 3.6 m in diameter, and gives a three-dimensional measurement for each track segment. It also provides a number of ionization measurements for each track to help in particle identification. Going further in, there is the inner tracking chamber (ITC), which is an axial wire drift chamber. It has inner and outer diameters of 13 cm and 29 cm, respectively, and a length of 2 m. It gives eight track coordinates and provides a trigger signal for charged particles emerging from the interaction point. Even further in, and closest to the beam pipe, is a silicon vertex detector. This records the two coordinates for particles along a 40 cm length of the beam line, one at 6.3 cm away and another at 11 cm away from the beam axis.

DELPHI Experiment

DELPHI collaboration, also a large group of scientists and engineers from different parts of the world, constructed the DELPHI detector for use at LEP. It consists of a central cylindrical section (called the *barrel*) and two end-caps (called the *forward sections*). Its overall length and diameter are each over 10 m, and it weighs about 3500 tons (see Figure 3.9).

The barrel part of the detector consists of the vertex detector, the inner detector containing JET chambers and trigger layers, the time projection chamber (TPC), the outer detector, and the muon chambers. The vertex detector is nearest the e^+e^- interaction point and provides very precise tracking information (to detect particles with very short life) by extrapolating the tracks back to the interaction point. Next, the JET chamber of the inner detector provides coordinate information for points on each track between 12 cm and 23 cm radii. The trigger system is made up of four levels, each of higher selectivity. The TPC consists of two cylinders of 1.3 m length each, and occupies the space between radii 29 cm and 122 cm. It provides the principal tracking information and measures ionization loss precisely to help with particle identification. The outer detector is composed of five layers of drift tubes between radii, 196 cm and 206 cm. Between the TPC and the outer detector is a ring imaging Cherenkov detector (RICH) detector, 3.5 m long, inner radius 1.23 m and outer radius 1.97 m, divided in half by a central support wall. Just outside the outer detector is the electromagnetic calorimeter, situated between radii 2.08 m and 2.60 m, mostly consisting of lead.

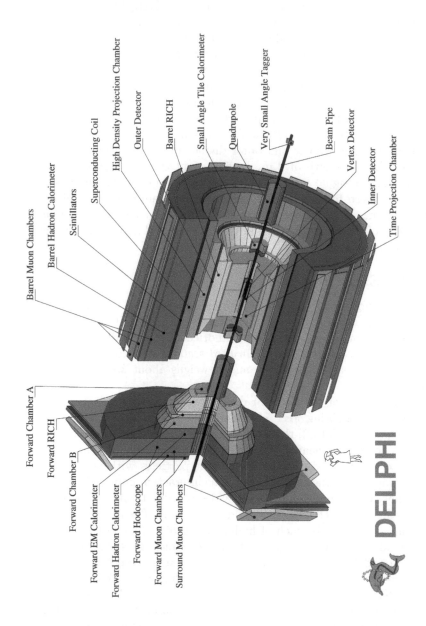

Barrel Muon Chambers

Barrel Hadron Calorimeter

Scintillators

Superconducting Coil

High Density Projection Chamber

Outer Detector

Barrel RICH

Small Angle Tile Calorimeter

Quadrupole

Very Small Angle Tagger

Beam Pipe

Vertex Detector

Inner Detector

Time Projection Chamber

Forward Chamber A

Forward RICH

Forward Chamber B

Forward EM Calorimeter

Forward Hadron Calorimeter

Forward Hodoscope

Forward Muon Chambers

Surround Muon Chambers

DELPHI

Figure 3.9 A cutaway view of the DELPHI detector showing its different components. (Courtesy DELPHI collaboration and CERN.)

Surrounding that is the superconducting coil which provides an axial magnetic field of 1.23 T. The hadron calorimeter is the next outer layer, consisting mainly of iron, which does energy measurements of neutral and charged hadrons. The muon chambers are the outermost part of the detector and the most distant from the collision point.

In the barrel part of the detector, the precision of trajectory measurements are 5 μm to 10 μm in the vertex detector, a fraction of 1 mm in the TPC, and 1 mm to 3 mm in the muon chambers.

The forward parts of the detector consist of the forward chambers, one on each end of the cylinder; the very forward tracker; the forward muon chambers; and the surrounding muon chambers. Components, similar to the components present in the barrel part of the detector, are also present in the forward parts of the detector. This ensures the provision of almost 4π solid angle coverage for the detector.

The component elements provide information in three-dimensional form. This information is read out by a number of dedicated microprocessors involving a large number of electronic channels. All the data are then joined together to form *events*, which are sent to central computers where they are stored for later analysis.

L3 Experiment

The L3 collaboration, again a large international one, constructed the L3 detector for work at LEP. As with the other detectors, it is also a multi-component cylindrical detector (see Figure 3.10). Going outward from the beam pipe, inside which the electrons and positrons collide, the silicon vertex detector is followed by a time expansion chamber (TEC). These give precise track information on the charged particles produced from the collision point. The next three cylindrical layers are the electromagnetic calorimeter called *BGO* (bismuth germanium oxide) calorimeter, the hadron calorimeter (HCAL), and the muon detector. The outermost layer contains the magnet which generates a magnetic field inside the detector for measuring the momenta of the charged particles created at the collision point.

The components of this detector function as in the other two detectors described above. All the information from a collision gathered by the different components of the detector is sent to computers where the event is reconstructed. The reconstruction gives a picture, identifying the particles and showing the paths taken by them and the energies they carry.

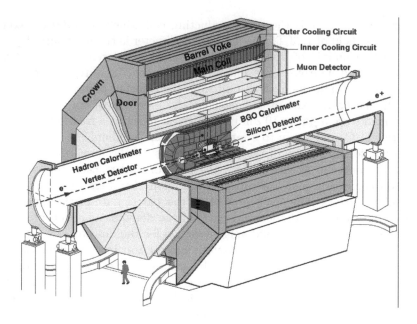

Figure 3.10: A cutaway view of the L3 detector showing its different components. (Courtesy L3 collaboration and CERN.)

OPAL Experiment

The OPAL collaboration also consists of a large consortium of scientists and engineers from different parts of the world and has constructed the OPAL detector. It is also a large, multi-purpose, multi-component detector (see Figure 3.11) and it measures about 12 m long, 12 m high, and 12 m wide.

There are three main layers in the detector arranged cylindrically about the beam axis and with the collision point at its center. These are the system of tracking detectors, electron and hadron shower calorimeters, and the muon detector system (see Figure 3.12). The central tracking system is made up of the silicon vertex detector, followed by the vertex detector, which is followed by the jet chamber, and then by the z-chambers. The tracking detectors detect the ionization caused by the outgoing charged particle. The measurement of the locations of the ionizations yields information to construct the path of the charged particle. The largest of the tracking chambers is the central jet chamber, where the ionization caused by the charged particle is measured at a large number of points along its path. These measurements allow a good de-

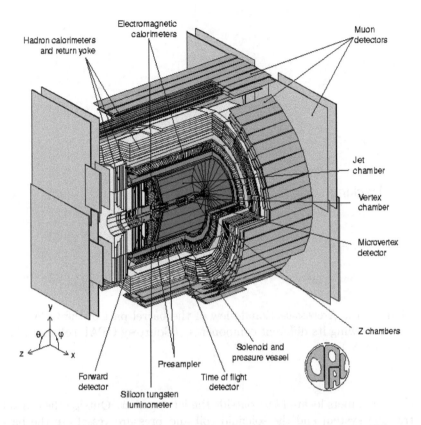

Figure 3.11: A cutaway view of the OPAL detector showing its different components. (Courtesy OPAL collaboration and CERN.)

termination of the particle's trajectory. There is a solenoid outside the jet chamber to provide an axial magnetic field in this detector just as in the other detectors described above. There is also a pressure vessel located here. A measurement of the curvature of the track of the charged particle allows a determination of its momentum. Data gathered on specific ionization loss along the track helps identify the particle as electron, pion, kaon, etc. The vertex chamber lies just inside the jet chamber and the silicon vertex detector is the innermost detector closest to the beam pipe. The information from these is used to find the decay vertices of short-lived particles, and also to improve the momentum resolution. The vertex and the jet chambers give very accurate measurements of the tracks in a plane perpendicular to the beam axis. Accurate information on the path along the beam axis is obtained by

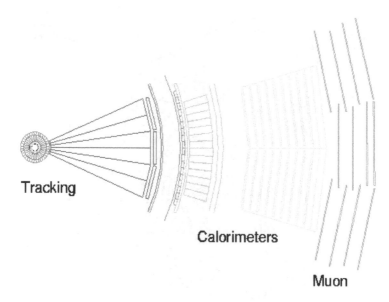

Tracking

Calorimeters

Muon

Figure 3.12: A cross-sectional view of the barrel part of the OPAL detector showing its different components. (Courtesy OPAL collaboration and CERN.)

the z-chambers located just outside the jet chamber. Outside the central tracking system and the solenoid coil and pressure vessel are the barrel electromagnetic calorimeters. These are made of lead-glass blocks. There are also lead-glass blocks in the end-caps. The barrel lead-glass blocks together with the lead-glass blocks in the end-caps cover 98% of the full solid angle. Most electromagnetic showers start before the lead-glass due to the matter already traversed, such as the magnet coil and the pressure vessel. Presamplers are used to measure this shower effect and to improve the energy resolution.

The iron of the magnet return yoke, outside the electromagnetic calorimeter, serves as the hadron calorimeter. All the particles which have cleared the electromagnetic calorimeter are either hadrons or muons. The hadron calorimeter serves to measure the energy of the hadrons coming through the electromagnetic calorimeter and also helps in identifying the muons. The thickness of the iron is four or more interaction lengths and covers 97% of the full solid angle. The yoke is divided into layers, and detectors are introduced in between the layers. These form a cylindrical sampling calorimeter of 1 m thickness. The end-caps also have hadron calorimeters in them to give full solid angle coverage.

Finally, the outermost part of the detector, whether in the barrel or at the end-caps, contains the muon detectors. The barrel part has 110 large area drift chambers, each chamber being 1.2 m wide and 9 cm deep. There are 44 of these on each side of the barrel, and ten on the top and twelve on the bottom of the barrel. At either end of the detector there are four layers of streamer tubes laid perpendicular to the axis of the beam and covering about 150 square meters area. The end-cap muon detector on either side consists of 8 quadrant chambers and 4 patch chambers, and each chamber has two layers of streamer tubes, spaced 19 mm apart. One layer of these has vertical wires, while the other layer has horizontal wires.

An example of the display of an event in OPAL detector, in which e^+e^- annihilate into a quark antiquark pair at 204 GeV, the quark and the antiquark subsequently turning into jets of hadrons, is shown in Figure 3.13 below.

Centre-of-mass energy 204 GeV

Figure 3.13: OPAL display of e^+e^- annihilation into q and $\bar{q}$, each of which gives jets of hadrons. (Courtesy OPAL collaboration and CERN.)

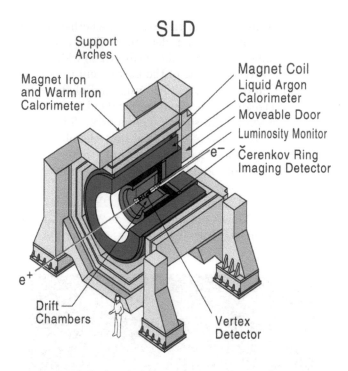

Figure 3.14: Cutaway view of the SLD detector at the SLC in SLAC. SLAC is a high energy physics research facility operated on behalf of the U. S. Department of Energy by Stanford University. (Courtesy SLAC.)

SLD Detector

The SLD detector was constructed at SLAC to work with SLC. Figure 3.14 above shows a cutaway view of the detector exposing its various components to view. Many of the components in this detector function in much the same way as in the detectors at LEP. A new feature of this detector is a 3D silicon CCD vertexing capability. Combined with beam spot sizes provided by SLC, this detector provides high precision tracking information, fine-grained calorimetry, and excellent particle identification using the RICH counters. The production of polarized electron beam at SLC with substantial polarization combined with the special capabilities of the SLD detector enabled the SLD experiment to make the world's best determination of the weak mixing angle with corresponding implications for the Higgs boson mass in the standard model.

GLOSSARY

Alphabetical Listing of Topics

In the part of the book that follows, is presented a glossary of terms, using key words of commonly used terms in particle physics listed alphabetically. Brief explanations of the topics are provided, maintaining a balance between depth and range, in each topic. A given topic is sometimes repeated under different alphabetical listings providing somewhat different perspectives and different emphases. Cross references between related topics are given throughout the listing. In as much as possible, the explanations under the different listings are self-contained. The history of the evolution and development of ideas is also discussed in the topics, providing elaboration of material contained in the early part of the book on the historical overview of the field of particle physics. For those interested in pursuing further details, references to the original sources are given for the material covered under each listing in the glossary.

Abelian Gauge Theories
See in section under "Gauge Theories".

Adler-Bell-Jackiw Anomaly
Let us consider a classical field theory described by a Lagrangian which
has certain continuous symmetries. According to Noether's theorem,
for each generator of symmetry, there exists a current that is conserved.
If we quantize the field theory, two questions arise: (1) Do the symmetries of the classical theory survive after quantization? (2) Do the
corresponding currents still remain conserved? Renormalization effects
in quantum field theory play a subtle role in the answer to these questions, and the answers are not always *yes* as one might think. There
are cases where such symmetries are preserved and other cases in which
they are not. In the latter cases, the corresponding currents are not
conserved. Such cases are said to possess *anomalies*. They were first discovered by Adler [24] and by Bell and Jackiw [25] and hence are referred
to as *Adler-Bell-Jackiw anomalies* or sometimes as *Triangle anomalies*.
Strictly speaking, theories possessing anomalies cannot be classified as
renormalizable. Theories with anomalies have to be put together in such
a way that, if there are several fields participating, each giving rise to
anomaly, the anomalies have to cancel among themselves.

Examples of theories where anomalies appear involve chiral symmetries with massless fermions. Consider the massless Dirac Lagrangian for
free fermions: $\mathcal{L} = \bar{\psi}\gamma^\mu \partial_\mu \psi$. It is invariant under the transformations
$\psi \to e^{ia}\psi$ and $\psi \to e^{i\gamma^5 a}\psi$, where the first is a phase transformation,
and the second is a chiral transformation. There are two currents J^μ
and $J^{\mu 5}$ corresponding to these symmetries, which are conserved according to Noether's theorem. It is easy to verify by direct calculation,
using the massless Dirac equations for ψ and $\bar{\psi}$, that $\partial_\mu J^\mu(x) = 0$ and
$\partial_\mu J^{\mu 5}(x) = 0$. Thus, for massless fermions, both the vector and the axial
vector currents are conserved. One can form linear combinations of these
currents which have left and right chiralities: $J_L^\mu = (1/2)(J^\mu - J^{\mu 5})$ and
$J_R^\mu = (1/2)(J^\mu + J^{\mu 5})$, respectively, and these are separately conserved.
When we extend the considerations to the Dirac fermion interacting with
gauge fields, calculations similar to the above seem to suggest that both
the vector and the axial vector currents should be conserved. However,
a careful examination shows that this is not quite the case, and the axial
vector current is not conserved although the vector current is. The axial
vector current has an anomaly.

In quantum field theories involving chiral fermions and gauge bosons,
the gauge bosons have different interaction strengths when coupled to
left- or right-handed fermions. This is typical of the electroweak theory

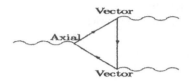

Figure 4.1: Triangle diagram leading to chiral anomaly.

where the coupling strength of the gauge boson to the fermion depends on its chirality. In such theories, if one calculates the interaction of one gauge boson with two other gauge bosons through an intermediate state involving a triangle loop of fermions as shown in Figure 4.1 above, where the gauge bosons have different couplings to left-handed and right-handed chiral states, one obtains a divergent contribution, which is not removable by the usual rules of renormalization theory. In other words, the theory is non-renormalizable and is said to possess a *chiral anomaly*. The only way theories involving such interactions can be made meaningful and consistent is to introduce a suitable multiplet of fermions such that the individual fermion anomaly contributions cancel one another exactly.

Alpha Particle Scattering

When a collimated beam of α particles from a radioactive source is made to pass through matter, some of the particles suffer deviations from their original direction. They are said to be *scattered*. The scattering must be due to the interaction between the particles of the beam and the atoms of the material. Detailed study of this scattering process is capable of giving information on both the scattered particles and the atoms of matter. The scattering of α particles in various materials was studied in detail by Rutherford in 1911 [26].

At that time there was evidence to suggest that atoms of matter consisted of electrically neutral collection of positive and negative charges and could be pictured as spheres of radius about 10^{-8}cm. Thomson proposed a simple model of the atom: a sphere of uniformly distributed positive charge of radius 10^{-8}cm throughout which was distributed an equal negative charge in the form of electrons. Given a model of the distribution of charges in the atom, the scattering of the positively charged alpha particle by the atom can be calculated quantitatively and compared with experimental findings. With the Thomson model, it was shown that the average deflection suffered by the alpha particle on a single atom should be very small. Even for matter in the form of a foil of thickness t, assuming that the alpha particle suffers multiple deflections in the foil, the average deflection is still found to be small. The probability of a large angle deflection is extremely small in Thomson model. For example, the number of alpha particles scattered through an angle of $10°$ or more in a gold foil of thickness 4×10^{-5}cm was calculated to be about 10^{-43} relative to the unscattered particles [27]. Experimentally, Geiger and Marsden [28] found that 1 in every 8,000 particles was scattered by an angle greater $90°$, a rate which is completely incompatible with the Thomson model predictions.

Rutherford [26] proposed a new atomic model to explain the results of alpha particle scattering in gold foils. In his model the positive charge in the atom, instead of being distributed in a sphere of radius 10^{-8}cm, is concentrated in a much smaller sphere, the *nucleus*, and the negatively charged electrons are distributed in the much larger sphere (radius 10^{-8}cm) outside the nucleus. Further, the nucleus was assumed to carry all the mass of the atom, since it was known that the electrons had very small mass. In this model, the alpha particle suffers very little scattering from the electrons, approaching the nucleus at very small distances where the electrostatic repulsion between the alpha particle and the nucleus is very large. Treating the nucleus and the alpha particle as point charges, and with the nucleus fixed, the calculation of the orbit of the alpha particle is a simple problem in classical mechanics. Using this theory, Rutherford calculated the fraction of alpha particles scattered through a given angle θ. Geiger and Marsden [29] undertook to test Rutherford's model and found satisfactory agreement with experimental results. This established the Rutherford model of the atom as the correct one.

Alpha Radioactivity
In 1896, Becquerel was the first person to notice that crystals of uranium salt emitted certain rays which could affect photographic plates. It was also established that the emitted radiation induced electrical conductivity in gases. Salts such as these were called *radioactive*. Further properties of the rays were gathered by examining a collimated beam of these rays obtained by placing the salt at the bottom of a long and narrow cavity in a lead block. At some distance above the lead block a photographic plate was placed. This whole system was enclosed in a chamber and evacuated. A strong magnetic field perpendicular to the channel of the rays was established. When the photographic plate was exposed to the rays and developed, two spots were found on the plate. One of these spots was directly above the long channel indicating that

this component of the radiation from the salt was not affected by the magnetic field. It must be electrically neutral. The other was a spot not in line with the channel but displaced from it, say to the left of it. If the magnetic field is reversed in direction, the deviated spot is found on the right of the channel line. This component of the rays is thus not electrically neutral. From the orientation of the magnetic field and the bending these rays experienced in the magnetic field, it could be established that these rays were made up of positively charged particles. The electrically neutral component was given the name *gamma rays* (γ), and the positively charged component was given the name *alpha rays* (α). If the strength of the magnetic field was reduced, one also found another deviated spot in a position on the side opposite to that of the alpha ray. Thus, this component must also be charged but oppositely to the alpha rays. This component was given the name *beta rays*. Detailed studies of the properties of the α and β radiations from uranium were carried out by Rutherford [30]. It was eventually established that the α rays consist of doubly charged helium ions, the β rays consist of nothing but electrons, and the γ rays are electromagnetic radiation of high energy.

AMANDA Neutrino Detector

The name of this detector stands for *Antarctic Muon And Neutrino Detector Array*. It is a detector that has been constructed at the South Pole to observe high energy neutrinos from astrophysical sources. The energies of the neutrinos will be in the range of 1 TeV. Strings of PMT's (PhotoMultiplier Tubes) are located in water-drilled holes deep under the South polar ice cap. Neutrinos coming through the earth will interact with the ice or other particles and give rise to muons. The PMT array will detect the Cherenkov light emitted by the muons. The tracking of the muons is done by measuring the arrival times of the Cherenkov photons at the PMT's. Located at a depth of 1500 m to 2000 m under the Antarctic ice, the detector array for AMANDA-B consists of 302 PMT's on 10 strings and has a detection area of about 10,000 square meters. The detector has been collecting data for more than a year and the data stored on tape are in the analysis phase. The detector collaboration is a large one, consisting of 116 scientists from 15 institutions in the U.S. and Europe.

In a further development of the operation, construction of AMANDA II with an effective area several times that of AMANDA-B has been completed. Initially, three strings of PMT's were installed in depths ranging between 1300 and 2400 meters. With these, information on the optical properties of polar ice in this range of depths could be obtained. In the early part of the year 2000, six further strings of PMT's were added, thus completing the construction of AMANDA-II. Data taking is in progress with the new system.

Annihilation of e^+e^-

Huge progress has been made in the last three decades in developing high luminosity colliders of electron and positron beams. These colliders provide a much higher center of mass energies than fixed target machines and consequently have a much higher discovery potential for new particles. Significant advances have been made in studying the properties of new particles produced with electron-positron colliders at Orsay, Frascati, Hamburg, Novosibirsk, Stanford, Japan, China, and LEP at CERN. The fact that electrons and positrons belong to the lepton family, and seem to behave like points even at the highest energies examined, makes the theoretical and experimental analysis of the production of new particles much simpler than with proton-proton or proton-antiproton colliders which involve composite particles. Z^0 and $W^\pm$, and a very rich spectrum of spin-1 mesons, have been produced and studied at these machines. Much of our knowledge of these particles comes from such studies. One can also look for heavy leptons with these colliders.

First evidence for two-jet structure in e^+e^- annihilation reaction leading to hadron production was found in 1975 at the SPEAR ring at SLAC. The jet structure is interpreted as evidence for quark-antiquark production from the e^+e^- annihilation, with subsequent formation of hadrons from the quark and the antiquark as they separate from the production point [31]. Evidence for three-jet structure in e^+e^- annihilation reaction leading to hadron production has also been abundant. The three-jet structure is interpreted as evidence for quark, antiquark, and gluon production with subsequent formation of hadrons from these particles as they separate from one another [32–34].

The linear electron-positron collider at SLAC, called the SLC, with unique capabilities to produce polarized electrons and positrons, has helped us greatly in understanding various physics questions in the region of the Z^0. A number of new proposals for constructing linear colliders of electrons and positrons is under active consideration for reaching center of mass energies of 500 GeV or higher with high luminosity (NLC—the Next Linear Collider). Colliders which involve muon beams are also being contemplated.

Anomalous Magnetic Moment

Dirac theory of spin 1/2 point particles attributes an intrinsic magnetic moment associated with the spin, equal to μ_B, which is called the *Bohr magneton.* The magnetic moment $\vec{\mu}$ and the spin vector $\vec{s}$ are related by $\vec{\mu} = g\mu_B\vec{s}$, where g is the Landé g-factor, and $g\mu_B$ is the gyromagnetic ratio (equal in magnitude to μ/s). Since the electron has spin $1/2$, g according to Dirac theory should have the value of exactly 2.

The g values for the electron and the muon have been determined to great precision experimentally and are found to be different from 2 by about 0.2%. This difference is called the *anomalous magnetic moment.* The existence of this difference suggests that the Dirac theory applies only approximately to the electron and the muon. The proton and the neutron are also experimentally found to have large anomalous magnetic moments. The large deviations observed for the proton and the neutron (compared with those for the electron and the muon) are attributed to the fact that these particles are hadrons and composite in structure, while the electron and the muon are essentially structureless points.

The deviation of g from the value 2 for spin $1/2$ particles has its origin in what are generally called the radiative corrections and, at least for the electrons and muons, are calculable from the theory known as *quantum electrodynamics* (QED). For the protons and neutrons, which are hadrons and hence participate in strong interactions, there are at present no reliable calculations of these corrections available (see further under "Quantum Electrodynamics (QED)" and "Quantum Chromodynamics (QCD)").

Antideuteron
First evidence for the production of antideuterons came from an experiment performed by Dorfan et al. [35] at the Brookhaven AGS proton synchrotron. They bombarded beryllium with protons of 30 GeV energy and used counters as detectors. This was further confirmed in an experiment using 43 GeV, 52 GeV, and 70 GeV protons on an aluminum target at the Serpukhov proton synchrotron [36]. The experiment by Dorfan et al. [35] also found evidence for antitriton. Based on these results, it might be said that antimatter was produced for the first time.

Antineutron
Direct experimental confirmation of the existence of the antineutron, the antiparticle of the neutron, came from a study of the charge exchange scattering in antiproton-proton scattering reactions using a heavy liquid bubble chamber [37] at the Berkeley Bevatron. Reactions that were studied included $\bar{p}p \rightarrow \bar{p}p, \bar{p}p \rightarrow \bar{n}n, \bar{n}n \rightarrow \pi$'s. These reactions established the existence of the antineutron.

Antiproton

Dirac's theory of the electron requires the existence of antiparticle to
the electron coming from the re-interpretation of the negative energy
solutions to the equation. Dirac's theory extended to the proton requires
the existence of the antiproton. Experimental finding of the antiproton
would affirm the correctness of using Dirac's theory for the proton. Such
a particle was discovered in 1955 in an experiment performed at the
Berkeley Bevatron by Chamberlain, Segrè, Wiegand, and Ypsilantis [38].
Somewhat earlier, in studies of cosmic rays, several events were observed
which could have been due to antiprotons [39–42]. No definitive proof of
their existence could be established based on those earlier observations.

The experiment performed at the Bevatron was specifically designed
to produce and detect the antiproton and ascertain that this negatively
charged particle has a mass equal to that of the proton. Protons from
the Bevatron were made to impinge on a target, and the momentum and
velocity β (in units of the velocity of light) of the negatively charged particles originating from the target were measured simultaneously. There
was a large contamination by negative pions (100,000 pions for every
antiproton) which were copiously produced from the target. In this
huge background of pions, detecting antiprotons was a real challenge.
The experimental setup in schematic form for the antiproton search is
shown in Figure 4.2 on the following page. The negatively charged particles produced from the target T were bent and focused onto a scintillation counter S_1 by a system of dipole and quadrupole magnets,
and on to a second scintillation counter S_2 by another set of bending
and focusing magnets. Then the beam passed through (1) a threshold
Cherenkov counter C_1 which selected velocities $\beta > 0.79$, (2) a second
differential Cherenkov counter C_2 which selected velocities in the range
$0.75 < \beta < 0.78$, then (3) a scintillation counter S_3, and finally (4) a
total absorption Cherenkov counter C_3 (not shown in the figure). They
estimated that the negative pions in the beam would have $\beta = 0.99$,
while the antiprotons (assuming protonic mass) would have $\beta = 0.76$.
Thus, a signal for pions would be a coincidence in the counters S_1, S_2,
C_1, and S_3 with no signal in C_2, while for antiprotons, there would be
coincident signals in S_1, S_2, C_2, and S_3 and no signal in C_1. The latter
events would be possible antiproton candidates. To further ensure that
they were indeed antiprotons, time of flight measurements were made
between the scintillation counters S_1 and S_2 which were separated by a
distance of 41 ft. For antiprotons this time of flight was expected to be
51 ns. Only those particles which gave counts in the counter and passed
this additional requirement of time of flight of 51 ns were accepted as
true antiprotons. Only 60 of these negatively charged particles passed

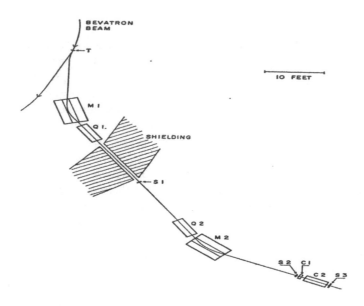

Figure 4.2: Experimental arrangement to detect antiprotons. (Figure from O. Chamberlain, E. Segrè, C. Wiegand, T. Ypsilantis, *Physical Review* 100, 947, 1955. Copyright 1955 by the American Physical Society.)

these cuts. A further check on the mass spectrum of these particles was made by putting the target in a different position, transporting positive pions and protons down the same beam (reversing currents in the bending magnets), and measuring the distribution of protons. There was excellent agreement between the mass distributions of protons and antiprotons, and it was established that the observed antiprotons had a mass equal to that of the proton within about 1%.

Antiproton Annihilations
One of the features of the Dirac theory of spin 1/2 particles (electron or proton) is that particles and antiparticles must be generated in pairs and annihilated in pairs. Unlike electron-positron annihilations in which photons are produced, in antiproton-nucleon annihilations, pions are the usual annihilation product, and products such as photons and/or electron-positron pairs are rare. Kaons are also only rarely produced. Examples of antiproton-nucleon annihilation were observed in a nuclear emulsion stack exposed to the antiproton beam from the Bevatron by the same group which discovered the antiproton [43]. In this paper there

is a beautiful example of an antiproton coming to rest and annihilating with a proton bound in a nucleus in the emulsion. The energy released in the annihilation process breaks up the nucleus in addition to producing a number of pions. Tracks of the nuclear fragments as well as those due to the charged pions are seen in the nuclear emulsion. Neutral pions are not visible in the emulsion. The total energy measured from the visible tracks is 1300 MeV, which is more than the energy equivalent of the mass of the incoming antiproton proving that annihilation has occurred. A determination of the mass of the incident antiproton could be made, and it gave a value within few percent of the protonic mass.

Certain selection rules can be formulated for antiprotons annihilating at rest. For example, it can be shown that if the antiproton-nucleon pair is in a $^3S_1$ state, it cannot annihilate into three pions. (See under "G-parity" for a discussion of the selection rules leading to final states of definite number of pions in nucleon-antinucleon annihilations).

Associated Production

The concept of associated production of new particles was put forward by Pais in 1952 in an attempt [44] to explain how it is possible to have a copious production of these particles and yet have long lifetimes for decay. A brief history of the discovery of these particles will explain how this concept came to be formulated.

Cosmic rays investigations [45] with cloud chambers, for a number of years since 1947, had found evidence for the existence of unstable particles with masses between the muon and the proton. Estimates of their lifetimes placed them somewhere around 10^{-10}s. These particles were called V particles as they left V shaped tracks in the cloud chamber. Two types of V particles were found, those that decayed into a proton and a negative pion with a Q value of about 40 MeV, and those that decayed into a pair of charged pions. The problem these particles posed is that, if the interactions responsible for their long decay times are also responsible for their production, then they must be produced only rarely. The associated production hypothesis put forward by Pais to explain this feature was as follows. If the V particle is produced in association with another unstable particle by an interaction which is strong, the production rate for the associated production will be high because of the strong interactions. If the V particle and the other unstable particles were to decay by means of a different interaction—the weak interaction—they would have a long lifetime.

Substantial progress could be made with the construction and operation of the Brookhaven Cosmotron. Using the 1.5 GeV negative pion beam from this machine, a number of V particles were produced and could be studied quantitatively. It was possible to establish that these particles were readily produced and yet decayed very slowly confirming the applicability of Pais' ideas. V particles heavier than the proton were called hyperons and the V particles that decayed into pions were a new kind of meson. (See also under "Lambda Hyperon", "Σ Particles", "Ξ Particles", "Kaons: The τ-θ Puzzle", "Kaons—Neutral $K_1^0(K_S^0)$, $K_2^0(K_L^0)$", and "Hyperons—Decay Selection Rules".)

Asymptotic Freedom

This is a property possessed by the non-Abelian gauge theory of quantum chromodynamics (QCD). Simply put, it states that the coupling constant, which characterizes the quark gluon interaction in QCD, becomes weak at large relative momenta or at short distances. A qualitative picture of how this comes about is explained below.

First, let us note that even in electrodynamics, the charge, which is a measure of the coupling between a charged particle and the electromagnetic field, is not an absolute constant. It is an effective coupling, which depends upon how far one is from the charge. To understand this, consider a positive charge placed at some point in a charge neutral medium, which is polarizable. The presence of the positive charge causes the neutral medium to be polarized; that is, it causes a separation of positive and negative charges of the medium in the near vicinity of the positive charge. The negative charges from the medium will be attracted to the positive charge and will be close to it all around, while the positive charges of the medium will be repelled away from it to large distances. The negative charges of the medium partially screen the positive charge, the amount of screening depending upon how far we are from the location of the positive charge. The scale of distance at which screening effect arises is set by the interparticle spacing in the medium. For distances small compared to this scale, one sees the unscreened positive charge, while at larger distances the amount of screening will vary. This screening effect of the medium is taken into account by introducing a dielectric constant for the medium according to which the effective charge of the particle is obtained by dividing the charge by the dielectric constant. In quantum electrodynamics, it turns out that the vacuum itself behaves as a polarizable medium, because of a process called *vacuum polarization* which arises due to the possibility of the creation of virtual electron-positron pairs from the vacuum. A positive charge placed in the vacuum causes it to be polarized, and the measured effective charge is a function of how close we are to the positive charge. In vacuum the scale of distance at which one starts seeing the unscreened charge is set by the Compton wavelength of the electron, 3.867×10^{-11} cm. Only in experiments where we are probing the charge at closer distances will we see the full charge. If the experiment probes the charge at greater distances, we will see the effect of the screening, and the measured charge is a function of the distance at which we measure it.

Now let us consider QCD in which one considers the color interaction of quarks with the gluons of the chromodynamic fields. The interaction between quarks is due to the exchange of gluons between the quarks. An analogous process of vacuum polarization occurs here in which colored virtual quark-antiquark pairs are produced by the gluon during its passage between the quarks. As in electrodynamics this produces a screening of the color charge. This is not the whole story however. There is another effect that comes into play. The gluon itself carries color charge and can produce virtual gluon pairs which will also contribute to the vacuum polarization. A calculation of this further effect shows that it leads to an "anti-screening" effect. The total effect depends upon the number of quarks of different flavors and on the number of gluons of different colors. Detailed calculation shows that the combination that enters the theory is $(2f - 11n)$, where f is the number of flavors and n is the number of quark colors. If this quantity is positive, the theory will be like QED, and the effective coupling will increase at short distances (scale $\simeq$ Compton wavelength of quark), while it will decrease at short distances for negative values. Because there are 6 flavors of quarks and three colors, this quantity is negative in QCD. This results in the effective coupling decreasing at short distances or high momenta which we term *Asymptotic Freedom*.

Atmospheric Neutrinos

These are neutrinos produced when high energy cosmic rays impinge on the earth's atmosphere and produce a shower of pions which decay and eventually produce electron and muon neutrinos. A π^+ produces $\mu^+ + \nu_\mu$, the positive muon then subsequently decays into $e^+ + \bar{\nu}_\mu + \nu_e$. Thus for each pion that decays, one gets a ν_μ, $\bar{\nu}_\mu$, and ν_e. On detecting the flavor of the neutrinos by experiments, one expects to see a ratio of two to one for the muon type of neutrinos relative to the electron type. If this ratio is different from 2:1, a new phenomenon must be coming into play. Experiment at the Super-Kamiokande (Super-K) project has found the ratio to be 1:1 rather than 2:1.

The detector for the neutrinos in Super-K consists of a cylinder containing 12.5 million gallons of very pure water, weighing about 50 kilotons, and looked at by some 13,000 photomultiplier tubes (PMT's) spread all around the cylinder surface. When a neutrino interacts in the water, it produces secondary particles, which move with velocities greater than the velocity of light in the water. This results in the emission of Cerenkov radiation which is seen by the PMT's as a ring shaped pulse of blue light of very short duration (nanoseconds). From knowledge of the position, timing, and the amplitude of the light signals, one can construct the track of the secondary particles. It is also possible to say in which direction the initial neutrinos are going. If the secondary particle is a muon, the ring of Cerenkov light tends to be rather sharp, while if it is an electron, the ring tends to be more diffuse. Thus it is possible to know whether electrons or muons are produced by the neutrino interactions, and the numbers of muons and electrons produced must reflect the numbers of the corresponding neutrino species. The number of muons relative to electrons is less than the expected ratio of 2:1.

Many explanations, such as, an anomalous external source of electron neutrinos, faults with the water detector, and incorrect theoretical estimates for the relative neutrino fluxes, were considered to explain the deficit of muon neutrinos and rejected. The most plausible explanation seems to be that we are seeing neutrino oscillations here; the muon neutrinos may be oscillating into tau neutrinos or into electron neutrinos. The argument for this explanation is further strengthened in the Super-K experiment by their measurement of up-down asymmetry for muon-type and electron-type events as a function of the observed charged particle momentum. These measurements involve selecting neutrinos coming into the detector from directly above the detector and those that enter the bottom of the detector from below. The neutrinos entering from the bottom of the detector have traveled through the earth, an extra 10,000 km, during which time they have had a greater chance to change their flavor through oscillation. The measurement of the up-down asymmetry by Super-K shows that there is a difference in the numbers of muon neutrinos, depending upon whether they have taken the shorter path or the longer path before they reach the detector. It was found the electron-type events are up-down symmetric (consistent with the geometry and no oscillations), while muon-type events have considerable asymmetry. The deficit in the muon-type events was nearly one half for the upward going muons. The shape of the asymmetry as a function of the charged particle momentum is also what one expects in a scenario involving oscillations. The favored oscillation parameters which are in accord with the atmospheric neutrino measurements are (see under "Neutrino Oscillations") $\sin^2 2\theta \sim 1$; $10^{-4} \leq \Delta m^2 \leq 10^{-2}$ eV^2.

The atmospheric neutrino experiment by Super-K makes it very plausible that neutrino oscillations are occurring. However, before we can claim that neutrino oscillations have definitely been found, we must look for corroborations from other experiments. Plans are afoot to mount experiments using neutrinos from accelerators, exploring the same range of neutrino oscillation parameters as in Super-K. These experiments are expected to produce results in the next few years.

Atomic Nucleus—Structure
With the discovery of the neutron by Chadwick, the question was posed as to the role neutrons play in the building of the atomic nucleus. The suggestion was put forward independently by Ivanenko [46] and Heisenberg [47] that the atomic nucleus consists of protons and neutrons (rather than protons and electrons). This suggestion had the immediate merit of resolving a couple of outstanding problems of the time. One of these had to do with the fact that, on account of the uncertainty principle, electrons could not be confined in a space of nuclear dimensions ($\simeq 10^{-13}$cm) without having extremely high kinetic energies. The second had to do with the fact that the study of the rotational spectrum of diatomic nitrogen molecule shows that the nuclei obeyed Bose-Einstein statistics rather than Fermi-Dirac statistics. The atomic number of nitrogen is 7, its atomic weight 14. This can be made up in the proton-electron model of the nucleus by having 14 protons and 7 electrons to give net 7 units of positive charge to the nucleus and atomic weight of 14. This requires a total of 21 particles. In this model, since each constituent particle obeys Fermi-Dirac statistics, the system with an odd total number of particles would also obey Fermi-Dirac statistics in conflict with requirements from the rotational spectrum. The proton-neutron model of the nucleus, on the other hand, could be made up of 7 protons and 7 neutrons for a total of 14 particles. The neutrons being nearly as massive as the protons, does not pose any problem with the uncertainty principle in confining it to a space of nuclear dimensions. If the neutrons also obey Fermi-Dirac statistics, this system of even total number of particles would obey Bose-Einstein statistics in conformity with the requirement from the rotational spectrum. Thus the model, in which an atom X with a nucleus of atomic weight A contains Z protons and $N = A - Z$ neutrons in its nucleus, became established. Such a nucleus is represented as ${}^{A}_{Z}X$.

Atomic Number
Studies of large angle α particle scattering by thin gold foils under the direction of Rutherford led to the formulation of the nuclear model of the atom [48]. According to this model, the electrically neutral atom consists of a net positively charged nucleus carrying almost all the mass of the atom, of size about 10^{-13} cm, surrounded by electrons in shells of

much larger radius (10^{-8} cm), such that the whole system is electrically neutral. The number of units of charge (in units of the magnitude of the electron charge $|e|$) carried by the nucleus could be deduced from the data on the scattering of α particles to be about $(1/2)A|e|$ where A is the atomic weight. The resultant electron charge in the atom must then also be $(1/2)A|e|$ for the atom to be electrically neutral.

Independently of the α particle scattering experiments, information on the number of electrons in the atom was obtained by the entirely different method of scattering of X-rays from light atoms by Barkla [49]. Barkla had shown that this number was equal to about half the atomic weight, a conclusion supported by the work on alpha particle scattering.

If a number called *atomic number*, which is simply the number of the atom when arranged according to increasing atomic weight, is introduced, the charge of the nucleus is about equal to the atomic number (times $|e|$). The importance of the atomic number lies in the fact that the frequencies associated with the X-ray spectra of elements studied by Moseley [50] were found to vary as the square of a number which differed by unity in going from one atom to the next. If this number is identified with the atomic number, the properties of the atom would be determined by a number which increases by unity as one goes from one atom to the next. The atomic nucleus would then be characterized by the atomic number of the nucleus Z and another number which represents the nearest whole number to the atomic weight A.

Atomic Structure

The first quantum theory of the spectra of atoms was proposed by N. Bohr [51] based on Rutherford's model of the atom. According to this model, the atom consists of a massive positively charged nucleus of small dimensions and negatively charged electrons going around the nucleus, with Coulomb interaction between the nucleus and the electrons. Classically such a system cannot be stable, as the electron in accelerated motion in its orbits will emit radiation, lose energy, and eventually spiral into the nucleus. To get around this difficulty, Bohr proposed that classical mechanics and classical electrodynamics principles may not be applicable to systems of atomic size. As support for this argument, he pointed to the difficulty with behavior of specific heats, photoelectric effect, black body radiation, etc., which for their resolution needed new ideas to be introduced, such as the relation between energy E and the frequency of the radiation ν, $E = h\nu$, h being Planck's constant. He proposed that the only stable states of an electron in motion around the nucleus must have angular momenta which are integral multiples of Planck's constant. With this one modification, he was able to derive an expression for the allowed energy states of the hydrogen atom. He also proposed that the atom emits radiation when it makes a transition from a higher discrete energy state to a discrete state with lower energy. Such radiation leads to a discrete spectrum and in fact provided an explanation of the Balmer formula for the hydrogen spectrum.

Axiomatic Field Theory

Lehmann, Symanzik, and Zimmermann [52] introduced a new formulation of quantized field theories in 1955. The matrix elements of field operators and of the S matrix are determined from a set of equations based on some very general principles of relativity and quantum mechanics. These equations contain only physical masses and physical coupling constants. The advantage these equations possess over the standard method of doing S matrix calculations is that this method does not introduce any divergences in the basic equations. Although the equations as set up are not restricted to perturbation theory, the solutions of the equations are obtained by a power series expansion in a coupling parameter. The results obtained with the new formulation are identical with the results of renormalized perturbation theory using Feynman diagrams.

Another approach to axiomatic field theory was developed by Wightman [53]. In this approach the focus of attention is on the vacuum expectation values of field operators and the properties they acquire due to very general requirements of Lorentz invariance, absence of negative energy states, and positive definiteness of the scalar product. It is shown that they are boundary values of analytic functions, and local commutativity of fields becomes equivalent to certain symmetry requirements on the analytic functions. Given the vacuum expectation values, the problem of determining the neutral scalar field in a neutral scalar field theory was solved by Wightman.

Axions

In our discussions of conservation of CP for strong interactions we show that, within the context of QCD, certain constraints on field configurations lead to modifications of the QCD Lagrangian density by the addition of a term, which gives rise to violation of CP in strong interactions. As there seems to be no experimental evidence for the violation of CP in strong interactions, such a term had better have a zero coefficient, if QCD is to be the theory for strong interactions.

The added term depends on a parameter θ (see under "Peccei-Quinn Symmetry") which gets modified into an effective θ_{eff} through inclusion of quark mass effects, etc. The presence of this term can be shown to give rise to an electric dipole moment for the neutron (whose magnitude depends on the size of θ_{eff}), which can be looked for experimentally. No nonzero electric dipole moment of the neutron has been measured within the accuracy of the experiments. The experiments serve to put limits on the magnitude of the θ_{eff}, of the order of 10^{-9}. A value of zero for this parameter is a natural result of Peccei-Quinn symmetry, which is exact at the classical level and which will give rise to a massless Goldstone boson. Such a particle is called the *Axion*. Quantum mechanical effects, such as the triangle anomaly with gluons, give a nonzero value to the mass of the axion arising from the spontaneous breaking of Peccei-Quinn symmetry. Thus, the axion is a pseudo-Goldstone boson.

The effect of adding the triangle anomaly contribution modifies the added term in the Lagrangian density to the form

$$\mathcal{L} = (\theta_{eff} - \frac{\phi}{f}) \frac{g^2}{32\pi^2} F^a_{\mu\nu} \tilde{F}^{\mu\nu}_a$$

where ϕ is the axion field and f is its decay constant. Including non-perturbative QCD effects, it is found that the potential for the axion field ϕ has a minimum when ϕ is equal to $f\theta_{eff}$ and the added term vanishes. Thus the existence of the axion would solve the problem of CP violation in strong interactions.

The question that naturally arises from these considerations is whether axions do exist in nature. To look for them experimentally, one needs to have some idea of their mass and the size of their couplings to the various standard model particles. The mass of the axion turns out to be inversely proportional to f.

Early axion models chose f to be of the order of the electroweak scale, about 250 GeV, and have two Higgs doublets in them. The axion masses and couplings are known once the ratio of the vacuum expectation values of the Higgs doublets is specified and flavor conservation at the tree level is imposed. Such an axion would have a mass of about 1.8 MeV and would couple weakly to electrons and positrons. It has been looked for and not found. Does this mean that one is brought back to the problem of strong interaction CP violation? Not necessarily, because it is possible that f may be much larger than the electroweak scale, 250 GeV.

Models in which f is much greater than the electroweak scale, are called *invisible axion models*. The axion couplings become so weak that they escape observation in laboratory experiments. However, even such invisible axions do have astrophysical consequences. In particular, they could be candidates for cold dark matter in the universe. Their existence will have an effect on time scales of evolution of stars. Such effects, which depend upon the strength of the interaction of axions with photons, electrons, and nucleons, should not be so large as to lead to conflict with observational limits. For example, globular-cluster stars, provide evidence that the strength of the photon-axion coupling must not be larger than about 0.6×10^{-10} GeV^{-1} [54]. Another astrophysical observation is the duration of Supernova SN1987A neutrino signal. This signal is of a few seconds duration. This establishes that the cooling of the neutron star has proceeded through emission of neutrinos rather than via the invisible axions [55]. Such considerations are useful in providing regions of exclusion for the axion nucleon couplings and provide direction for future searches for axions. Further experimental searches for axions, both laboratory and astrophysical, are in progress. At present it is inconclusive whether axions exist.

B Baryon

The first indication of a b quark bound in a baryon came from the CERN Intersecting Storage Ring pp collider [56]. Its mass as measured by this group is about 5.43 GeV. It was found to be electrically neutral, decayed into a proton, D_0 meson and π^-, and was produced in association with another particle whose decay product contained a positron. The interpretation given was that this is an example of associated production of naked "b" states in pp interactions with a quark composition (udb), that is, a Λ_b^0. This was further confirmed by observation at the $p\bar{p}$ collider by the UA1 collaboration [57].

B Meson

If $b\bar{b}$ bound states in the form of $\Upsilon(1S)$ states exist, there should be other bound states in which a b quark or antiquark is bound with other quarks, such as one of the (u, d) pair or one of the (c, s) pair. Such quark-antiquark bound states should exist as mesons, while a bound state of three quark combinations should be baryons. For example, the $B^\pm$ mesons are the bound states $(\bar{b}u)$ and $(b\bar{u})$, respectively. Particles such as these were found in a systematic study of one of the excited Υ states called the $\Upsilon(4S)$ by the CLEO collaboration at the Cornell e^+e^- ring [58]. At the energy of the production of $\Upsilon(4S)$ in this machine, a strong enhancement of single electron production was observed. This observation was interpreted as the fact that the energy of the $\Upsilon(4S)$ is above the energy required to produce a B meson and its antiparticle $\bar{B}$, and the observation of the single electrons came from the weak decay of the B meson or the $\bar{B}$. Thus, the interpretation of this process is: $e^+e^- \to \Upsilon(4S) \to B\bar{B}$ and $B \to e^- +$ anything and $\bar{B} \to e^+ +$ anything.

B_s Meson

The B_s meson is the bound state formed from $b\bar{s}$, is electrically neutral, and may be called the strange meson B_s. Its antiparticle $\bar{B}_s$ would be the bound state $\bar{b}s$. The first direct observation and measurement of its properties were done at the CERN LEP e^+e^- collider by the ALEPH and OPAL collaborations [59].

b Quark

The first indication of the existence of a heavy quark with mass around 4.5 GeV came from the observation of a dimuon resonance in 400 GeV proton-nucleus collisions carried out at the Fermilab proton synchrotron [60]. This strong enhancement of the dimuon signal at 9.5 GeV is interpreted as being due to the production of the $\Upsilon(1S)$ state, which is thought to be a bound state of a new quark and its antiparticle (the first of a third generation family) called *beauty quark, bottom quark,* or just *b quark* and which decays into a pair of muons. Other excited Υ states called the $\Upsilon(2S)$, $\Upsilon(3S)$, and $\Upsilon(4S)$ have also been found and their modes of decay have also been studied [61].

Details of their masses and decay widths and other properties can be found in the "Review of Particle Physics" produced by the Particle Data Group [62].

$B^0\bar{B}^0$ Mixing

The B mesons contain b quark as one of their constituents. The B^0 meson's quark composition is $d\bar{b}$ while that of the $\bar{B}^0$ is $\bar{d}b$. These are also referred to as B_d^0 and $\bar{B}_d^0$, respectively. If the d quark in these is replaced by the s quark, we have the B_s^0 and the $\bar{B}_s^0$ mesons, respectively. These mesons are eigenstates of the strong interaction Hamiltonian and have the same values of mass and other quantum numbers, such as total angular momentum J, parity P, and the charge conjugation parity C. In the production processes of the B mesons, the appropriate eigenstates are those of the strong interaction Hamiltonian. Once they are produced, however, the weak interactions do not preserve quark flavours and, therefore, lead to a mixing of the B^0 and $\bar{B}^0$ states by second order weak interactions. The mixings result in B^0–$\bar{B}^0$ oscillations. Such mixing was first found in the neutral kaon system, $(K^0$–$\bar{K}^0)$ mixing (see under "Kaons—Neutral"), and led to K^0–$\bar{K}^0$ oscillations.

A natural question that arises is whether oscillations occur in the B^0–$\bar{B}^0$ system similar to those in the neutral kaon system. First evidence for such oscillations occurring was obtained by Albajar et al. [63], working with the UA1 collaboration at the CERN proton-antiproton collider. If B^0–$\bar{B}^0$ oscillations do not occur, the two muons coming from the decay of the B^0 and the $\bar{B}^0$ have opposite signs: $B^0 \to \mu^- X$, $\bar{B}^0 \to \mu^+ X'$. The appearance of same sign dimuons would signal the presence of such oscillations. Unfortunately the situation is not that simple. Same sign dimuons come also from a background process in which the B^0 decays directly to $\mu^- X$, while the other decays through a cascade process $\bar{B}^0 \to DX' \to \mu^- Y$, where X, X', and Y are some hadronic states. If one takes care to eliminate this background signal experimentally, in the absence of oscillations, there should be no dimuon signal of the same sign. In the experiment such background elimination was carried out, but still a signal of same sign dimuon was left over. They interpreted this excess signal as arising from the presence of oscillations in the B^0–$\bar{B}^0$ system. They found that the fraction of primary decays of the B's that give opposite sign dimuons from that expected without mixing was 0.121 ± 0.047.

The presence of B^0–$\bar{B}^0$ mixing was also found at electron-positron colliders by the Argus collaboration [64] at DESY working with the Doris II storage ring. It has been confirmed by the CLEO collaboration [65] at the Cornell $e^+ e^-$ storage ring. Both these groups have measured $B^0 \bar{B}^0$ mixing by looking for like sign dilepton events coming from the decay of $\Upsilon(4S)$. The Argus collaboration had a total of 50 like sign dileptons, of which 25.2 were determined to be background and 24.8$\pm$7.6 were signal. Comparable numbers for the CLEO collaboration were: total 71 dilepton events, 38.5 background, and 32.5$\pm$8.4 signal. CLEO calculates a value for the mixing parameter r, which is defined as

$$ r = \frac{N_{++} + N_{--}}{N_\pm - N_\pm(\text{from } B^+ B^- \text{ decay})}, $$

where in the denominator, the number of opposite sign dileptons from $B^+ B^-$ decays have been subtracted. Since the detector efficiencies for ee, $\mu\mu$, and $e\mu$ are not the same, CLEO first calculates from the signal data, $r_{ee} = 0.158 \pm 0.085 \pm 0.056$, $r_{\mu\mu} = 0.036 \pm 0.098 \pm 0.062$, and $r_{e\mu} = 0.262 \pm 0.088 \pm 0.051$, and then calculates a weighted average, $r = w_{ee} r_{ee} + w_{\mu\mu} r_{\mu\mu} + w_{e\mu} r_{e\mu} = 0.0188 \pm 0.055 \pm 0.056$, with $w_{ee} = 0.38, w_{\mu\mu} = 0.15, w_{e\mu} = 0.47$. These weights are proportional to the expected number of dileptons from the single lepton rates. In all these r values, the first error is statistical, the second one systematic. The conclusion of these results is that the $B^0 \bar{B}^0$ mixing is substantial.

More recently, observations of $B^0 \bar{B}^0$ mixing in Z^0 decays to $b\bar{b}$, have been done at LEP ring at CERN by the L3 [66] and the ALEPH [67] collaborations. The L3 group found a mixing parameter, $0.178^{+0.049}_{-0.040}$, while the ALEPH group found a value, $0.132^{+0.027}_{-0.026}$.

BaBar Experiment

This is an experiment that is being undertaken to study the properties of B mesons and conducted at a facility dedicated to producing large numbers of B mesons, called a B-factory. This B-factory is located at the Stanford Linear Accelerator Center (SLAC). The B mesons are produced in an electron-positron ring called *PEP-II*, which is an upgrading of the older $\beta0$ GeV, 800 meter diameter, colliding-beam storage ring called *PEP*. PEP-II is an asymmetric electron-positron collider, colliding 9 GeV electrons with 3.1 GeV positrons.

A new detector called *BaBar detector* has been built, by a large international collaboration of scientists and engineers from 72 institutions in 9 countries, to study the properties of the produced B mesons from the collision region. (The acronym for the name comes from the fact that it will study B and $\bar{B}$ system of mesons.) This detector, just as other colliding beam detectors, consists of a silicon vertex detector in its innermost part, followed by a drift chamber assembly and a particle identification system. The electromagnetic calorimeter is made up of CsI. There is a magnet to provide measurements of momentum, and the return iron yoke is instrumented.

Apart from studying in detail the properties of B mesons, BaBar will also be used for studying CP violation in the B meson system where the effects are expected to be large (see further under "CP Violation— Neutral B Mesons").

BAIKAL Neutrino Telescope

This is a neutrino detector that is being constructed in Lake Baikal. It is a unique telescope with an effective area of 2 times 11,000 square meters. The water volume is controlled at about 200,000 cubic meters. This detector will also be able to look at high energy neutrinos coming from astrophysical objects.

BAKSAN Neutrino Detector

This neutrino observatory is situated in Prielbrusye in the Caucasus mountains. It consists of a complex of detectors comprising a Gallium-Germanium neutrino telescope, Lithium-Beryllium and Chlorine-Argon telescopes (under construction), and an underground scintillation telescope, all with the object of further observations on neutrinos, solar as well as of astrophysical origin.

Baryon Number Conservation

Baryons are particles having a mass equal to or greater than protonic mass. The proton is the lowest state among the baryons; all the other baryons are heavier and decay to lower mass states but the decays stop when they reach the state of the proton. The neutron, is only about 1.29 MeV heavier than the proton and undergoes β decay emitting a proton, an electron, and an (anti)neutrino. Protons have not been observed to decay into any lower mass states despite the fact that there are many lower mass particles, such as various mesons, muons, and electrons. This lack of decay is supposed to be due to the operation of a selection rule forbidding the decay. This selection rule goes under the name of *the law of conservation of baryons*. Particles lighter than the proton are assigned a baryon number zero. Particles heavier than the proton which ultimately decay into the proton are all assigned a baryon number 1. Antibaryons carry the opposite baryon number, -1. The specific statement of the conservation law for baryons is attributed to Stückelberg [68]. This was brought into further prominence by Wigner [69]. There are also earlier discussions on this subject by Weyl [70].

Unlike some of the other conservation laws, this law is not directly related to a symmetry principle. In fact, in theories of grand unification of forces, the proton has a finite probability to decay, with violation of conservation of baryon number. Many experiments have been set up to look for proton decay; so far no such decays have been found.

Baryonic Resonance
The first among a very large number of baryonic resonances to be discovered was what is now called the $\Delta(1232)P_{33}$. This was discovered in the scattering of positive pions on protons, with pions having a kinetic energy between 80 MeV and 150 MeV, from the Chicago cyclotron. The cross section for this process was found to increase rapidly with energy in this energy range. In these first experiments it was not clear whether the cross sections go through a peak and come down to smaller values as would be necessary if it was a resonance. Subsequent work [71] using the pion beams of higher energies from Brookhaven Cosmotron showed that it is indeed a resonance in the isospin 3/2, angular momentum $P_{3/2}$ state of the pion-nucleon system.

Baryonic States
Many low lying baryonic states with excitation energies of the order of several hundred MeV above the proton state were discovered in accelerator experiments in the late 1950's and 1960's. These states group themselves systematically into families, with specific values of spin and parity for all members of the family. Among these, two families, the baryon octet and the decuplet, are specially worthy of note, because they provided clues which led to classifying them in terms of a sym-

metry scheme higher than isospin symmetry. The baryon octet has spin $J = 1/2$, has positive parity, and contains both non-strange and strange baryon states. It contains the proton and the neutron (mass 939 MeV); three strangeness -1 particles, $\Sigma^{\pm,0}$ (mean mass 1190 MeV); the Λ^0 particle (mass 1115 MeV); and two strangeness -2 particles, $\Xi^{-,0}$ (mean mass 1318 MeV), making a total of eight particles. The baryon decuplet has spin $J = 3/2$ and positive parity and also contains both non-strange and strange baryon states. There are ten members in this family, four of these with a mean mass of 1230 MeV, three particles with mean mass about 1380 MeV, two particles with a mean mass of about 1530 MeV, and one particle with mass about 1675 MeV. Of these ten particles, through a series of detailed experiments, it has been established that the first 4 states carry no strangeness, while the others carry strangeness quantum numbers which are nonzero. The three of mass $\simeq$ 1380 MeV have strangeness -1, the next two (mass $\simeq$ 1530 MeV) strangeness -2, and the last one strangeness -3. The mean mass spacing of the different strangeness states in these baryon multiplets is about 120 MeV to 150 MeV, while the mass differences between members of a family with same strangeness are very small. This suggests a grouping of the particles with given strangeness into multiplets of isotopic spin. For the octet, the proton and neutron form an isospin doublet, the Σ's an isospin triplet, the Λ an isospin singlet, the Ξ's an isospin doublet. For the decuplet, the quartet of states with mass around 1230 MeV form an isospin $(3/2)$ multiplet, the triplet of states with mass around 1380 MeV are assigned an isospin 1, the doublet of states with mass around 1530 MeV are assigned an isospin $(1/2)$, and finally the single state at about 1675 MeV is assigned an isospin 0. The eight members of the octet multiplet can be assigned to the eight dimensional representation of SU_3 and the ten members of the decuplet can be assigned to the ten dimensional representation of SU_3 (see further under "Eightfold Way"). In the decuplet, the first four are referred to as the "Δ" particles, $\Delta^{++}, \Delta^+, \Delta^0, \Delta^-$, the next three "$\Sigma$"-like particles, $\Sigma^+, \Sigma^0, \Sigma^-$, the next two "Ξ"-like particles Ξ^0, Ξ^-, and the last is called Ω^-. Further work on the SU_3 symmetry scheme showed that all these states can be understood as the excitations in a bound state of three quarks, called the u, d, and s quarks, in a model called the constituent quark model (see further under "Constituent Quark Model of Hadrons").

At higher masses there are other multiplets of baryons, called *charm baryons* and *beauty (or bottom) baryons*. The charm baryon contains a "charm" quark, while the beauty (or bottom) baryon contains a "beauty" (or "bottom") quark in addition to two other quarks.

Beta Decay

Radioactive sources emit radiations which have been classified as α, β, and γ rays which are supposed to arise when the nucleus of the atom undergoes a transformation. Of these, the β rays have been found to be electrons. Early observations of the energy spectrum of the emitted electrons showed the existence of a discrete energy spectrum on top of which was superposed a continuous energy spectrum. Further studies revealed that the discrete spectrum is associated with a process in which the nucleus emits γ radiation. Sometimes, instead of the γ ray being actually emitted by the nucleus, the energy of the nuclear transformation is transferred to one of the electrons bound in the atom, and ejection of the electron occurs with an energy which corresponds to the nuclear transition energy, minus the binding energy of the electron in the atom. Such electrons are called *conversion electrons* and are closely associated with the emission of γ radiation. The electrons which have a continuous energy distribution and which are emitted in nuclear transformations are called β *rays*.

Before we discuss the continuous energy distribution and its implications, we digress a little and discuss some of the features of the energy changes in β decay in general. The atomic mass, the nuclear mass, and the electron mass all play a role here. Let us denote the bare nuclear mass with atomic number Z as N_Z. The mass of the atom M_Z is made up of the bare nuclear mass together with the masses of the Z electrons in the atom, less the sum total of their binding energies B_Z: $M_Z = N_Z + Zm - B_Z$, where m is the mass of the electron (recall that we are taking c, the velocity of light, equal to one). For the bare nucleus undergoing β (electron) decay, we may write $N_Z = N_{Z+1} + m + Q$, where Q is sum total of the kinetic energy of the electron, the recoil kinetic energy given to the nucleus, and the kinetic energy given to any other particles that may be emitted in the process, and is simply called the Q value for the decay. (We will see immediately below that at least one other particle, called the *neutrino*, is indeed emitted.) Rewriting this in terms of atomic masses, we have

$$M_Z = M_{Z+1} + (B_{Z+1} - B_Z) + Q \qquad \text{(electron emission)}$$

The second term in the round brackets, namely, the difference in the binding energies, is usually very small compared to the beta ray energies. We may write similar equations for the case of positron emission. Here the atomic number changes from Z to $Z-1$: hence, $N_Z = N_{Z-1} + m + Q$, or in terms of atomic masses,

$$M_Z = M_{Z-1} + 2m + (B_{Z-1} - B_Z) + Q \qquad \text{(positron emission)}$$

Another process that occurs, which is included under the term β decay, is orbital electron capture. This usually occurs in heavy nuclei, from the K-shell of the atom because these electrons have an appreciable probability of spending a lot of time inside the nuclear volume. Here, if the binding energy of the electron is represented by B_Z^K, then we have $N_Z + m - B_Z^K = N_{Z-1} + Q$, or equivalently

$$M_Z = M_{Z-1} + (B_{Z-1} - B_Z) + B_Z^K + Q \qquad \text{(electron capture)}$$

In electron capture, Q represents the sum total of the energy carried away by the neutrino and the recoil energy given to the nucleus. If the atomic masses are such that K-shell capture is not possible, capture of electrons may occur from higher shells, such as L-shell or M-shell.

Let us get back to the continuous energy distribution of the β particles. The continuous energy distribution for the electrons poses a serious problem with energy conservation. The electrons are emitted when the nucleus makes a transition between two definite energy states and therefore must have a unique energy. It is found, however, that the continuous energy distribution has a definite upper limit, and that only this upper limit (plus the rest mass energy of the electron) equals the energy difference between the nuclear states involved.

Another problem has to do with angular momentum conservation. Consider, for example, β decay of tritium, $_1^3\text{H} \rightarrow {}_2^3\text{He} + e^-$. The spin angular momentum (measured in units of $\hbar$, the unit of angular momentum which we take to be 1) of $_1^3\text{H}$ on the left-hand side and that of $_2^3\text{He}$ on the right-hand side are known to be $(1/2)$ each. On the right-hand side, the total angular momentum carried is made up of the total intrinsic spin carried by the two particles, which can thus have the values 0 or 1, plus any orbital angular momentum between the two. Since the orbital angular momentum can take on values which are only integral, the total angular momentum on the right-hand side can be only integral. There is a discrepancy in the value of the angular momentum on the left and right sides leading one to question angular momentum conservation.

Both these problems disappear, if one assumes that along with the electron, an additional electrically neutral particle carrying half a unit of intrinsic angular momentum is emitted undetected. This particle must also have very little rest mass energy associated with it so that the maximum total electron energy (kinetic and rest mass energy) will be equal to the energy difference of the nuclear states, as is observed. Such a solution of this problem was proposed by Pauli [72], and this particle has been called the *neutrino* (see also under "Neutrino"). It is a massless, chargeless particle, carrying $(1/2)$ unit of intrinsic angular momentum.

Beta Decay Leading to Bound Electrons

Normally when β decay occurs, the decay electron or positron is emitted and escapes from the atom to the outside. Since the *beta* energy spectrum has all kinetic energies ranging from zero to some maximum value, those electrons at the low energy end of the spectrum, as they escape, can be strongly influenced by the attractive electric force of the nucleus. If the attractive force they feel is strong enough, it is possible that the electron might end up in a bound state in the atom corresponding to the daughter nucleus. A precise calculation of the probability of this bound state decay has been done quantum mechanically [73]. These calculations indicate that there is a small probability for the decay electron to end up in the bound state, typically of the order of 10^{-4} to 10^{-5} relative to the normal decay probability.

The first direct observation of β-decay electrons ending up in a bound state was carried out at the Darmstadt heavy ion facility by Jung et al. [74]. They observed the β decay of $^{163}_{66}\text{Dy}^{66+}$ ions stored in the ring of the heavy ion facility. The number of daughter ions, $^{163}_{67}\text{Ho}^{66+}$, produced from the decay of the parent ions was measured as a function of time spent in the ring. From this, a half-life for the decay could be determined. It was found to be 47^{+5}_{-4} days. This result, taken in conjunction with the measured half-lives for electron capture from the M_1 and M_2 shells of neutral $^{163}_{67}\text{Ho}$, provided the information necessary to put a limit on both the Q value for electron capture and the electron neutrino mass.

Beta Decay of the Λ

The Λ particle is an electrically neutral baryon having strangeness -1. It has been observed to decay into a proton and π^- meson. The first observation of the β decay of this baryon, $\Lambda \rightarrow \text{proton} + \text{electron} + \bar{\nu}_e$, was reported by Crawford et al. [75] using the Berkeley Bevatron and a liquid hydrogen bubble chamber. The sequence of the reactions which led to this decay was established to be:

$$\pi^- p \rightarrow \Sigma^0 + K^0; \quad \Sigma^0 \rightarrow \Lambda + \gamma; \quad \Lambda \rightarrow p + e^- + \bar{\nu}_e.$$

Another observation which confirmed this finding was due to Nordin et al. [76]. It was done with a different initial beam. A beam of separated K^- was allowed to impinge on a target which was a liquid hydrogen bubble chamber. The K^- reacted with the protons in the liquid hydrogen in the bubble chamber, by the reaction $K^- + p \rightarrow \Lambda + \pi^0$, the Λ from which underwent a β decay. Since these initial observations many more such β decays of the Λ have been observed.

Beta Decay of the Neutron

The neutron, being heavier than the proton by 1.29 MeV in energy units, can undergo decay into a proton: $n \rightarrow p + e^+ + \nu_e$. The Q-value in this reaction is rather small, and hence one would expect a rather small decay probability and therefore a long lifetime. A first evidence that such a decay of the free neutron was occurring came from experiments performed at a nuclear reactor by Snell and Miller [77]. This was confirmed in further works by Robson and by Snell et al. [78], who also measured the lifetime to be in the range 9 to 25 min. The lifetime is now known much better. It has the value (886.7 ± 1.9) s.

Beta Decay of π^+

The positive pion is heavier than the neutral pion by a small amount, the difference in energy units being about 4.59 MeV. From theoretical considerations, Gershtein and Zeldovich [79] and, independently, Feynman and Gell-Mann [80] predicted that the charged pion ought to suffer β decay into the neutral pion state: $\pi^+ \rightarrow \pi^0 + e^+ + \nu_e$. The first observation of this decay was made by Dunaitsev et al. [81] at the Dubna synchrocyclotron using spectrometers.

Beta Decay—Strong Interaction Corrections

In the introduction to the theory of beta decay, we mentioned that the interaction Hamiltonian in this case is formed from the product of two currents, one from the proton-neutron system and the other from the electron-neutrino system. Of these, the proton-neutron system belongs to the family of particles called *hadrons*, which are subject to strong interactions, while the electron-neutrino system is not. The latter system belongs to the family of particles called *leptons*. The question, then, is whether the strong interactions that the hadrons feel require corrections to the β decay theory which have to be taken into account. In general the answer is yes. However, for the *vector* hadronic current, it can be shown that the strong interaction corrections do not modify the coupling involved. This was first shown by Gershtein and Zeldovich [79], and also by Feynman and Gell-Mann [80], and goes under the name of *Conserved Vector Current Hypothesis*.

The answer comes from analogy with a similar question in electrodynamics about the equality of the observed electric charge of the proton and the positron. This equality of the observed charges is a statement about the electric charges after all renormalization effects due to strong interactions for the proton are taken into account. It can be shown that, if one starts with equal "bare" electric charges for the proton and the positron, and the divergence of the vector currents of the particles vanish (that is, conservation of vector current holds), then the renormalized electric charges are also equal.

Renormalization effects arise from strong interactions which have the property of being charge independent. Description of this is simple in terms of the concept of isotopic spin: the proton and neutron are considered as two substates of a particle, the nucleon, with isospin vector $\vec{I}$, (components (I_1, I_2, I_3)), with eigenvalues $\vec{I}^2 = (1/2)((1/2) + 1)$ and $I_3 = +(1/2)$ for the proton and $I_3 = -(1/2)$ for the neutron state (see further under "Isotopic Spin"). In analogy with ordinary spin, $I_1 + iI_2$ and $I_1 - iI_2$ will play the role of raising and lowering operators, changing neutron to proton and proton to neutron, respectively. Statement of charge independence becomes a statement of symmetry under rotations in the isotopic spin space. One can introduce an isospin (vector) current for the nucleon $\vec{I}_\mu(x)$, whose isospin "3" component $I_{3,\mu}(x)$ is related to the electromagnetic current. The other components, $A_\mu^\pm(x) = I_{1,\mu}(x) \pm iI_{2,\mu}(x)$, can be associated with currents, which change a neutron into a proton (upper sign) and proton into a neutron (lower sign), respectively, and which play the role of currents in beta decay. If the currents $\vec{I}_\mu(x)$ have divergence zero (that is, a conserved vector current (CVC) exists), and the "bare" weak coupling constants are equal, then the renormalized weak coupling constants will also be equal in analogy with the renormalized electric charge. Thus, the consequence of CVC is that the weak coupling constant for vector interactions in beta decay is not renormalized.

Beta Decay—Theory

The quantum field theory of β decay was first given by Fermi [82] in 1934. Assuming the existence of the neutrino, he formulated the theory for the emission of the electron and an antineutrino by a method similar to that used for treating the emission of radiation by an excited atom. (An antineutrino is emitted along with the electron rather than a neutrino, because of the association of a lepton number with these particles; see discussion under "Leptons".) In radiation emission by the atom, the interaction Hamiltonian density is formed by the scalar product of the electromagnetic vector current of the electron, $\bar{\psi}(x)\gamma_\mu\psi(x)$, and the vector potential of the electromagnetic field, $A^\mu(x)$,

$$H(x) = e\bar{\psi}(x)\gamma_\mu\psi(x)A^\mu(x).$$

By analogy, Fermi assumed that in beta decay, the interaction Hamiltonian density is the scalar product of the vector current formed from the proton-neutron system with a similar vector current formed from the electron-neutrino system,

$$H(x) = G_F \bar{\psi}_p(x)\gamma_\mu \psi_n(x)\bar{\psi}_e(x)\gamma^\mu \psi_\nu(x).$$

In radiation emission, the coupling constant is the electric charge e of the electron. In beta decay, a corresponding coupling constant G_F called the *Fermi weak coupling constant* was introduced, which has the dimensions of energy times volume (or dimension of length squared, in units where $\hbar = c = 1$). The calculation of the probability of transition proceeds in first order perturbation theory using the expression,

$$P = 2\pi|M|^2 \rho_f,$$

where M is the matrix element for the transition, and ρ_f is the number of final (f) states per unit energy interval accessible to the products of the decay.

The electron and the antineutrino, each have intrinsic angular momentum (spin) 1/2, so the total spin angular momentum carried by these is 0 or 1. If no orbital angular momentum is carried by them with respect to the nucleus, then the total change in the angular momentum ΔJ of the nuclear states can only be 0 or 1, and the parity of the nuclear state does not change. These are called *allowed transitions*. Of these, the transitions in which $\Delta J = 0$ are called *Fermi transitions* (no spin flip in $n \to p$), the ones in which $\Delta J = 1$ are called *Gamow-Teller transitions* (spin flip in $n \to p$). It should be noted that in the Fermi theory with vector currents alone, no spin flip can occur in the transformation of a neutron into a proton or vice versa, and hence it cannot give rise to allowed Gamow-Teller transitions. To accommodate allowed Gamow-Teller transitions, modification of the simple Fermi theory above is necessary (More on this below). Other transitions, in which nonzero orbital angular momentum is involved in the nuclear transition, so that $\Delta J > 1$, are called *forbidden transitions*. In forbidden transitions, in addition to spin flip or no spin flip, there is a change in the orbital angular momentum in the transformation $n \leftrightarrow p$.

Fermi succeeded in deriving expressions for the decay rate (or the mean lifetime) and for the energy distributions of the β-particles in allowed Fermi transitions. As mentioned above, the vector interaction used by Fermi did not allow spin flip in the $n \to p$ transformation, hence it did not allow Gamow-Teller transitions, and to account for these a modification of the theory was necessary. It was soon found that, within the requirements of special relativity, five possible forms of the currents are allowed, a linear combination of whose scalar products occurs in the interaction Hamiltonian for β decay. These forms are named according to their transformation properties under Lorentz transformations and space reflections in Dirac theory: (S) $(\bar{\psi}\psi)$ scalar, (V)$(\bar{\psi}\gamma_\mu\psi)$ vector, (T) $(\bar{\psi}\gamma_\mu\gamma_\nu\psi)$ tensor, (A) $(\bar{\psi}\gamma_5\gamma_\mu\psi)$ axial vector, and (P) $(\bar{\psi}\gamma_5\psi)$ pseudoscalar. Of these, the (P) pseudoscalar form can be safely ignored in nuclear beta decay processes because it produces effects only of order $\beta^2 \simeq 10^{-3}$, where β is the typical velocity of a nuclear particle (in units of the velocity of light). Of the rest, (S) scalar and (V) vector forms do not allow spin-flip, and hence can account for Fermi transitions, while (T) tensor and (A) axial vector forms allow spin flip, and hence can account for Gamow-Teller transitions. Many years of very detailed work had to be done to determine which of these combinations best described all the data in the field including the discovery that neutrinos and electrons have left-handed helicities -1 and $-\beta$, respectively, and antineutrinos and positrons have right-handed helicities $+1$ and $+\beta$, respectively, where helicity measures the projection of the spin on the direction of the momentum (longitudinal polarization). (See also under "Parity—Nonconservation in Nuclear β Decays".)

In an allowed Fermi transition, since the total angular momentum carried by the electron and the antineutrino is zero, the right-handed antineutrino (helicity $+1$) is accompanied by an electron of opposite helicity. Theoretical calculations based on the vector interaction (V) tend to favor the configuration in which the electron and the antineutrino have opposite helicities and, hence, tend to go in the same direction, whereas the scalar interaction (S) favors same helicities and, hence, tend to go in opposite directions. Experimentally, the antineutrino direction is determined by momentum conservation, if one knows the electron momentum and the recoil momentum given to the nucleus. When the electron and the antineutrino go in the same direction, large values of nuclear recoil will occur, while when they go in opposite directions, small values of nuclear recoil will occur. Thus, an experiment determining the recoil distribution in allowed Fermi decays can decide whether (V) or (S) interaction is favored, because for (V) one expects large nuclear recoil and for (S) small nuclear recoil. Experimentally one finds large nuclear-recoil momenta are favored, hence (V) rather than (S) is selected.

In an allowed Gamow-Teller transition, the total angular momentum carried by the electron and the antineutrino is 1. Thus, they have parallel spin. Again, the right-handed antineutrino (helicity $= +1$) is accompanied by an electron of opposite helicity. It is found that the (A) interaction favors opposite helicities, whereas the (T) interaction favors the same helicities. Hence one expects to see large nuclear recoil momenta only for (A) interactions and small recoil nuclear momenta only for (T) interactions. Experimentally one finds again large nuclear recoil momenta are favored, thus picking (A) over (T). Further work has shown that the combination (V-A) best describes the entire set of data available in the field [80].

Details of the theoretical derivation of the expression for probability of β decay per unit time, and hence an expression for the half-life for β decay can be found in any book on nuclear physics, for example, in the book by Segrè [83]. We will only quote the result here and not present the details. The probability, $w(p_e)dp_e$, of a β decay, giving an electron with momentum in the interval $(p_e, p_e + dp_e)$ is derived to be (assuming zero neutrino mass)

$$w(p_e)dp_e = \frac{G_F^2}{2\pi^3}|M|^2 F(Z, E_e)(W - E_e)^2 p_e^2 dp_e.$$

(If a finite neutrino mass, $m_\nu \neq 0$ is assumed, the factor $(W - E_e)^2$ in this expression, should be replaced by $[W - E_e]\sqrt{[W - E_e]^2 - m_\nu^2}$.) Here, G_F is the Fermi weak coupling constant, characterizing the β decay coupling, W is the total disintegration energy (that is, the energy difference between the nuclear states), E_e is the total energy of the electron (including its rest mass energy), $p_e = \sqrt{E_e^2 - m_e^2}$ is the momentum of the electron, and E_ν is the energy of the antineutrino. Energy conservation (neglecting the small nuclear recoil energy) gives $W = E_e + E_\nu$. $|M|^2$ is the square of the matrix element for the nuclear transition. The matrix element M is reduced to an integral over the nuclear volume of the electron wave function, the antineutrino wave function, and the wave functions of the transforming neutron and the produced proton, summed over all the particles of the nucleus. The function $F(Z, E_e)$ is called the *Fermi function* and corrects for the fact that the wave function of the electron is not a plane wave but affected by the Coulomb field of the nucleus. This function has the approximate form $F(Z, E_e) = 2\pi\eta_1[1 - \exp -2\pi\eta_1]^{-1}$, where $\eta_1 = \frac{Z}{137\beta_e}$ with β_e being the velocity of the electron (in units of the velocity of light) far from the nucleus. This factor is nearly unity, except for very low energy electrons and/or high Z nuclei. For allowed transitions, M is independent of p_e and p_ν, and may be considered a constant for the discussion of the energy distribution of the electrons in β decay. The above expression represents the form of the β electron spectrum. An exactly similar expression holds for the positron spectrum, the only modification occuring in the Fermi function in which the η_1 is opposite in sign for positrons. Using $\eta = (p_e/m_e)$, $\epsilon = (E_e/m_e)$, $W_0 = (W/m_e)$.

We see that a plot of

$$\sqrt{\frac{w(\eta)}{\eta^2 F(Z, E_e)}} \quad \text{versus } (W - E_e)$$

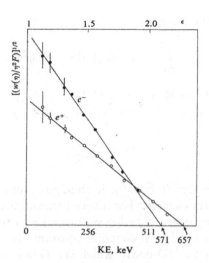

Figure 4.3: Kurie plot for ^{64}Cu electrons and positrons. (Figure from G. E. Owen and C. S. Cook, *Physical Review* 76, 1726, 1949. Copyright 1949 by the American Physical Society.)

must give a straight line. Such a plot is called a *Kurie plot*. This plot is shown in Figure 4.3 above, for electrons and positrons from ^{64}Cu. Deviations from a straight line, for energies E_e away from the end point W in this plot, indicates forbidden transitions. (For energies very near the end point W, the result of assuming $m_\nu \neq 0$ can be shown to lead to a Kurie plot with a vertical tangent at the end point. There is thus a possibility, at least theoretically, to find whether m_ν is zero or not. Experimentally, however, this seems very difficult for reasons of detector efficiency and resolution.)

The total decay rate λ is obtained from the above expression for the probability by integrating it over all electron momenta up to the maximum allowable, $p_{e,max} = \sqrt{(W^2 - m_e^2)}$. The expression is

$$\lambda = \frac{G_F^2}{2\pi^3}|M|^2 \int_0^{p_{e,max}} (W - E_e)^2 p_e^2 dp_e.$$

In evaluating the integral over p_e, introducing $\eta = p_e/m_e$, we get

$$\lambda = G_F^2|M|^2 \frac{m_e^5}{2\pi^3} f,$$

and f stands for the integral

$$f = \int_0^{\eta_{max}} F(Z,\eta)(W_0 - \epsilon)^2\eta^2 d\eta,$$

where we recall $W_0 = W/m_e, \epsilon = E_e/m_e$, and $E_e = \sqrt{p_e^2 + m_e^2}$. The half-life τ is given by $\tau = \frac{\ln_e 2}{\lambda}$. Thus the product $f\tau$ can be written as

$$f\tau = \frac{\text{const}}{|M|^2},$$

where the const $= [2\pi^3/(G_F^2 m_e^5)]$; it thus gives information about the square of the matrix element. For allowed transitions, $|M|^2$ is nearly 1. Using the measured value of $f\tau$ for various allowed transitions, one can deduce a value for the Fermi coupling constant G_F. For example, the measured $f\tau$-value for ^{14}O-decay, which is a *Fermi transition*, is 3100 ± 20 s. Putting in the values of the various quantities in the "const" above, one gets for G_F the numerical value 1.4×10^{-49} erg cm^3. Analysis of many β decays gives values for G_F which are close to this value thus giving credibility to this quantity being the universal coupling constant for β decay interactions. The value for G_F can be rewritten as

$$G_F = 1.0 \times 10^{-5}(\frac{1}{M_p})^2,$$

where M_p is the mass of the proton. In a more refined theory, where one takes into account the structure of the proton and neutron one gets the numerical value of G_F for Fermi transitions modified to

$$G_F = 1.02 \times 10^{-5}(\frac{1}{M_p})^2.$$

One can also get an idea of the coupling constant for a decay, which is a mixture of Fermi and Gamow-Teller transitions, from its measured $f\tau$-value. An example is the free neutron decay, which has an $f\tau$-value $= 1080 \pm 16$ s. From this, the ratio of the Gamow-Teller to Fermi couplings can be evaluated (except for the sign), and it turns out to be about 20% to 25% larger than the Fermi coupling in magnitude.

Usually, the value of $\ln f\tau$, rather than $f\tau$, is quoted from experiments on β decays. The $\ln f\tau$-values range from about 3 for allowed transitions to larger values, such as 6, 9, 12, 15, for forbidden transitions. The largest $\ln f\tau$-value known is 23 for ^{115}In, which has a half-life of 6×10^{14} years.

Bhabha Scattering

The elastic scattering process $e^+e^- \rightarrow e^+e^-$ was first calculated by Bhabha [84]. The scattering process is due to photon exchange between the particles involved and can be calculated, using QED, to the order in perturbation theory necessary to give the desired accuracy. Because the electrons and positrons are found to be points (with no structure) even at the highest energies investigated, the theoretically calculated Bhabha scattering cross section can be used to measure luminosity of electrons and positrons in colliders. All LEP experiments have luminosity monitors installed as close to the beam direction as possible and measure Bhabha scattering continuously, along with any other measurements that may be going on at the same time. The continuous monitoring of the luminosity is very important for high precision measurements.

Big-Bang Cosmology

In the early stages of the evolution of our universe after the big bang, particle physics has a significant role to play in answering some questions of cosmology. This interrelation between particle physics and cosmology has emerged as a very fruitful area of study and we briefly summarize here the highlights of these endeavors.

The space-time metric for our universe at early times after the big bang is usually taken to be described by the so called Robertson-Walker-Friedmann metric,

$$ds^2 = dt^2 - R^2(t) \left[\frac{dr^2}{1 - kr^2} + r^2 d\Omega^2 \right] \qquad d\Omega^2 = d\theta^2 + \sin^2\theta d\phi^2$$

In this equation, $R(t)$ may be loosely referred to as the radius, or the "size" of the universe. The constant k can be chosen to be either $+1$, 0, or -1 by appropriate scaling of the coordinates. These values of k give closed universe, open universe, or spatially flat universe, respectively. Einstein's general relativity theory gives the time rate of change of the radius of the Universe, called the *Hubble parameter*, $H = \frac{\dot{R}(t)}{R(t)}$, where

$$\left[\frac{\dot{R}(t)}{R(t)} \right]^2 = H^2 = \frac{8\pi}{3} G\rho - \frac{k}{R^2(t)} + \frac{\Lambda}{3},$$

where G is Newtonian gravitational constant, ρ is the total energy density, and Λ is the cosmological constant. We should note that, for a system of relativistic particles at temperature T, the radiation energy density varies like T^4, and the number density varies like T^3. The total number of relativistic particles $T^3 R^3$, which is essentially the entropy, remains constant if the expansion is adiabatic, which implies $T \propto (1/R)$.

If we divide both sides of the equation above by H^2, and introduce $\rho_c = [3H^2/(8\pi G)]$, called the *critical density*, we may rewrite the preceding equation as

$$1 = \frac{\rho}{\rho_c} - \frac{k}{H^2R^2} + \frac{\Lambda}{3H^2}.$$

Energy conservation is expressed by equating the time rate of change of energy in a sphere of radius R to the rate of flow of energy through the bounding surface,

$$\frac{d}{dt}\left(\frac{4\pi}{3}R^3\rho\right) = -p\dot{R} \cdot 4\pi R^2,$$

where p is the pressure. From this we can immediately derive, $\dot{\rho} = -3(\rho + p)(\dot{R}/R)$. Differentiating the equation given above for $(\dot{R}/R)^2$ with respect to time, we get,

$$\frac{\ddot{R}}{R} = \frac{\Lambda}{3} - \frac{4\pi G}{3}(\rho + 3p).$$

We shall take $\Lambda = 0$ hereafter. Putting a suffix 0 to denote present day values, we have

$$\frac{k}{R_0^2} = H_0^2(\Omega_0 - 1), \qquad \Omega_0 = (\rho_0/\rho_c).$$

Here R_0, H_0, and ρ_0 are the present day values of radius, Hubble parameter, and total energy density, respectively. The parameter Ω_0 is called the *density parameter*. The ultimate fate of the universe depends on this parameter; $\Omega_0 > 1, < 1, = 1$ define whether $k = +1, -1, 0$, respectively, and determine whether the universe is closed, open, or spatially flat.

Now we refer to some observational facts.

- The present day value of the Hubble parameter is not known precisely even after half a century of observations. It is usually expressed as $H_0 = 10^2 h_0$ km s^{-1} Mpc^{-1} (Mpc=megaparsec=3.26 × 10^6 light years), with h_0 lying somewhere in the range 0.4 to 1.

- The presence of a cosmic microwave background radiation (CMB) has been established [85]. It is found that the entire universe is pervaded by photons having a black body distribution with temperature $T = 2.7°$K. Within a "horizon" distance of 10^{28} cm we find 10^{87} photons.

- Knowledge of Ω_0 is obtained from several different measurements. A very important input is the observationally determined baryon to photon number densities (n_B/n_γ), in the range $(0.3-10) \times 10^{-10}$. This fact is used in deducing, from the observation of visible matter in the form of stars and galaxies, an estimate of Ω_0(visible) ≤ 0.01. Other observations on rotational curves of galaxies, use of virial theorem to determine cluster masses, gravitational lensing effects, etc., and all give estimates of Ω_0 in the range 0.1 to 2. These estimates show that there is a lot of invisible or "dark matter" out there. Particle physics may provide an answer as to what the dark matter may be (see under "Dark Matter, Machos, Wimps").

- Until recent observations with high resolution angular scales, it was thought that the CMB is completely isotropic (see further under "Cosmic Microwave Background Radiation").

- Measurements of the relative abundance of primordial light elements, hydrogen, and helium (^{4}He) have been done by astronomers for a long time. It is found that the abundance of hydrogen is about 70%, and of ^{4}He about 25%. A very interesting fact is that, based on our knowledge of particle physics and nuclear processes, all the primordial relative abundances observed can be determined in terms of one parameter, n_B/n_γ in the range $(1.5 - 6.3) \times 10^{-10}$.

The scenario for the evolution of the universe is as follows. In the early stages, the energy density is dominated by radiation, the temperature is very high, and the universe is populated by relativistic particles. Taking Boltzmann's constant equal to unity (1 eV $= 1.2 \times 10^4$ K), for $T > 10^{15}$ GeV, the total number of degrees of freedom g of relativistic paticles could be as much as 100 (such as in grand unified theories). In the radiation dominated regime, the k/R^2 term may be neglected in comparison with the other term and by integration we can get a "temperature clock":

$$t \simeq (gG)^{-1/2}/T^2 \simeq g^{-1/2}(2.4 \text{ s})(1 \text{ MeV}/T)^2.$$

Thus instead of the time, it is convenient to specify the temperature at which various events occur. However, in astrophysics, the events in the history of the universe, and are said to occur at certain red shifts. The red shift parameter z is related to $R(t)$ by $1 + z = (R(t_0)/R(t))$, where $R(t_0)$ is the present size of the universe, and $1 + z$ measures the red shift as the ratio of the observed wavelength of some particular spectral line to the wavelength that line has in the laboratory.

At a temperature of tens of MeV (time $\simeq$ a few millisec), the universe is populated by protons, neutrons, electrons, positrons, photons, and different neutrino species. The baryons are nonrelativistic while the other particles are all relativistic. The baryons contribute very little to the energy density, since $(n_B/n_\gamma)(m_N/T) \simeq 10^{-8}$, where m_N is the nucleon mass. The particles are kept in equilibrium by various processes, such as $\nu\bar{\nu} \leftrightarrow e^+e^-$, $\nu n \leftrightarrow pe^+$, $\gamma\gamma \leftrightarrow e^+e^-$. At this stage, these reaction rates are faster than the expansion rate of the universe. The characteristic expansion rate drops as the temperature drops. The weak interaction rates drop even faster, so a stage will be reached at which the neutrinos will no longer be in equilibrium. This happens at a temperature of about 3 MeV. The stronger interactions, such as electromagnetic interactions and strong interactions, keep the protons, neutrons, positrons, and photons in kinetic equilibrium. For $T > m_e$ (m_e is the rest mass energy of the electron), the numbers of electrons, positrons, and photons are comparable. Electrical charge neutrality gives $n(e^-) - n(e^+) = n(\text{protons})$ and hence there is a very slight excess (about 10^{-10}), of electrons over positrons. When T starts to drop below m_e (time $\simeq 1$ s) the process $\gamma\gamma \rightarrow e^+e^-$ is highly suppressed, and positrons annihilate with electrons and are not replenished. The annihilation heats up the photons relative to the neutrinos. Following this, a small number of electrons (and an equal number of protons), about 10^{-10} per photon, are left over.

As time progresses, since the radiation energy density falls like T^4 (or R^{-4}), and the energy density of nonrelativistic matter falls like R^{-3}, the universe will become matter dominated at some stage. The time at which this occurs can be found by equating radiation and matter energy densities. It is found that the temperature T_{eq} at which this happens is $T_{eq} = 5.6\Omega_0 h_0^2$ eV, and the corresponding red shift parameter z_{eq} is obtained from $1 + z_{eq} = 2.4 \times 10^4 \Omega_0 h_0^2$. Prior to these values, radiation dominates, while after these values, matter dominates.

The big achievement of big-bang cosmology is nucleosynthesis. Going back to the stage when rates of weak processes were still fast compared to the expansion rate, neutrons $\leftrightarrow$ protons reactions were rapid, and the neutron to proton ratio followed the Boltzmann factor, $\exp[-(m_n - m_p)/T]$. As temperature falls, there are fewer neutrons. The weak interaction rates are still significant until a temperature of a few MeV. When the temperature drops further, neutrons decay, and these decays become significant. The first step in making heavier nuclei is the production of deuterons, represented as d. The binding energy of the deuteron is small, about 2.2 MeV. If the temperature is above a certain critical value, any deuterons formed will be photo-dissociated right away. Since the photon to nucleon ratio is so large, the temperature has to fall substantially below the binding energy of 2.2 MeV so that the deuterons that are formed do not get dissociated. It turns out this temperature is about 0.1 MeV, and this is the temperature below which nucleosynthesis may be said to begin. There are not enough nucleons at this stage to have production of ^{4}He by many body reactions; nuclei must be

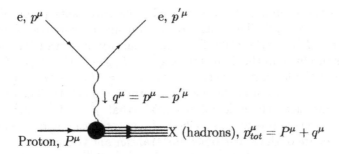

Figure 4.4: Kinematics of electron-nucleon deep inelastic scattering.

built step by step by strong and electromagnetic interactions. Reactions which play a role are: $pn \rightarrow d\gamma$, $nd \rightarrow {}^3\text{H}\gamma$, $dd \rightarrow {}^3\text{He } n$, ${}^3\text{He } n \rightarrow {}^4\text{He}\gamma$, ${}^3\text{H}d \rightarrow {}^4\text{He } n$, etc. Once the bottleneck of deuteron production is crossed at $T \simeq 0.1$ MeV, further nucleosynthesis up to ${}^7\text{Li}$ proceeds rapidly and all the neutrons are rapidly used up in these processes. The primordial abundances calculated depend on the ratio $n_B/n_\gamma \simeq 10^{-10}$ and are in reasonable agreement with observed values. Elements heavier than these primordial ones are synthesized later in stars, after stars and galaxies are formed.

Bjorken Scaling

This concept was developed in 1969 by J. D. Bjorken [86] in connection with the theory of deep inelastic scattering of leptons on nucleons. It plays a very significant role in elucidating the substructure of the nucleon, a role as important as that played by the Rutherford scattering of alpha particles for understanding the structure of the atom and its nucleus.

To explain Bjorken scaling, let us consider the kinematics involved in deep inelastic scattering of leptons (electrons or neutrinos and their antiparticles) on nucleons. Here we consider only the scattering of electrons. In Figure 4.4 above, we have the diagram in which an incident electron with a well defined four-momentum $p^\mu = (E, \vec{p})$ is incident on a nucleon at rest in the laboratory with four-momentum $P^\mu = (M, \vec{0})$. The nucleon suffers a highly inelastic scattering process, producing a whole slew of particles in the final state X, the total of the four-momenta of these produced particles being, $p^\mu_{tot} = (E_{tot}, \vec{p}_{tot})$. The electron acquires a final four-momentum $p'^\mu = (E', \vec{p}')$, which is measured. We specify the invariant mass W of the products of the nucleon breakup by $W^2 = E^2_{tot} - \vec{p}^2_{tot}$. The four-momentum q^μ transferred by the electron to the proton is known from the measurements to be $q^\mu (q^0 = E - E', \vec{q} = \vec{p} - \vec{p}')$. The square of the four-momentum transfer is then $q^2 = (E - E')^2 - (\vec{p} - \vec{p}')^2$. The same quantity in terms of the nucleon and its products in the final state is $q^2 = (E_{tot} - M)^2 - (\vec{p}_{tot} - \vec{0})^2$. We introduce the laboratory energy transfer, $\nu = (E_{tot} - M) = (E - E')$, which can be written in the invariant form $\nu = q \cdot P/M$. Then a simple calculation using these definitions shows that $q^2 = W^2 - M^2 - 2M\nu$. Thus, the whole process is specified, if one gets just the two quantities, q^2 the square of the four-momentum transfer and ν the energy transfer. The invariant mass of the produced hadrons W^2 can be calculated to be $W^2 = M^2 + 2M\nu + q^2$. Such a reaction in which one measures only the outgoing electron's energy and momentum, (the initial electron's energy and momentum being known), and in which no specific hadrons are detected in the final state, is called an *inclusive reaction*. Incidentally, we notice that if we were to treat elastic scattering, we would set $W = M$, and we would have $q^2 = -2M\nu$. Let $Q^2 = -q^2$, and if we define a variable, $x = \frac{Q^2}{2M\nu}$, then we have $1 - x = (W^2 - M^2)/(2M\nu)$, and we see that x is restricted to the range $0 < x \leq 1$. The value $x = 1$ holds for elastic scattering, whereas $x \to 0$ is the region of deep inelastic scattering. The smaller the x value, the higher the energy transfer, and the higher the resolution with which the nucleon is seen. Another useful variable is $y = [1 - (E'/E)]$, which has the invariant form $y = q \cdot P/p \cdot P$ and represents the fraction of the energy lost by the incident electron in the laboratory.

The cross section for the inclusive deep inelastic scattering of the electron by the nucleon has been worked out. In Figure 4.4, if we suppose that the momentum transfer that occurs between the electron and the nucleon is by a single photon exchange, the differential cross section for the process is proportional to the product of two tensors, one coming from the Lepton vertex $L_{\mu\nu}$ and the other coming from the hadron vertex $H^{\mu\nu}$,

$$d\sigma = (\frac{4\pi\alpha}{q^2})^2 \frac{1}{4[(p \cdot P)^2 - m_e^2 M^2]^{(1/2)}} 4\pi M \, L_{\mu\nu} H^{\mu\nu} \frac{d^3 p'}{2E'(2\pi)^3},$$

where α is the fine structure constant, M the nucleon mass, m_e the electron mass, the second factor is the flux factor, and the last factor is the final electron phase space factor [87]. For the electron, using the rules of quantum electrodynamics, $L_{\mu\nu}$ has the form

$$L_{\mu\nu} = 2[p'_\mu p_\nu + p_\mu p'_\nu + (q^2/2)g_{\mu\nu}].$$

The hadronic tensor, obtained by summing over all accessible hadronic states, has a form dictated by general invariance requirements and is written in terms of two structure functions, $W_1(Q^2, \nu)$ and $W_2(Q^2, \nu)$, which are functions of the two variables Q^2 and ν as

$$
\begin{aligned}
H^{\mu\nu} &= (-g^{\mu\nu} + q^\mu q^\nu / q^2) W_1(q^2, \nu) \\
&+ [P^\mu - (P \cdot q/q^2) q^\mu][P^\nu - (P \cdot q/q^2) q^\nu] W_2(Q^2, \nu).
\end{aligned}
$$

In terms of laboratory coordinates, neglecting electron mass, one can write,

$$
\frac{d^2\sigma}{dQ^2 d\nu} = \frac{\pi\alpha^2}{4p^2 \sin^4 \theta/2} \frac{1}{pp'} \left[W_2(Q^2, \nu) \cos^2 \theta/2 + 2W_1(Q^2, \nu) \sin^2 \theta/2 \right].
$$

Bjorken showed that the structure functions W_1 and W_2 are related to matrix elements of commutators of hadronic currents at almost equal times in the infinite momentum limit. He showed that this infinite momentum limit is not divergent. If this limit is nonzero, he predicted that the structure functions, when $Q^2 \to \infty$ and $\nu \to \infty$, but Q^2/ν is finite, can depend on Q^2 and ν only through the ratio Q^2/ν. Thus Bjorken scaling for the functions W_1 and W_2 are the statements that, for $Q^2 \to \infty$ and $\nu \to \infty$, but $x = Q^2/(2M\nu)$ remaining fixed,

$$
\begin{aligned}
MW_1(Q^2, \nu) &\to F_1(x) \\
\nu W_2(Q^2, \nu) &\to F_2(x)
\end{aligned}
$$

and the limiting functions $F_1(x)$ and $F_2(x)$ are finite.

In terms of the variables x and y, we may cast the expression for the deep inelastic scattering of electron on nucleon as

$$
\frac{d^2\sigma}{dxdy} = \frac{2\pi\alpha^2 s}{Q^4} [xy^2 F_1(x) + 2(1-y)F_2(x)].
$$

Bjorken Scaling—Experimental Confirmation
The first experimental evidence for Bjorken scaling behavior was reported by Bloom et al. [88] and confirmed by Breidenbach et al. [89]. Figures, adapted from the experimental paper of Miller et al. [90] and from Friedman and Kendall [91] are reproduced here to illustrate Bjorken scaling in Figure 4.5 and Figure 4.6, respectively.

Bjorken Scaling—Explanation
Feynman [92] showed that the scaling behavior is what one gets if the proton was composed of "pointlike" constituents (which he called *partons*, from which the incident electrons suffer scattering, and in the limit

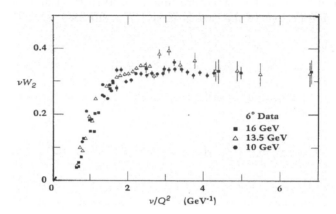

Figure 4.5: Bjorken scaling of νW_2, shown as a function of $(\frac{\nu}{Q^2} \sim \frac{1}{x})$ for different Q^2. (Reprinted from the *Beam Line*, Vol. 20, No. 3, 1990.)

when $Q^2 \to \infty$ and $\nu \to \infty$, but Q^2/ν remains finite, the deep inelastic electron-proton cross section is given by the incoherent sum of the parton cross sections, summed over all the partons of the proton. For the case of *elastic scattering* of the electron from a particular pointlike constituent i, with charge e_i, and mass m_i, the result for the cross section can be derived making use of the result for the elastic scattering of electrons on muons [87]. We may write

$$
\begin{aligned}
\frac{d^2\sigma^i}{dQ^2 d\nu} &= \frac{\pi\alpha^2}{4p^2 \sin^4 \theta/2} \frac{1}{pp'} \left(e_i^2 \cos^2 \theta/2 + e_i^2 \frac{Q^2}{4m_i^2} 2\sin^2 \theta/2 \right) \\
&\times \quad \delta(\nu - Q^2/(2m_i)),
\end{aligned}
$$

where the result is written in terms of laboratory coordinates, the incident electron four-momentum being p_μ, the final momentum being p'_μ, $q^2 = -Q^2$ being the square of the four-momentum transfer $q_\mu = p_\mu - p'_\mu$, and θ the angle of scattering of the electron. We also suppose that the parton's initial four-momentum is $p_i^\mu = xP^\mu$, and its mass $m_i = xM$, where P^μ and M are the four-momentum of the proton and its mass, respectively. This expression may be compared with the previously derived result for the inclusive deep inelastic cross section,

$$
\frac{d^2\sigma}{dQ^2 d\nu} = \frac{\pi\alpha^2}{4p^2 \sin^4 \theta/2} \frac{1}{pp'} \left[W_2(Q^2, \nu) \cos^2 \theta/2 + 2W_1(Q^2, \nu) \sin^2 \theta/2 \right].
$$

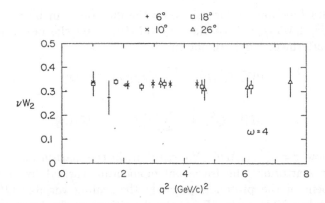

Figure 4.6: Bjorken scaling of νW_2, shown as a function of Q^2 for a single value of $\omega = \frac{1}{x} = 4$. (Reproduced with permission, from the *Annual Reviews of Nuclear Science*, Volume 22 ©1972 by Annual Reviews www.AnnualReviews.org.)

This shows that, if each parton i contributes incoherently to W_1^i and W_2^i (we assume this to be the case for $Q^2 \to \infty$ and $\nu \to \infty$), then comparison gives

$$
\begin{aligned}
W_1^i &= e_i^2 \frac{Q^2}{4M^2 x^2} \delta(\nu - Q^2/(2Mx)), \\
W_2^i &= e_i^2 \delta(\nu - Q^2/(2Mx)).
\end{aligned}
$$

The total contribution to W_1 and W_2 from all partons can be obtained by summing these expressions over i. The variable x, the fraction of the momentum of the proton P carried by the parton i, is not a discrete quantity but actually is continuously distributed over the range 0 to 1 for each parton. Thus the sum over partons is replaced by an integral over x with a weighting function, $f_i(x)$, which represents the probability that the parton of type i carries a fraction x of the momentum of the proton. This probability function is called the *parton distribution function* and parameterizes the proton in the parton model. (It is not calculable in the model but has to be obtained from fits to the experimental data.) Then one has

$$
\begin{aligned}
W_1 &= \sum_i \int_0^1 dx \, f_i(x) e_i^2 \frac{Q^2}{4M^2 x^2} \delta(\nu - Q^2/(2Mx)), \\
W_2 &= \sum_i \int_0^1 dx \, f_i(x) e_i^2 \delta(\nu - Q^2/(2Mx)).
\end{aligned}
$$

The delta functions occurring inside the integrals can be rewritten as $\delta(\nu - Q^2/(2Mx)) = (x/\nu)\delta(x - Q^2/(2M\nu))$, and the integrals over x easily performed. Then we obtain

$$\nu W_2(Q^2, \nu) = \sum_i e_i^2 x f_i(x) \equiv F_2(x),$$

$$2M W_1(Q^2, \nu) = \sum_i e_i^2 f_i(x) \equiv 2F_1(x),$$

where now $x = Q^2/(2M\nu)$. We see that, in this parton model of deep inelastic scattering, the fractional momentum carried by the parton constituent of the proton is precisely the scaling variable $Q^2/(2M\nu)$ that Bjorken introduced. From the definitions of the functions $F_1(x)$ and $F_2(x)$ defined in the parton model, we see that they are related: $2xF_1(x) = F_2(x)$. This relation between F_1 and F_2 is called the *Callan-Gross relation* [93]. Parametrizations of the parton distribution functions can be obtained by imposing the requirement that they reproduce the experimentally measured values of F_1 and F_2 as a function of x.

BooNE—Booster Neutrino Experiment

This is an experiment being set up to detect neutrino oscillations at Fermilab using the neutrinos created by the protons from the 8 GeV booster. The goals of BooNE are, first, to confirm or refute the observations that have already been made by the LSND experiment at Los Alamos on the oscillation of ν_μ neutrinos and to a much better statistical precision (thousands of events), and second, determine the oscillation parameters accurately.

Borexino Experiment

This is an experiment designed to study solar neutrinos. It is being set up in the Gran Sasso laboratories, situated between Rome and Terramo in Italy, and is designed to detect the monoenergetic ^{7}Be neutrinos in the solar nuclear cycle. Three hundred tons of liquid scintillator with a fiducial volume of 100 tons is expected to produce about 50 events per day due mostly to ^{7}Be neutrinos. Two thousand PMT's (Phototmultiplier Tubes), each of 20 cm diameter, will detect scintillation light in real time.

Bottomonium

The first evidence of a new vector meson, called the $\Upsilon(1S)$, was found by Herb et al. [60]. With the Fermilab proton synchrotron, this group studied production of pairs of muons in 400 GeV proton-nucleus collisions. The reactions studied were $p + \text{Cu} \rightarrow \mu^+ \mu^- + X$ and $p + \text{Pt} \rightarrow \mu^+ \mu^- + X$.

They found a strong enhancement of the muon pairs at an invariant mass of 9.5 GeV in a sample of 9000 dimuon events. This strong enhancement was interpreted as being due to the production of a neutral vector meson state, called the $\Upsilon(1S)$, which subsequently decayed into a pair of muons. The $\Upsilon(1S)$ meson is interpreted as the bound state of a new quark-antiquark pair, $(b\bar{b})$, of the b-quark (and its antiquark), called by some the *bottom quark* and by others as the *beauty quark*. This quark represents the first discovery of the bottom member of a third quark family $(-, b)$, after the (u, d) and the (c, s) families. (The "$-$" in the third quark family represents a vacancy, because, at that time, the top member of the family, called the *top quark*, was yet to be found.) The electric charge carried by the b-quark was found to be (-1/3) times the unit of charge $(|e|)$ just like the d and the s quarks of the first two generations. This was accomplished by measuring the decay width of the $\Upsilon(1S)$ state to electron pairs and fitting to theoretical calculations which favored a charge assignment of $(-1/3)|e|$ for the b-quark. The production of the $\Upsilon(1S)$ state was independently confirmed in electron-positron annihilations by the Pluto detector collaboration [60] and the DASP detector collaboration [60] working at DESY in Hamburg, Germany.

Evidence for the existence of higher excited states of the Υ family started coming in shortly after the discovery of the $\Upsilon(1S)$. First evidence for further structure in the Υ region, now called the $\Upsilon(2S)$ state (with an invariant mass around 10.0 GeV) and the $\Upsilon(3S)$ state, was announced by Innes et al. [94] working with the Fermilab proton synchrotron. The existence of the $\Upsilon(2S)$ state at an invariant mass of around 10.01 to 10.02 GeV was also confirmed independently in electron-positron annihilations by Bienlein et al. [61], and by the DASP detector collaboration [61] at DESY. Confirmation of the existence of the $\Upsilon(3S)$ and the $\Upsilon(4S)$ states of the $b\bar{b}$ system came from the work of the CLEO detector collaboration [61] and from the CUSB detector collaboration [61]. The CLEO collaboration pointed out, from the observed characteristics of the $\Upsilon(4S)$ state, that this state lies in energy above the threshold for the production of B mesons. The values of the masses and widths of these states as reported in the most recent "Review of Particle Physics" [62] are: $\Upsilon(1S)$, 9.46037 ± 0.00021 GeV, $\Gamma = 52.5 \pm 1.8$ keV; $\Upsilon(2S)$, 10.02330 ± 0.00031 GeV, $\Gamma = 44 \pm 7$ keV; $\Upsilon(3S)$, 10.3553 ± 0.0005 GeV, $\Gamma = 26.3 \pm 3.5$ keV; and $\Upsilon(4S)$, 10.5800 ± 0.0035 GeV, $\Gamma = 10 \pm 4$ MeV. The spectroscopy of the states of the $b\bar{b}$ system reveals a lot of information about the strong forces that are responsible for these bound states, hence, the great interest in their study.

Bremsstrahlung

When a high energy charged particle passes through a medium with an energy greater than its rest energy, it radiates energy in the form of electromagnetic radiation, due to collisions with the atoms of the medium. This radiation is called *Bremsstrahlung*. The cross section for this process has been derived in QED. (See section on "Energy Loss by Radiation: Bremsstrahlung" in Chapter 3, and in the Glossary under "Infrared Divergence".)

Cabibbo Angle—Cabibbo Mixing Matrix

In our earlier discussions of β decay processes, we concerned ourselves with nuclear β decays only. It was soon found from other experiments that many of the other elementary particles also decayed with lifetimes in the range from 10^{-6} s to 10^{-10} s. The muon was discovered with a mass about 206 times the electron mass and a lifetime of 2 μs and was found to decay into an electron and two neutrinos. Also the pion, with a mass of about 276 times the mass of the electron, was found, which decayed with a lifetime of about 2.7×10^{-8} s into a muon and a neutrino. Particles such as the Λ with strangeness quantum number $S = -1$ were found to undergo β decay, decay products of which had strangeness $S = 0$ associated with them. The question soon arose whether the Fermi theory developed for β decay could be extended to describe these and other decays. In other words, the question arose as to how "universal" the applicability of Fermi theory was.

The question was soon answered in the affirmative. When Fermi theory was applied to negative muon decay with (V-A) coupling of two currents, one current formed from the electron and its (anti-)neutrino system, and the other current from the muon and its neutrino system, the energy distribution of the decay electrons and the lifetime could be fitted with *about* the same value of the coupling G_F as was the case with neutron decay. Actually a very careful comparison of muon decay and neutron decay revealed that the two G_F values differed from one another by a small amount, the value obtained from neutron decay being somewhat smaller, about 97% of the one obtained from muon decay. Similar calculation for Λ beta decay revealed that this process required a much smaller value of G_F. Of these, the neutron decay involves the decay of a non-strange particle into non-strange particles, so that the strangenesss change is zero, while Λ decay involves a change in the strangeness quantum number. Faced with this situation, Cabibbo [95] tried to see if the data on strangeness conserving hadronic currents with $\Delta S = 0$ and the strangeness changing hadronic currrents with $\Delta S = 1$ could be brought into line with the same value of G_F if the hadronic currents were somehow shared in the decay interactions. Specifically, he proposed that strangeness conserving hadronic currents coupled with the lepton currents with a further factor $\cos\theta_C$, while the strangeness changing hadronic currents coupled with leptonic currents with a factor $\sin\theta_C$, where θ_C is called the *Cabibbo angle*. It was soon found that with a value of $\sin\theta_C$ about 0.23, one could fit these decays with the same value of G_F. This Cabibbo modification of the Fermi theory has been found very successful in bringing a number of decay processes involving elementary particles in line with the same value of G_F. In other words, there is universality in such decay processes. When β decay theory is thus, generalised to apply to decays of other elementary particles, one calls the generalised theory, a theory of *weak interactions of elementary particles*. Thus, the theory of weak interactions includes the theory of β decay as part of it.

In terms of the quark picture, Cabibbo's modifications can be accommodated by introducing a mixing between the d and s quarks through a matrix U which has the form,

$$\begin{pmatrix} d' \\ s' \end{pmatrix} = U \begin{pmatrix} d \\ s \end{pmatrix},$$

where U is the matrix

$$U = \begin{pmatrix} \cos\theta_C & \sin\theta_C \\ -\sin\theta_C & \cos\theta_C \end{pmatrix}.$$

The Cabibbo rotated quark states d' and s' are the ones which are involved in forming the weak $\Delta S = 0$ and $\Delta S = 1$ currents.

Cabibbo-Kobayashi-Maskawa (CKM) Matrix

The extension of Cabibbo mixing involving two quark families, d and s, to one involving Cabibbo-like mixing in three quark families, d, s, and b was done by Kobayashi and Maskawa [96] in order to accommodate the phenomenon of CP violation. It can be shown that the phenomenon of CP violation cannot be accommodated with just two quark families; it requires at least three families. By convention the mixing is introduced in terms of the quarks d, s, and b, all of the same charge, $-(1/3)|e|$, in the form,

$$\begin{pmatrix} d' \\ s' \\ b' \end{pmatrix} = V \begin{pmatrix} d \\ s \\ b \end{pmatrix},$$

where the matrix V has the form,

$$V = \begin{pmatrix} V_{ud} & V_{us} & V_{ub} \\ V_{cd} & V_{cs} & V_{cb} \\ V_{td} & V_{ts} & V_{tb} \end{pmatrix}.$$

Introducing the generation labels $i, j = 1, 2, 3$, $c_{ij} = \cos\theta_{ij}$, and $s_{ij} = \sin\theta_{ij}$, the matrix U is usually parametrized, involving the angles, θ_{12}, θ_{23}, θ_{13}, and the phase angle δ_{13}, in the form,

$$V = \begin{pmatrix} c_{12}c_{13} & s_{12}c_{13} & s_{13}e^{-i\delta_{13}} \\ -s_{12}c_{23} - c_{12}s_{23}s_{13}e^{i\delta_{13}} & c_{12}c_{23} - s_{12}s_{23}s_{13}e^{i\delta_{13}} & c_{13}s_{23} \\ s_{12}s_{23} - c_{12}c_{23}s_{13}e^{i\delta_{13}} & -c_{12}s_{23} - c_{23}s_{12}s_{13}e^{i\delta_{13}} & c_{13}c_{23} \end{pmatrix}.$$

This particular parametrization has the advantage that if one of the inter-generational mixing angles is zero, the mixing between those generations vanishes. Further, when $\theta_{23} = \theta_{13} = 0$, the mixing matrix reduces to the Cabibbo form of the mixing matrix, allowing the identification of $\theta_{12} = \theta_C$. By suitably choosing the phases of the quark fields, the three angles, θ_{12}, θ_{23}, and θ_{13}, may each be restricted to the range between 0 and $\pi/2$, and δ_{13} to the range 0 to 2π.

There is experimental information on the size of the different elements of the V matrix. $|V_{ud}|$ is the best known element; it is obtained by comparing superallowed nuclear beta decay with muon decay, including radiative corrections and isospin corrections. Its value is $|V_{ud}| = 0.9740 \pm 0.0010$. $|V_{us}|$ is determined from the analysis of the decay $K \to \pi e\nu$, and is found to be $|V_{us}| = 0.2196 \pm 0.0023$. To find $|V_{cd}|$ one can use data on production of charm in neutrino nucleon collisions: $\nu_\mu N \to \mu + \text{charm} + X$. From these one gets, $|V_{cd}| = 0.224 \pm 0.016$. The data on charm production also allows one to extract $|V_{cs}|$ depending on assumptions one makes about the strange quark content in the sea of partons in the nucleon. From these considerations $|V_{cs}| = 1.04 \pm 0.16$ is found. $|V_{cb}|$ is found from the measurements on decays of B mesons (which occur predominantly through $b \to c$ quark transitions), $B \to \bar{D} + l^+ + \nu_l$, where l is a lepton and ν_l is its corresponding neutrino. By these measurements one gets $|V_{cb}| = 0.0395 \pm 0.0017$. $|V_{ub}|$ is obtained by looking for the semi-leptonic decay of B mesons produced at the $\Upsilon(4S)$ resonance. These decays are due to $b \to u + l + \bar{\nu}_l$ and its charge conjugate; one gets at these by measuring the lepton energy spectrum above the end point of the lepton energy spectrum in $b \to c + l + \bar{\nu}_l$. In this manner, one gets the ratio $(|V_{ub}|/|V_{cb}|) = 0.08 \pm 0.02$. From the additional constraint provided by the fact that the mixing matrix V is unitary, with just three generations of quarks, one can derive bounds on the remaining matrix elements. In this manner, the modulii of the matrix elements of V are found to be in the ranges as shown below [62]:

$$V = \begin{pmatrix} 0.9745 - 0.9760 & 0.217 - 0.224 & 0.0018 - 0.0045 \\ 0.217 - 0.224 & 0.9737 - 0.9753 & 0.036 - 0.042 \\ 0.004 - 0.013 & 0.035 - 0.042 & 0.9991 - 0.9994 \end{pmatrix}$$

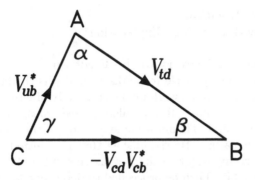

Figure 4.7: The unitarity condition on the CKM matrix elements, represented geometrically as a triangle. Correct estimates of the matrix elements should lead to closure of the triangle.

The phase δ_{13} can be obtained from measurements on CP violation. In the case of CP violation in the neutral Kaon system (see "CP Violation—Neutral Kaons"), the calculation of the ϵ parameter involves δ_{13}.

The information provided by the unitarity of the V matrix and the direct measurements of the modulii of some of the matrix elements, allows one to construct what is called the *unitarity triangle*. Applying unitarity to the first and third columns of the V matrix, we have

$$V_{ud}V_{ub}^* + V_{cd}V_{cb}^* + V_{td}V_{tb}^* = 0$$

A geometrical representation of this equation in the complex plane provides us with the unitarity triangle.

In the parametrization we have chosen for V, we note from the above that V_{cb} is real, and V_{cd} is very nearly real and equal to $-s_{12}$. V_{ud} and V_{tb} are both real and nearly 1, so that the unitarity condition written above may be reduced to

$$V_{ub}^* + V_{td} = -V_{cd}V_{cb}^*.$$

This is represented by a triangle ABC as shown in Figure 4.7 on the preceding page.

Let us choose to orient the triangle such that the side CB of the triangle is horizontal with length $-V_{cd}V_{cb}^*$. The other sides are CA representing V_{ub}^* and AB representing V_{td}. The side CB is the resultant of the addition of the two complex vectors CA and AB. If the relevant matrix elements are correctly estimated from the measurements, the unitarity triangle should close.

Callan-Gross Relation

See under "Bjorken Scaling—Explanation".

Cathode Rays—Discovery of the Electron

The first elementary particle to be discovered was the electron in 1897 by J.J. Thomson [97]. His work involved a study of electrical discharges in gases at low pressures. Rays, called *cathode rays*, were produced in these discharges. J. J. Thomson made a study of these cathode rays to determine their nature. He established that the cathode rays consisted of particles carrying negative charges, as they were affected by electric and magnetic fields. Then he proceeded to determine the ratio of charge e to mass m of these particles, e/m, by subjecting them to deflections in electric and magnetic fields. The cathode ray particles were made to pass through a region between two plates of length l (say, in the x direction) between which a constant electric field E_y was maintained (say, in the y direction). The particles suffered a deflection in the y direction transverse to their motion, and, from the known direction of the electric field and the direction of deflection, it could be established that these particles carried a negative charge. If the particles had a velocity v and traversed a distance l in the x direction between the plates, it is easy to show that the magnitude of the deflection D suffered in the y direction is $D = \frac{1}{2}\frac{eE_y}{m}(l/v)^2$. To determine the velocity v of the particles, Thomson imposed a magnetic field B_z in a direction z, mutually perpendicular to the x and y directions. The direction of the magnetic field $(+z$ or $-z)$ was so chosen that the deflection it caused tended to oppose the deflection caused by the electric field. The magnitude of the magnetic field was adjusted until it produced a deflection which cancelled the deflection due to the electric field. It is easy to show that under this circumstance, the velocity $v = (E_y/B_z)$. Having measured the velocity of these particles, a measurement of the deflection D in the absence of the magnetic field, can be used to determine the value of e/m for the cathode rays. This gave a value for $(|e|/m) = 1.76 \times 10^{11}$ Coulombs/kg. The negatively charged particles in the cathode rays were given the name *electron*. In 1911 Millikan, by an ingenious method, determined the magnitude of the charge of the electron $|e|$ to be $|e| = 1.6 \times 10^{-19}$ Coulomb. Combining this with $(|e|/m)$ measured earlier by J. J. Thomson, the mass of the electron could be determined: $m = 9 \times 10^{-31}$ kg. The electron is the first of a family of elementary particles to be called the *lepton*; its charge and mass have been measured much more accurately since those early measurements, and the modern values are: $e = -(1.60217733 \pm 0.00000044) \times 10^{-19}$ Coulomb and $m = (9.109389 \pm 0.000054) \times 10^{-31}$ kg.

Causality Condition—Quantum Field Theory

Causality condition is a recognition of the fact that no *effect* can occur which precedes the *cause*. In quantum field theory, it is a statement that measurements of a field at one point in space-time do not affect the measurements of the field at another point in space-time, if the two space-time points are separated by a space-like interval. For an understanding of this concept and its implications, an introduction to some of the basic ideas of relativity and quantum field theory is necessary. This will be done in brief.

In special theory of relativity, the maximum speed with which a particle can propagate between two space-time points is the velocity of light c (which in our units is 1, since we work in units in which $\hbar = c = 1$). This implies that if we have two space-time points, x^μ with components $(x^0, \vec{x})$ and y^μ with components $(y^0, \vec{y})$ in the four-dimensional continuum, a disturbance occurring at y^μ can be felt at x^μ, only if the space-time interval between these points, defined in terms of their components as $(x - y)^\mu (x - y)_\mu = (x^0 - y^0)^2 - (\vec{x} - \vec{y})^2$, is zero or positive. If it is zero, the interval is called *light-like*, and if it is positive, the interval is called *time-like*. In the case of light-like interval, it is easy to understand why this is zero: the spatial distance between the two points is $|\vec{x} - \vec{y}|$, and the time taken for light (photons), starting at time y^0 at $\vec{y}$, will reach the point $\vec{x}$ at time x^0, if $x^0 = y^0 + |\vec{x} - \vec{y}|$ or $x^0 - y^0 = |\vec{x} - \vec{y}|$. Any material particle other than a photon travels with a velocity less than 1, hence the time taken by it to traverse the distance between the two points will be more, in other words, $x^0 - y^0 > |\vec{x} - \vec{y}|$. For points in the four-dimensional continuum where the coordinate components are such that $x^0 - y^0 < |\vec{x} - \vec{y}|$, no physical particle will be able to propagate between such points. The space-time interval for such points will satisfy $(x^0 - y^0)^2 < (\vec{x} - \vec{y})^2$, and such intervals are called *space-like intervals*. We say no causal communication is possible between points separated by space-like intervals.

It can be shown that in a relativistic theory of *single* particles, described by wave functions obeying relativistically invariant wave equations, one has violations of causal connections between points separated by space-like intervals. If one calculates the probability amplitude for a particle to propagate between such points, it is found to be non-vanishing (although small). It also turns out other difficulties are encountered; the equations have solutions for positive as well as negative energies, making the interpretations of negative energy solutions problematical in such a theory [98]. In a relativistic situation, one cannot restrict attention to the single particle only, for when its energy becomes large enough, it can create other particle-antiparticle pairs. In fact, even when there is not sufficient energy to create real particles, quantum mechanics via the uncertainty principle, allows for arbitrarily large fluctuations in energy, ΔE, for arbitrarily small time intervals, Δt, provided $\Delta E \Delta t = 1$ (since in our units $\hbar = 1$). Thus intermediate states can occur in which arbitrarily large number of particles can exist for arbitrarily small intervals of time. Such particles are called *virtual*. Thus in a relativistic theory, a single particle theory is untenable and a many particle description becomes necessary. The many particle theory is constructed by quantizing a field theory, thus leading to quantum field theory.

In classical relativistic field theory, the fundamental dynamical variables are fields, described by functions of space-time coordinates. One starts with a relativistically invariant expression for the Lagrangian constructed with the field functions and their space and time derivatives. Following well-known procedures, the Hamiltonian is constructed in terms of the field variables $\phi(x)$ and the canonically conjugate variables $\pi(x)$. (For details refer to [98].) Quantization is carried out by elevating $\phi(x)$ and its canonical conjugate to operator status and introducing commutation relations between them. In quantizing the field theory, the Fourier expansions of the field operators with positive frequency component have as their coefficient an operator $a(\vec{p})$ that *annihilates* a particle with momentum $\vec{p}$ and energy $\omega(\vec{p}) = p_0 = \sqrt{\vec{p}^2 + m^2}$, while the negative frequency component, being the Hermitian conjugate of the positive frequency solution, will have as its coefficent an operator $a^\dagger(\vec{p})$ that *creates* a particle with momentum $\vec{p}$ and energy $\omega(\vec{p})$. The Hamiltonian is expressible as an integral over all momenta of the product of the number operator, $n_p = a^\dagger(\vec{p})a(\vec{p})$ and $\omega(\vec{p})$. $\phi(x)$ which are real are suitable for describing electrically neutral particles, annihilation and creation operators are Hermitian conjugates of one another, and a particle is its own antiparticle. If $\phi(x)$ is complex, it will annihilate negatively charged particles and create positively charged particles, and the Hermitian conjugate operator $\phi^\dagger(x)$ will create negatively charged particles and annihilate positively charged particles. Thus this quantum field theory leads to a multiparticle theory with both particles and antiparticles. All the states of the system are constructed from the vacuum state $|0\rangle$, defined by $a(\vec{p})|0\rangle = 0$, applying creation operators repeatedly to the vacuum state.

Now we return to causality condition in quantum field theory. This condition is stated in the form that the commutator of field operators at space-like separations vanishes, that is, $[\phi(x), \phi(y)] = 0$, for $(x - y)^\mu (x - y)_\mu < 0$. With this condition, if one calculates the amplitude for a particle to propagate from x to y, it also obtains a contribution for the antiparticle to go from y to x, the two contributions cancelling each other exactly. Thus for space-like separations, measurement of the field at x does not affect the measurement of the field at y and causality is preserved in quantum field theory.

Using the causality conditions in quantum field theory, Gell-Mann, Goldberger, and Thirring [99] treated the scattering of spin zero particles by a force center and the scattering of photons by a quantized matter field. The dispersion relations of Kramers and Kronig for applications in optics were derived from field theory. Many other applications of dispersion relations appeared over the next several years. (See also under "Dispersion Relations".)

Charge Conjugation (*C*) Operation

If one examines the classical Maxwell's equations for electromagnetism, one notices that these equations remain invariant under the operation of change of sign of charge and magnetic moments provided the electric and the magnetic fields also are changed in sign. Such a change in the sign of the charge and magnetic moment of a particle (without changing any other of its properties) is called the operation of *charge conjugation*. At high energies, one has to use concepts of quantum field theory to describe the behavior of particles and their interactions, and the situation is somewhat different. In such a regime, particles are accompanied by antiparticles, where the antiparticles, besides being distinguished by the sign of their charge and magnetic moment, may have other quantum numbers, such as baryon number or lepton number, which have to be reversed, too. Thus, charge conjugation operation in relativistic quantum field theory involves not only the change of sign of charge and magnetic moment of the particle but also the change of sign of baryon number, lepton number, etc., if these are nonvanishing. Mesons have neither baryon nor lepton number.

A question naturally arises whether the four fundamental interactions observed in nature—gravitational, electromagnetic, weak, and strong—are symmetric under the operation of charge conjugation. Experimental evidence is available on this subject. The available data indicate that three of these interactions—gravitational, electromagentic, and strong—respect this symmetry, while weak interactions do not. Before we present these evidences, we present some details on the operation of charge conjugation.

The mesons π^+ and π^- are antiparticles of one another with baryon and lepton numbers zero. If the $\pi^\pm$ meson is represented by the ket $|\pi^\pm\rangle$, then the charge conjugation operation represented by C gives $C|\pi^\pm\rangle \rightarrow \alpha|\pi^\mp\rangle$, where α is a possible phase factor. It is clear from this operation, the states $|\pi^\pm\rangle$ are not eigenstates of the operator C. On the other hand, a charge neutral state may be an eigenstate with some eigenvalue λ: $C|\pi^0\rangle = \lambda|\pi^0\rangle$ [100]. Repeating the operation a second time we get back to the original state, so we must have $\lambda^2 = 1$ or $\lambda = \pm 1$. To see whether we should assign the eigenvalue $+1$ or -1 to the $|\pi^0\rangle$ state, we recall the fact that the π^0 meson decays into two photons through electromagnetic interactions. Now the photon is a quantum of the electromagnetic field. Suppose we represent the photon state by $|\gamma\rangle$. As this is also electrically neutral, $C|\gamma\rangle = \lambda_\gamma|\gamma\rangle$, where λ_γ is the eigenvalue. Repeating the operation again we get back the initial state, so $\lambda_\gamma^2 = 1$ or $\lambda_\gamma = \pm 1$. This eigenvalue must be chosen to be -1, because the electromagnetic field changes sign when the charge-currents producing the electromagnetic field change sign, and the photon is a quantum of the electromagnetic field. Thus for each photon, the charge conjugation eigenvalue is -1. A system of n photons will have eigenvalue $(-1)^n$, which for $n = 2$ gives the value $+1$. Thus, if in π^0, decaying through electromagnetic interactions, charge conjugation symmetry is respected, the $|\pi^0\rangle$ state must be one with eigenvalue $\lambda = \lambda_\gamma^2 = +1$.

If we assign the charge conjugation eigenvalue $+1$ to π^0, it has the further consequence that π^0 should not decay into 3 photons if charge conjugation symmetry is valid in electromagnetic interactions. This has been tested experimentally. The branching ratio for $\pi^0 \to 3\gamma$ to $\pi^0 \to 2\gamma$ is less than 3×10^{-8}. Another test comes from the decay of the η^0. It also decays 39% of the time into 2γ's, so that it can be assigned C eigenvalue $+1$ like the π^0. The branching ratio for $\eta^0 \to 3\gamma$ decay is less than 5×10^{-4} and for $\eta^0 \to \pi^0 e^+ e^-$ is less than 4×10^{-5}. Thus, electromagnetic interactions are invariant under the C operation.

To test C invariance for strong interactions, one studies reactions in which certain particles are produced, their rates of production, their energy distribution etc., and compares them with reactions in which all the particles are replaced by the antiparticles. Experiments performed show that they give identical results at the level of much better than 1%.

Charge conjugation parity can be defined for a neutral fermion-antifermion system, such as $p\bar{p}$, $n\bar{n}$, or quark-antiquark system. If the fermion-antifermion system is in a state of orbital angular momentum l and the total spin is s, it can be shown that the charge conjugation parity has the eigenvalue $C = (-1)^{l+s}$. (See section under "Positronium" where this is explicitly shown.)

Weak interactions do not respect charge conjugation symmetry. They also violate reflection symmetry or parity operation (P). Neutrinos are products of decays through weak interactions, for example, in meson decays. Experimentally they are found to be left-handed, while antineutrinos are found to be only right-handed. (See further under "Neutrino".) There are no left-handed antineutrinos, nor any right-handed neutrinos. If C operation applied to these particles, right-handed neutrinos and left-handed antineutrinos would have to be present. The combined operation of charge conjugation and reflection (CP) seems to hold for weak interactions. Under the combined operation of CP, a left-handed neutrino would go into a right-handed antineutrino. This is in conformity with what is observed. Actually a small violation of CP is observed in weak interaction processes involving neutral K mesons. (See more under "Conservation/Violation of CP".)

Chargino
These are mixtures of the hypothetical w-inos and charged higgsinos, which are the supersymmetric partners of the W and the charged Higgs bosons, respectively. (See under "Supersymmetry".)

Charm Particles—Charmonium
The existence of charm quark was established through the observation of a vector meson resonance called the J/ψ in e^+e^- collisions at SLAC, and simultaneously, in the production of massive e^+e^- pairs in proton collisions with a beryllium target at Brookhaven National Laboratory [101]. The resonance peak in both cases was at 3.1 GeV. The J/ψ resonance was interpreted as the bound state of a new quark, called the *charm quark* c, and its antiparticle, the *charm antiquark* $\bar{c}$. Higher excited states of this bound system exist, too. The family of bound states of $c\bar{c}$ are called *Charmonium* states. From measurements on charmonium, the mass of the charm quark was estimated to be between 1.1 and 1.4 GeV. In the quark model of elementary particles, the "charm" quark along with the "strange" quark form the second generation of quarks after the first generation of "up" and "down" quarks. The electric charge carried by the charm quark is $+(2/3)|e|$ while that carried by the strange quark is $-(1/3)|e|$ (where $|e|$ is the magnitude of the electron charge), just like the up and down quarks.

According to the constituent quark model of elementary particles, mesons are made up from bound states of quark-antiquark pairs, while the baryons are made up from bound states of three quarks. The nonstrange mesons and baryons are made from the first generation quarks, u and d. The strange mesons and baryons contain at least one strange quark, s. Similarly, the charm mesons and baryons will contain at least one charm quark, c. The first proposal for the existence of such hadrons was made by Bjorken and Glashow [102]. Many of the charm hadrons have been found.

First evidence for charm baryons $\Lambda_c^+ = (udc)$ and $\Sigma_c^{++} = (uuc)$ was reported by Cazzoli et al. [103] who observed the production of charm baryons with neutrinos at the Brookhaven proton synchrotron, detectors being hydrogen and deuterium bubble chambers. The mass of the Σ_c^{++} was found to be 2426±12 MeV. Confirmation of this discovery came from the work of Baltay et al. [104] working with the broad band neutrino beam at the Fermilab proton synchrotron, detector being heavy liquid bubble chamber. From their measurements they came up with a mass for the Λ_c^+ state at 2257 ± 10 MeV, and the mass difference between the Σ_c^{++} and the Λ_c^+ state to be 168 ± 3 MeV, giving the Σ_c^{++} a mass about 2425 MeV. First evidence for the doubly strange baryon $\Omega_c = (ssc)$ came from the work of Biagi et al. [105] who studied the production of charm-strange baryon states produced in the collision of Σ^- with nucleus from the hyperon beam at SPS in CERN. They reported a mass of 2.74 GeV for this state. First evidence for a charm-strange baryon Ξ_c^+ was also presented by Biagi et al. [106] from a study of the reaction Σ^- from the hyperon beam at CERN SPS impinging on a Be target which led to the production of $\Xi_c^+ + X$ with subsequent decay of the Ξ_c^+ to $\Lambda K^- 2\pi^+$. The mass of this state was given by them to be 2.46 GeV. The neutral counterpart of this baryon Ξ_c^0 was subsequently found by Avery et al. [107] working with the CLEO detector at the Cornell e^+e^- ring. The quark content of Ξ_c^0 is (dsc), and its mass was quoted to be about 2471 MeV. A measurement of the mass difference between the Ξ_c^+ and the Ξ_c^0 states was done by Alam et al. [108] also working with the CLEO detector at the Cornell electron-positron ring. They quoted a value for the mass difference, $M(\Xi_c^+) - M(\Xi_c^0) = (-5 \pm 4 \pm 1)$ MeV. Observation of a narrow anti-baryon state, interpreted as $\bar{\Lambda}_c^- = (\bar{u}\bar{d}\bar{c})$, at 2.26 GeV with a decay width less than 75 MeV, decaying to $\bar{\Lambda}\pi^-\pi^-\pi^+$, was reported by Knapp et al. [109].

First evidence for the production of charm mesons $D^+ = (c\bar{d})$ and $D^- = (\bar{c}d)$ was presented by Peruzzi et al. [110] working with the SLAC-SPEAR electron-positon storage ring. They looked for the production of a new narrow charged resonance in electron-positron annihilation at a center-of-mass energy of 4.03 GeV. They found a state at a mass of 1876±15 MeV in the $K^\pm 2\pi^\mp$ channel, but not in the channel $K^\mp\pi^+\pi^-$. They interpreted these events in terms of production of D^+ and D^- mesons which then decayed to $D^+ \rightarrow K^- 2\pi^+$ and $D^- \rightarrow K^+ 2\pi^-$. The neutral counterparts $D^0 = (c\bar{u})$ and $\bar{D}^0 = (\bar{c}u)$ were also found in a SLAC-SPEAR experiment in multihadronic final neutral states produced in electron-positron annihilation at center-of-mass energies in the range between 3.90 GeV and 4.60 GeV. The data indicated new narrow neutral states with mass 1865 ± 15 MeV and a decay width less than 40 MeV, which decayed into $K^\pm\pi^\mp$ and $K^\pm\pi^\mp\pi^\pm\pi^\mp$. The interpretation of the data was that D^0 and $\bar{D}^0$ were produced in the annihilation reaction and these particles decayed to $D^0 \to K^-2\pi^+\pi^-$ and $\bar{D}^0 \to K^+\pi^+2\pi^-$. Charm-strange mesons have also been found. These are the $D_s^+ = (c\bar{s})$ and $D_s^- = (\bar{c}s)$, first evidence of which came from the DASP collaboration [111]. Evidence for excited states, D_s^{*+} as well as D_s^{*-}, was also presented in this work. The mass of the lower state was given as 2.03 ± 0.06 GeV while that of the excited state was given as 2.14 ± 0.06 GeV.

Continuing work in these studies has culminated in a lot of information on lifetimes, branching ratios for decay into various modes, and checks on assignment of various quantum numbers given to these particles, be they baryons or mesons. Details can be found in the "Review of Particle Physics" [62].

Chew-Frautschi Plot
A number of Regge trajectories (see under "Regge Poles") are known which are nicely exhibited in the *Chew-Frautschi Plot*. If one plots along the abscissa M^2 where M is the mass of the particle or the resonance, and along the ordinate J, the spin of the particle or resonance, one finds that the known particles and resonances fall on straight lines in the plot (see Figure 4.8).

Chiral Symmetry
To introduce the idea of chiral symmetry, we consider the Lagrangian for QCD for very light u and d quarks. Writing only the fermionic part, in terms of the isotopic spin doublet $q = \begin{pmatrix} u \\ d \end{pmatrix}$, we have

$$\mathcal{L} = \bar{q}(i\gamma^\mu D_\mu)q - \bar{q}\mathbf{m}q,$$

where D represents the covariant derivative, and $\mathbf{m}$ is the (diagonal) mass matrix of the u and the d quarks. This Lagrangian has isospin SU_2 symmetry which transforms the q doublet. If we introduce quark chiral components q_L and q_R by

$$q_L = \left(\frac{1-\gamma_5}{2}\right)q; \qquad q_R = \left(\frac{1+\gamma_5}{2}\right)q,$$

and set the mass terms to zero in the above Lagrangian, we see that the left (L) and right (R) chirality projections of q do not mix, and the Lagrangian is symmetric under separate unitary (U_L and U_R ($U_L \neq U_R$)) transformations of the L and R chirality projections. Associated with

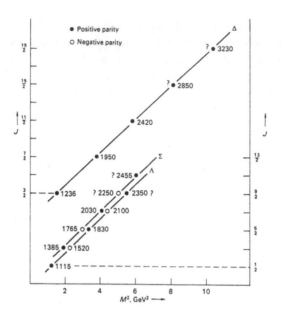

Figure 4.8: Chew-Frautschi plot of Regge trajectories for baryons and baryonic resonances. (Figure from *Introduction to High Energy Physics* by Donald H. Perkins. Copyright © 1982 by Addison-Wesley Publishing Company, Inc. Reprinted by permission of Addison Wesley Longman Publishers, Inc.)

these symmetries are four currents, which have both isosinglet U_1 and isovector (SU_2) parts,

$$\bar{q}_L\gamma^\mu q_L, \qquad\qquad \bar{q}_R\gamma^\mu q_R$$

$$\bar{q}_L\gamma^\mu T^i q_L, \qquad\qquad \bar{q}_R\gamma^\mu T^i q_R$$

where T^i, $i = 1, 2, 3$ are the generators of SU_2. If the chiral symmetry is exact, these chiral currents will be conserved. From these chiral currents, we can form combinations which are four-vector (V) and axial four-vector currents (A)

$$J^{\mu(V)} = \bar{q}\gamma^\mu q, \qquad J^{i\mu(V)} = \bar{q}\gamma^\mu T^i q$$

$$J^{\mu(A)} = \bar{q}\gamma^\mu\gamma^5 q, \qquad J^{i\mu(A)} = \bar{q}\gamma^\mu\gamma^5 T^i q.$$

Chiral symmetry is said to hold if these vector and axial vector currents are conserved. It is known that the vector currents are exactly conserved corresponding to transformations with $U_L = U_R$. What about the axial vector currents? Their conservation does not seem to correspond with any obvious symmetry of strong interactions, and it is possible these symmetries are broken spontaneously. Spontaneous breaking of the continuous symmetries associated with these axial curents will lead to the appearance of massless, spinless Goldstone bosons. The only particles in nature which seem to come close to satisfying this requirement seem to be isospin triplets of pions which have a rather low mass, are pseudoscalar particles, and are capable of being created by axial vector isospin currents.

Chirality States of Dirac Particles

Violations of parity (P) and charge conjugation (C) occur in the electroweak interactions of the Standard Model. To describe this concisely, it is useful to introduce the notion of *chirality states*. Let us introduce the matrix $\gamma_5 = i\gamma^0\gamma^1\gamma^2\gamma^3$ from the Dirac theory for the electron. States for which the eigenvalue of γ_5 is $+1$ are called states of *right-handed chirality*, and those with eigenvalue -1 are states of *left-handed chirality*. The Dirac spinor ψ, using projection operators constructed with γ_5, can be written as a sum of right- and left-handed chiral components:

$$\psi = \psi_R + \psi_L; \ \psi_R = [(1 + \gamma_5)/2]\psi, \ \psi_L = [(1 - \gamma_5)/2]\psi.$$

Maximal violations of parity and charge conjugation appear in the theory, if the coupling favors one chirality of the particle over another. For massless particles, chirality is the same as helicity, the projection of spin along the direction of motion.

CHORUS Experiment

The acronym CHORUS stands for *Cern Hybrid Oscillation Research apparatUS*. It is an experiment at CERN, designed to study the oscillations of the muon neutrino into tau neutrinos from a pure muon neutrino beam produced by the CERN SPS accelerator. Muon neutrinos of high intensity (about 10^6 cm^{-2} s^{-1}) and energy 27 GeV will travel through 850 m before being registered in a detector which is sensitive to tau neutrinos. Scintillation fibres are used as sensitive elements in the tracker, and special CCD cameras and nuclear emulsion techniques will be used for track recognitions and scanning.

Colored Quarks and Gluons

The need for color as an extra quantum number for the quarks comes about as follows. In the constituent quark model, where baryons are bound states of quarks, the ten states of the decuplet can be considered as quark composites of u, d, s quark flavors as follws: $\Delta^{++} = (uuu)$, $\Delta^{+} = (uud)$, $\Delta^{0} = (udd)$, $\Delta^{-} = (ddd)$; $\Sigma^{+} = (uus)$, $\Sigma^{0} = (uds)$, $\Sigma^{-} = (dds)$; $\Xi^{0} = (uss)$, $\Xi^{-} = (dss)$; $\Omega^{-} = (sss)$. The mass differences between the states of differing strangeness can be accommodated in the model by assuming that the mass of the strange quark m_s is about 150 MeV more than $m_u \simeq m_d$. To get a positive parity for the decuplet, we have to assume that the quarks move in a spatial state of orbital angular momentum zero. However, one immediately runs into a difficulty. The bound state, being made of three quarks (each of which is a fermion), has to be described by a wave function which is totally antisymmetric in all its variables. The wave function is a product of spatial part, flavor part, and spin part. If the orbital angular momentum is zero, the spatial part is symmetric, so the product of the flavor and the spin part must be antisymmetric. The Δ^{++}, Δ^{-}, and Ω^{-} have flavor symmetric quark combinations, being made of identical quarks (uuu), (ddd), and (sss), respectively. To get the total spin for the bound state to be $J = 3/2$ all three quarks will have to be in a spin symmetric state leading to a violation of Pauli principle. In order to get over this problem, another degree of freedom is attributed to quarks, together with the demand that the wave function be antisymmetric in this new degree of freedom. This new degree of freedom is called *color* and it is postulated that each flavor of quark comes in three colors, traditionally called *red, green,* and *blue,* and that baryons which are bound states of quarks have no net color, that is they are color singlets. Likewise mesons, which are bound states of quarks and antiquarks, must also have no net color (i.e., must be color singlets). The attribution of three colors to each quark flavor finds support in two other experimental facts: (a) three colors are required to get the decay rate for $\pi^0 \to 2\gamma$ right, and (b) at high energies in e^+e^- reactions, the ratio of the cross section for annihilation into hadrons to that of annihilation into a pair of muons also requires three colors to get it right. A formal theory called *quantum chromodynamics (QCD)*, has been formulated which uses the notion of color as a dynamical degree of freedom and which governs the color interactions of quarks (see under "QCD—Quantum Chromodynamics"). The color charge for quark interactions plays a role similar to that played by the electric charge in the electromagnetic interactions of charged particles (Quantum Electrodynamics (QED)). Electromagnetic interactions are mediated by a massless vector particle, the photon. Likewise, color interactions of the quarks are mediated by a new massless vector particle called the *gluon*. While the quantum of the electromagnetic field, the photon, is electrically neutral, the quantum of the color field, the gluon, is not color neutral but does carry color charge, leading to some very important differences between QED and QCD.

Composite Model of Pions

The first composite model of pions was explored in a very interesting paper by Fermi and Yang [112]. In this paper, the model for the pion as a bound state of nucleon and an antinucleon was shown to lead to properties of the meson required in the theory of Yukawa [113] to explain the short range nature of nuclear forces. It is interesting that replacing the nucleon and antinucleon of this model by quarks and antiquarks leads to the modern constituent quark model of mesons.

Compton Effect

A. H. Compton discovered in 1922 that the scattering of X-rays by thin foils of materials did not follow what was expected from classical electromagnetic theory [114]. The classical theory of the scattering was given by Thomson. According to this theory, the incident radiation sets free electrons of the material into oscillatory motion, which then re-radiate energy in the form of scattered radiation in all directions. He showed that the intensity of the scattered radiation at an angle θ to the incident radiation varies as $(1 + \cos^2 \theta)$ and that it is independent of the wavelength of the incident radiation.

Compton studied the scattering of molybdenum K_α-rays by graphite at various angles to the incident beam and compared the scattered spectrum with the incident spectrum. Compton found that at a given angle, the scattered radiation had two components. One of these had the same wavelength as the incident one, while the other component had a wavelength which was shifted with respect to the incident one by an amount which depended on the angle of scattering θ, clearly at variance with what is expected from Thomson's theory.

In order to understand this behavior, departing from Thomson's classical picture, Compton treated the incident radiation according to quantum theory. According to this theory, radiation consists of quanta, photons of energy $\hbar\omega$, and momentum $\hbar\omega/c$, where ω is the angular frequency of radiation, c the velocity of light, $\hbar = h/(2\pi)$, and h is Planck's constant. (In units in which $\hbar = c = 1$, energy and momentum of the photon are both equal to ω.) He treated the scattering as an elastic collision of the incident photons with the free electrons of the material in which the scattering is being studied. Using laws of energy and momentum conservation in the elastic scattering process, he was able to derive $\lambda' - \lambda = (2\pi/m)(1 - \cos\theta)$, where λ' is the wave length of the scattered radiation, λ is the wave length of the incident radiation, and m is the mass of the electron. The wavelength shifts measured from the experiment were in excellent accord with the quantum theory derivation of the collision. This experiment provided direct experimental verification that the photon is an elementary particle with energy and momentum. The intensity of the scattered radiation and the polarization properties of the scattered radiation have also been derived in quantum electrodynamics. The result for the intensity can be derived from an expression for the scattering cross section for photons on electrons known as the *Klein Nishina formula* [98] and plays a very important role in the energy loss of photons in materials.

Confinement of Color

It is believed that QCD does not allow asymptotically free particle states which carry a net color. In other words, colored quarks or gluons will not be seen as free particles. They are permanently confined within hadrons. As free particles we will see hadrons only because they are color singlets, that is, colorless objects, which are combinations of colored quarks and antiquarks.

We may understand how this property arises from the following qualitative considerations. Elsewhere in this book, we have mentioned another property *asymptotic freedom* possessed by QCD. According to this property, the effective quark gluon coupling vanishes at infinite energies or short distances, so that at high energies, the quarks and gluons inside a hadron can be treated approximately as non-interacting particles. As one proceeds to lower energies or longer distances, this effective coupling increases and is expected to become very large at very large distances. At high energies, one can develop a perturbation treatment of QCD, expanding in terms of the small effective coupling constant. Such a perturbative treatment has been found to give results which agree with experiments in a number of phenomena involving hadrons at high energies. As one proceeds toward low energies, such a perturbative treatment is not good because of the increase in the size of the expansion parameter. In the region of very strong coupling, an alternative method due to Wilson [115] may be employed. In this method, the continuum QCD is replaced by discretizing it on a lattice of four-dimensional Euclidean variables, and the statistical mechanics of this system is considered. Through such studies Wilson showed that color is confined, in the sense that if two colored quarks on the lattice are pulled to a large distance apart, the energy of such a configuration increases proportional to the length separating the quarks. Clearly the energy needed to pull the quark to infinite distance from the other quark will be infinite. He also showed in this lattice version of the theory, the only states that have finite energy for infinite separations are configurations in which the separating objects are color singlet objects. Color confinement in lattice version of QCD is thus established. It is believed, although there is no good formal proof, that the confinement property holds true in continuum QCD also.

Conservation/Violation of CP

It has been stated, under our discussion of charge conjugation operation, that the weak interactions are not invariant under the separate operations of spatial reflection (parity, P) and replacement of particles by antiparticles (charge conjugation, C). They seem to be invariant under the combined operation of CP. Experimental proofs of these statements come from the work of Wu et al. [116] on testing parity conservation in weak interactions. For violation of charge conjugation the evidence comes from the establishment of the property of the neutrino being only left-handed in electron beta decays [117]. Measurements of electron and positron polarizations in several beta decays have been carried out resulting in helicity $-\beta$ for electrons and $+\beta$ for positrons, where β is the velocity of the particle in units of the velocity of light. Just as left-handed electrons are accompanied by left-handed neutrinos in electron decays, right-handed positrons must be accompanied by right-handed antineutrinos. The attribution of helicity $+1$ to the antineutrino is consistent with the fact that when the mass can be ignored for high energy positrons ($v \simeq c$), the helicity of the positron tends to the value $+1$. The fact that one does not have left-handed antineutrinos is evidence for the violation of C in beta-decay reactions and weak interaction reactions generally. The operation of CP seems to be respected by beta decay processes in which left-handed neutrinos go over into right-handed antineutrinos.

The question has been raised whether CP is conserved in general in weak interaction reactions. Studies of decays of neutral K mesons have revealed a small violation of CP (see further under "Kaons—Neutral $K_1^0(K_S^0)$, $K_2^0(K_L^0)$). Violation of CP is also being looked for in other elementary particle decays, notably among the $B^0, \bar{B}^0$ mesons.

CP violation is being searched for because it has some very important consequences with respect to the current constitution of the universe—it is predominantly matter and no antimatter. In a very interesting paper, Sakharov [118] has pointed out that a solution to the long standing problem of the asymmetry between the numbers of baryons in the universe relative to the number of antibaryons may be linked to CP violation (see further under "Universe—Baryon Asymmetry").

Conserved Vector Current Hypothesis
See under "Beta Decay—Strong Interaction Corrections".

Constituent Quark Model of Hadrons
Gell-Mann [119] and Zweig [120] independently suggested in 1964, that use be made of the objects corresponding to the fundamental three dimensional representation of SU_3 as the basic building blocks of all hadrons. These objects were named *quarks* by Gell-Mann and *aces* by Zweig, but the name *quarks* has come to be accepted by the particle physics community. This triplet of quarks has a spin 1/2 for each and carries a baryon number 1/3. The triplet breaks up into an isotopic spin doublet with charges (in units of $|e|$) $+(2/3), -(1/3)$ and an isotopic spin singlet with charge $-(1/3)$. These are called the (u, d) and s quarks, respectively. The antiquarks will consist of $(\bar{u}, \bar{d})$ and $\bar{s}$, carrying opposite sign baryon number and opposite sign for charges from the corresponding quarks. SU_3 symmetry is assumed to be the symmetry of strong interactions, and interaction Lagrangians can be constructed in terms of the quarks. If the SU_3 symmetry is exact, all the hadrons formed from these building blocks must have the same mass. Because there are mass differences between the different hadrons, the SU_3 symmetry is not exact but broken. The breaking of this symmetry is assumed to be due to mass differences between the different quarks. Thus, there must be a lightest quark which must be stable. If one attributes a property called *strangeness* $S = -1$ to the s quark (and therefore, $S = +1$ to the $\bar{s}$ antiquark) and $S = 0$ to the isodoublet (u, d), one can form all the low lying meson and baryon states, non-strange as well as strange, with quark-antiquark combinations for the mesons, and three quark combinations for the baryons. If one of the quarks or the antiquarks possesses non-zero strangeness, then the hadron containing it will have non-zero strangeness and the hadron will be called a *strange hadron*. Thus, baryons can be formed which will have strangenesses in the range zero to -3 and mesons with strangenesses -1, 0, and $+1$. If the states which have been seen experimentally are put into correspondence with these theoretical states, a mass difference between the s and the (u, d) quarks of about 150 MeV is required to fit the data. The constituent quark model is quite successful in describing the low energy hadronic spectrum. It was subsequently found necessary to give an additional quantum number, called *color*, to the quarks (see under "Colored Quarks and Gluons").

Initially, only three types of quarks were introduced, the u, d, s ("up", "down", and "strange"). Now the standard model envisages the existence of six quarks: (u, d), (c, s), and (t, b), where c is the charm quark, t the top quark, and b the bottom quark. We briefly review the properties of these quarks: u, c, t carry charge $+(2/3)|e|$, while d, s, b carry charge $-(1/3)|e|$, and each quark carries a baryon number $(1/3)$. For mesons, bound state combinations can be formed from these and the antiquarks. As quarks carry a spin of $1/2$, the lowest bound states would be classified as $^1S_0$ and the $^3S_1$ states with total baryon number zero. These could be identified with the pseudoscalar and vector mesons. Higher excited states could come from higher orbital angular momenta between the quark and antiquark. The quarks also carry a quantum number called *color*, and the color dependent forces are such that only color neutral (color singlet) combinations manifest as mesons. Thus now, in addition to non-strange and strange mesons, we can have charm mesons or bottom mesons, if the quark in them is charm or bottom, respectively. We can also form three quark combinations whose states will represent the baryons. Again, just as in the case of mesons, we can have charm baryons or bottom baryons, if one of the quarks in them is charm or bottom. The "Review of Particle Physics" [62] lists many of these meson and baryon states and their quark constitutions and various detailed properties such as rates of transitions between various states. It is a rich source of information on mesons and baryons.

Cooling—Particle Beams

In storage ring colliders an important objective is to obtain the highest possible luminosities in the beams. There are methods known as *beam cooling* which achieve this end. The entire topic of techniques for handling of beams and enhancing their desired characteristics and minimizing the undesirable qualities is a highly specialized one. It is technically a complex field but very fascinating in the fantastic results that have been achieved and are still being developed. We cannot hope to do justice to this field in the short space that is being devoted to the topic of beam cooling here. We give only a brief account of the physics principles which are of importance in this field.

Cooling a beam of particles is a method by which one focuses the beam particles in space to a very small spot (small spatial volume), and also concentrate in that spot only particles with a very small spread in momentum (small momentum volume). The six coordinates, three spatial coordinates and three momentum coordinates specify a point in what is called *phase space*. The product of an element of spatial volume and an element of momentum volume is called *phase volume*, and the beam particles are represented in the phase volume by a distribution of points in it. Liouville's theorem states that for particles subject to conservative forces, the overall phase volume is an invariant. This seems to suggest that it may be a physical impossibility to achieve cooling. However, a deeper understanding of the applicability of this theorem to particle beams suggests that there are effective ways of circumventing the theorem; for example, the overall phase volume can be held constant to satisfy the theorem, while at the same time the phase volume associated with particles of the beam can be reduced.

Major landmarks in the field of cooling started with the invention of "electron cooling" by Budker in 1967 [121] and that of "stochastic cooling" by van der Meer in 1968 [122]. The experimental verification that electron cooling works came from Budker et al. in 1976 [123]. The experimental verification of stochastic cooling and its use in the Intersecting Storage Ring at CERN occurred in 1975. Since then these methods are incorporated in all storage ring colliders. For details on electron cooling and stochastic cooling, please see sections under "Electron Cooling" and "Stochastic Cooling".

Cosmic Microwave Background Radiation

One of the predictions of the big bang theory of the origin of the universe is the existence of a primordial, relic radiation left over from the big bang red shifted by the expansion of the universe [124]. Calculations suggest that such relic radiation will be isotropic and will have a black body temperature of about 3 degrees Kelvin at present. The detection of this relic radiation, referred to as *Cosmic Microwave Background (CMB)*, will provide very good evidence to establish the big bang theory. CMB was discovered by Penzias and Wilson in 1965 [125] thus establishing the big bang theory. They found that the power spectrum was isotropic and unpolarized within an accuracy of ten percent and fitted to a Planck distribution with a temperature of about 3 K.

Subsequent to the discovery, much more precise work has been done by many groups on the power spectrum, and the black body temperature has been determined more precisely to be 2.73 ± 0.01 K. These further studies have also shown that the radiation is unpolarized to the level of 10^{-5}. Small deviations from isotropy have also been found. Spherical harmonic analysis of the CMB $\Delta T(\theta, \phi)/T$, where T is the temperature and θ, ϕ are the spherical polar angles, reveals that there exists a dipole anisotropy at the level of 10^{-3}. The power at an angular scale θ is given by a term of multipole order l of order $(1/\theta)$. The dipole anisotropy seen can be understood in terms of the solar system moving relative to the isotropic CMB radiation. Using this interpretation, one can determine the velocity of the center of mass of the solar system with respect to the CMB and its direction. In precision studies of the CMB anisotropy, this effect and the effect of the Earth's velocity around the Sun are removed.

The interest in CMB anisotropy stems from the fact that theoretical considerations suggest that we should have temperature fluctuations in higher multipoles of order 10^{-5} in order to seed the formation of galaxies. COBE [126] indeed found such an anisotropy in 1992. The angular resolution of COBE was limited to 7 degrees. Even higher precision studies have been done by the Boomerang collaboration [127]. The Boomerang experiment studied the CMB with a microwave telescope, looking at 2.5% of the sky, with an angular resolution of (1/4) of a degree. The microwave telescope was flown on a balloon over Antarctica for a 10-day duration.

The higher order anisotropies are thought to arise from primordial perturbations in the energy density of early universe present in an epoch when the matter and radiation stopped to interact through photon scattering. Thus the detection of the anisotropies is proof for the existence of the density perturbations which are enhanced by the gravitational instability leading to the kind of structures we see at present. In the case of a flat universe, the peak of the power spectrum is expected to occur at an angular scale of 45 arcminutes. Boomerang has measured this peak with great precision, and the location and the magnitude of the peak agree with what is expected from a *flat* universe. From the location of the peak we can infer that the density of the matter (including dark matter) is within 10% of the critical value. However, this need not necessarily mean that the universe will expand forever, for "dark energy" may be present which provides repulsion. Dark energy may be looked upon as a manifestation of the cosmological constant. The lesson of the Boomerang experiment is that the universe has a large amount of dark energy.

Dark energy has the effect of accelerating the expansion of the universe. Measurements of distances to some type Ia supernovae show that the expansion of the universe is indeed accelerating. This effect is what one gets for a universe which is spatially flat and in which two thirds of the critical density arises from dark energy associated with the cosmological constant.

Data from Boomerang continue beyond the peak at 45 arcseconds up to an angular scale of 15 arcseconds. Theory predicts a second peak at 15 arcseconds. Boomerang data has such a peak but its height is less than theoretical expectations. At present there is no understanding of this discrepancy.

Cosmic Rays

Ever since the discovery by Hess in 1912 that the earth is being constantly bombarded by extremely energetic particles, called *cosmic rays*, a large amount of effort has been expended in understanding the nature of the radiation, its content, and its source or sources. Besides the inherent interest in the study of cosmic rays for its own sake, there is another practical reason to study them. In many high energy physics experiments, identification and suppression of the background effects due to cosmic rays is essential for the success of the experiment. We include here a brief summary of this field.

Cosmic rays reaching the earth from outer space consist of stable charged particles and stable nuclei, and are classified as *primary* or as *secondary*. Primary cosmic rays are those that are produced by special accelerating mechanisms operating in various astrophysical sources, while secondary cosmic rays originate from the collisions of primary cosmic ray particles with intestellar material. Electrons, protons, nuclei of helium, carbon, oxygen, iron, heavier nuclei, etc. are all produced in stars and constitute primary radiation. Nuclei of lithium, beryllium, and boron are classified as secondary because they are not produced abundantly in stars.

Measurements of the intensity spectrum of primary nucleons in the energy range from a few GeV to about 100 TeV are described very well by a power law [62]. It is approximately given by the expression $I_{nuc}(E) \sim 1.8E^{-\alpha}$, measured in units of nucleons $cm^{-2} \, s^{-1} \, sr^{-1} \, GeV^{-1}$, where the exponent α is about 2.7 and the energy E is the total energy of the particle including the rest mass energy. A quantity $\gamma = \alpha - 1$ is called the *integral spectral index*. It turns out that about 79% of primary nuclei are free protons. Another 15% are nucleons which are bound in helium nuclei. To get an idea of the actual numbers, we may mention that the primary oxygen flux at 10.6 GeV/nucleon is $3.26 \times 10^{-6}/(cm^2 \, s \, sr \, GeV/nucleon)$. At the same energy per nucleon, the proton flux and the helium flux are about 730 times and 34 times the oxygen flux, respectively. The differential flux of electrons and positrons incident at the top of the atmosphere as a function of the energy shows a steeper fall compared to the spectra of protons and nuclei (more like E^{-3}). Above 10 GeV energy, the proportion of antiprotons relative to protons is about 10^{-4}. These are all secondary, for at present there exists no evidence for a primary antiprotons component.

When the primary cosmic rays strike the atmosphere of the earth, many charged mesons and neutral mesons are produced by the collisions of primary cosmic rays with air atoms and molecules. Muons and neutrinos arise as decay products of the charged mesons, while electrons, positrons, and photons are produced from the decay of neutral mesons. As one goes down from the top of the atmosphere, the meson and nucleon components decrease due to decays or interactions. It is found that the vertically incident nucleon component follows an exponential attenuation law with an attenuation length of about 120 g cm^{-2}. The vertical intensity of $\pi^{\pm}$ with energy less than about 115 GeV reaches a maximum at a depth of 120 g cm^{-2}, which is at a height of about 15 km. The intensity of low energy pions is small because, for most of them, the decay time is shorter than the interaction time, and so these decay.

Let us first consider the neutrino component. Every pion decay gives rise to $\mu^{+} + \nu_{\mu}$ or $\mu^{-} + \bar{\nu}_{\mu}$, so a measurement of the flux of muons will give a good estimate for the flux of muon neutrinos (and antineutrinos) produced in the atmosphere. The value of the pion intensity at its maximum provides a good knowledge of the $\nu_{\mu}(\bar{\nu}_{\mu})$ fluxes to be expected in the atmosphere. The muons decay and give rise to an additional muon neutrino and an electron neutrino per muon. Thus from each pion decay (and subsequent muon decay), one expects two muon neutrinos and one electron neutrino, giving a ratio, $\nu_{\mu}/\nu_{e} = 2$ for the atmospheric neutrinos. Once produced, these weakly interacting particles, with energies of the order of several GeV, propagate for large distances without much hindrance.

The high energy muons produced in the top of the atmosphere reach sea level because of their low interactions. They typically lose about 2 GeV of energy through ionization loss before they get to ground level. The mean energy of muons at ground level is about 4 GeV. The energy spectrum is more or less flat for energies lower than 1 GeV and drops in the energy range 10 GeV–100 GeV, because the pions capable of giving rise to muons of this energy range do not decay but instead interact. A detector at sea level receives a flux, 1 cm^{-2} min^{-1} of muons. The angular distribution for muons of energy ~ 3 GeV at ground level follows a $\cos^2 \theta$ behavior, where θ is the angle from the vertical.

The electromagnetic component at sea level consists of electrons, positrons, and photons. These originate from cascades initiated by decay of neutral and charged pions. Most electrons of low energy at sea level have originated from muon decays. The flux of electrons and positrons of energies greater than 10 MeV, 100 MeV, and 1 GeV, are roughly about 30 m^{-2} s^{-1} sr^{-1}, 6 m^{-2} s^{-1} sr^{-1}, and 0.2 m^{-2} s^{-1} sr^{-1}, respectively.

Protons seen at sea level are essentially the energy degraded primary cosmic ray protons. About 30% of the vertically arriving nucleons at sea level are neutrons; the total flux of all protons of energy greater than 1 GeV arriving from a vertical direction is about 0.9 m^{-2} s^{-1} sr^{-1} at ground level.

Once one goes deep underground, the only particles that penetrate are the muons of high energy and neutrinos. The intensity of muons deep underground is estimated by taking the intensity at the top of the atmosphere and considering all energy losses they suffer. The average range of muons is given in terms of depth of water equivalent (1 km water equivalent $= 10^5$ g cm^{-2}). The average ranges for muons of 10 GeV, 100 GeV, 1 TeV, and 10 TeV, found from literature, are 0.05, 0.41, 2.42, and 6.30 km water equivalent, respectively.

Another phenomenon worthy of note is extensive air showers initiated by cosmic ray particles. Such showers occur when the shower initiated by a single cosmic ray particle of very high energy at the top of the atmosphere reaches the ground level. The shower has a hadron at its core, which subsequently develops an electromagnetic shower mainly through $\pi^0 \to \gamma\gamma$ processes in the core. Electrons and positrons are most numerous in these showers with muons an order of magnitude smaller in number. Such showers are found spread over a large area on ground and are detected by a large array of detectors in coincidence. Extensive air showers probe cosmic rays of energy greater than 100 TeV. An approximate relation between the shower size, as measured by the number of electrons n_e in the shower, and the energy of the primary cosmic ray E_0 is found to be $E_0 \sim 3.9 \times 10^6 (n_e/10^6)^{0.9}$ valid at a depth of 920 g cm^{-2}.

At the very highest energies, of the order of 10^{17} eV and above, there is intense interest at present in studying the cosmic ray particle spectrum. When the $E^{-2.7}$ factor is removed from the primary spectrum, the spectrum is still found to fall steeply between 10^{15} and 10^{16} eV. This has been called the *knee*. Between 10^{18} and 10^{19} eV, a rise in the spectrum has been noticed, called the *ankle*. The cosmic ray community is studying this region of energies quite intensively. It looks as though the ankle is caused by a higher energy cosmic ray population mixed in with a lower energy population at energies a couple of orders of magnitude below the ankle. The interpretation that is being placed at present on these results is that the spectrum below 10^{18} eV is of galactic origin, and the higher energy population is due to cosmic rays originating outside the galaxy.

Cosmological Bound on Neutrino Rest Mass

The first cosmological upper bound on neutrino rest mass was derived by Gershtein and Zeldovich [128]. They showed how one can greatly reduce the upper limit on the muon neutrino (ν_μ) mass by using cosmological considerations connected with the hot model of the universe. Their considerations are as follows. In the big bang theory of the creation of the universe, one envisages the universe as expanding from an initial very hot and extremely dense state. In the early stages of the universe, at temperatures T such that the energy was greater than ~ 1 MeV, neutrino generating processes occurred, producing various kinds of neutrinos. At thermal equilibrium, one can estimate the number densities of the different kinds of fermions (and antifermions) and bosons that are generated using equilibrium statistical mechanics. When the universe expands and cools down, the neutrinos survive because of the extremely low $\nu\bar{\nu}$ annihilation rates. Their density goes down because of the expansion in volume due to the expansion of the universe. Because the volume increases as Volume $\propto (1 + z)^{-3}$, where z is the red shift, one can estimate the number density of the neutrinos in the current epoch, characterized by $z = 0$, from that prevalent at the latest epoch when thermal equilibrium was maintained, $T(z_{eq}) \sim 1$ MeV, where z_{eq} is the red shift at that epoch. Such considerations give an estimate of the number density, $n_\nu + n_{\bar{\nu}}$, of 300 cm^{-3} for each species of neutrinos at the current epoch. This number density is enormous; by comparison, all the visible matter in the universe contributes to an average density of hydrogen atoms of only 2×10^{-8} cm^{-3}. If each neutrino has a mass, the number density derived can be translated into a mass density. Now, the mass density, ρ_{tot}, of all possible sources of gravitational potential in the universe has been ascertained to be $\rho_{tot} \leq 10^{-29}$ g cm^{-3}. The mass density provided by the different neutrino species has to be less than ρ_{tot}. This provides an upper limit on the neutrino mass. Assuming only electron and muon neutrinos (and their antiparticles) each having mass m_ν, Gershtein and Zeldovich derived this upper limit to be 400 eV. Subsequent independent works by Cowsik and McLelland [129] and by Marx and Szalay [130] have shown this limit to be about 8 eV. Comparing this with improved laboratory measurements, from which an upper bound for $m(\nu_\mu)$ of 0.17 MeV is found, we see that cosmological considerations reduce this upper bound by several orders of magnitude for the muon neutrino, while not doing very much for the electron neutrino.

CP Violation—Neutral *B* Mesons

CP violation was first observed in the neutral K meson system. With three families of quarks discovered since then, and mixings among the charge $-(1/3)|e|$ quarks of these families represented by the Cabibbo Kobayashi Maskawa (CKM) matrix, it came to be realized that the phenomenon of CP violation may not be restricted to the neutral Kaon system but may also be present in neutral mesons containing charm and bottom quarks. Investigations in the matter have shown that CP violation in mesons containing charm may not be observable, but it should be observable in neutral B mesons containing bottom (or beauty) quarks. As the violation is not expected to be large, one requires a large supply of neutral B mesons to make the observation possible. To meet these requirements and study other properties of B mesons, dedicated accelerators have been built to serve as B factories. One of these is the BaBar Experiment at SLAC (see further under "BaBar Experiment").

We present a brief description of CP violation in neutral B mesons [62]. To observe CP violation in neutral B meson systems, one has to observe B^0 and $\bar{B}^0$ meson decays and compare them. The quark content of $B^0(\bar{B}^0)$ is $d\bar{b}(\bar{d}b)$. Unlike in the neutral kaon system, the life time difference between the two B^0 eigenstates is expected to be negligibly small. These states will be distinguished only by the mass difference ΔM between the two states. If the decay of the neutral $B(\bar{B})$ meson occurs through the weak transformation of one b quark ($\bar{b}$ antiquark), it is found that the difference between B^0 and $\bar{B}^0$ decays is expressible in terms of an asymmetry parameter A_a. This parameter depends upon the particular C eigenstate a with eigenvalue η_a, the phase ϕ_M associated with B^0-$\bar{B}^0$ mixing, and the phase ϕ_D associated with the weak decay transitions according, to

$$A_a = \eta_a \sin\left[2(\phi_M + \phi_D)\right].$$

Considering the transition $b \to c\bar{c}s$ leading to $B^0(\bar{B}^0) \to \psi K_S$, the asymmetry in the Standard Model is a quantity without any uncertainty due to the hadronic matrix elements. The experimental constraints on the elements of the CKM matrix allow a prediction for this asymmetry $A_{\psi K_S}$ to be between -0.3 and -0.9. If there is any sizable deviation from this range, physics beyond the Standard Model may be required.

Another decay mode of interest is $B^0(\bar{B}^0) \to \pi\pi$ and involves the quark transition $b \to u\bar{u}d$. The asymmetry $A_{\pi\pi}$ can be worked out similarly to the other case. Both could involve $B^0\bar{B}^0$ mixing, but the difference between $A_{\pi\pi}$ and $A_{\psi K_S}$ would be a signal for CP violation arising from beyond the CKM matrix, called *direct CP violation*.

CP violation could also be looked for in charged B meson decays. B^+ and B^- decay to charge conjugate states. Difference in these rates signals CP violation. Here there are more theoretical uncertainties in evaluating what the difference should be. Predicted effects are small and not precise.

The B^0_s meson ($s\bar{b}$ quark content) affords another possibility for observing CP violation. Here the mass difference ΔM is much larger, although not yet measured. The width difference is also expected to be larger for B_s. Thus, there may be a possibility of finding CP violation just as in neutral K meson system by observing the B_s with different lifetimes decaying into the same CP eigenstates.

Current work in progress should soon enrich our knowledge of the details of *CP* violation, how much of it is due to the CKM mixing matrix, and how much is due to direct *CP* violation.

CP Violation—Neutral Kaons

Data on beta decays support the idea of *CP* conservation. *CP* violation was first observed in neutral Kaon decays by Christenson et al. [131]. Neutral Kaon systems consisting of K^0 and $\bar{K}^0$ exhibit some strange properties in their decays. These particles have opposite strangeness quantum numbers and are produced copiously in association with other particles also possessing strangeness, by collisions of non-strange particles involving strong interaction processes. The total strangeness of the particles produced in strong interaction must add up to zero because it was zero initially. Because strangeness is conserved in strong interactions, such interactions cannot induce $K^0 \to \bar{K}^0$ transitions. When the weak interactions are included, strangeness is no longer a good quantum number, and $K^0 \to \bar{K}^0$ transitions are possible, for example, via $K^0 \to \pi^+ + \pi^- \to \bar{K}^0$. Thus, the degenerate K^0 and $\bar{K}^0$ states cannot be eigenstates of the full Hamiltonian when the weak interactions are also included and linear combinations are needed. *CP* eigenstates can be formed, $|K_1^0\rangle = (|K^0\rangle + |\bar{K}^0\rangle)/2^{1/2}$ and $|K_2^0\rangle = (|K^0\rangle - |\bar{K}^0\rangle)/2^{1/2}$, with eigenvalues $+1$ and -1 respectively. The $|K_1^0\rangle$ and $|K_2^0\rangle$ states are distinguishable by their decay modes if *CP* is conserved during the decay.

Neutral K mesons have been observed to decay into 2π ($\pi^0\pi^0, \pi^+\pi^-$) and 3π ($\pi^+\pi^-\pi^0, 3\pi^0$) decay modes. By Bose symmetry, the total wave function of the 2π decay mode must be symmetric under the exchange of the two particles. As the pions are spinless, the exchange involves only the operation of C followed by P, so $CP = +1$ for the 2 pion state. For the three pion state, because the Q value for the decay is small (about 70 MeV), the three pions will mostly be in a relative orbital angular momentum zero state. The $\pi^+\pi^-$ in the three pion state will have $CP = +1$ (by argument similar to the one for the 2 pion state). The π^0 has $C = +1$ (because of its 2γ decay) and has $P = -1$ (because the pion has odd parity), so this 3 pion state has $CP = -1$. Contributions from higher relative orbital angular momenta will not allow one to have such a clean argument as to the value of *CP*, and both $+1$ and -1 may be possible. However, these contributions will be suppressed due to centrifugal barrier effects. The end result of these considerations is that the $|K_1^0\rangle$ with $CP = +1$ can decay only into 2 pions, and the $|K_2^0\rangle$ state with $CP = -1$ can decay only into 3 pions if *CP* conservation holds.

These decays involve quite different Q values, populate quite different regions of the phase space, and will have different disintegration rates. The measured lifetimes for these decay modes are 0.9×10^{-10} s and 0.5×10^{-7} s, respectively.

In a very interesting experiment performed by Christenson et al. [131], they demonstrated that $|K_2^0\rangle$ state also decays into 2 pions with a small branching ratio of 10^{-3}, thus indicating violation of CP. Hence, the nomenclature for describing the short-lived and the long-lived states has to be revised in view of the discovery that CP is violated. The short-lived state is called $|K_S\rangle$ and the long-lived state is called $|K_L\rangle$; the short-lived state is mostly $|K_1^0\rangle$ with a small admixture of $|K_2^0\rangle$, and the long-lived state is mostly $|K_2^0\rangle$ with a small admixture of $|K_1^0\rangle$. CP violation is characterized by the value η_{+-} for the ratio of amplitude for the long-lived state to decay into 2 pions to that for the short-lived state to decay into 2 pions. Experimental determination of this ratio was possible because of observable interference effects in the $\pi^+\pi^-$ signal due to the fact that both the long-lived and the short-lived states can decay into $\pi^+\pi^-$. Similar CP violation is observed in the $2\pi^0$ mode also, characterized by a value for η_{00}. η_{+-} and η_{00} are in general complex quantities with a modulus and a phase ϕ. Their experimental values are $|\eta_{+-}| = (2.274 \pm 0.022) \times 10^{-3}$, $\phi_{+-} = (44.6 \pm 1.2)$ deg and $|\eta_{00}| = (2.33 \pm 0.08) \times 10^{-3}$, $\phi_{00} = (54 \pm 5)$ deg. Two new parameters called ϵ and ϵ' are introduced related to η_{+-} and η_{00} by $\eta_{+-} = \epsilon - \epsilon'$, $\eta_{00} = \epsilon - 2\epsilon'$. CP violation has also been established in leptonic decay modes of the neutral K mesons: $K_L^0 \to e^+\nu_e\pi^-$ and $K_L^0 \to e^-\bar{\nu}_e\pi^+$; the final products are CP conjugates of each other and one would expect an asymmetry if CP is violated [132]. If the rate of decay to positrons is Γ_+ and to electrons is Γ_-, then the asymmetry they measure is

$$\delta = \frac{\Gamma_+ - \Gamma_-}{\Gamma_+ + \Gamma_-} = (+2.24 \pm 0.36) \times 10^{-3}.$$

The data available on CP violation in K^0 decays have been accommodated in the "superweak" model due to Wolfenstein [133]. In this model, CP violating effects are due to a new interaction much weaker than the usual weak interaction and seen only in the K^0 system. One of the predictions of this model is $\epsilon' = 0$ so that $|\eta_{+-}| = |\eta_{00}|$ and $\phi_{+-} = \phi_{00}$. It also leads to predictions for the asymmetry in leptonic decay modes of the K^0 which are in agreement with experiments.

There are other models of CP violation which involve six flavors of quarks and mixing among them, more general than the Cabibbo mixing discussed elsewhere. The mixing is characterized by the *Cabibbo Kobayashi Maskawa (CKM) Matrix* [96]. In this model of six flavors of quarks, a finite value of ϵ' is expected. Recent experiments have determined that ϵ' is different from zero and that CP violation may be a more general phenomenon than that observed in K^0 system only. Thus, for example, systems of neutral bottom (B^0) mesons may also exhibit CP violation, and this is being searched for in special accelerators, called B *factories*, designed for this purpose.

CPT Invariance

The CPT *theorem* due to Pauli [134] states that in relativistic field theories, all the interactions are invariant under the combined transformations C, P, and T, where C represents the particle-antiparticle conjugation, P the reflection operation, and T the time reversal operation, the operations being performed in any order. The theorem is based in local quantum field theory on very general assumptions and is not easy to circumvent. However, in view of the fact that many symmetries which were thought to be good turned out experimentally not to be borne out (such as parity violation and CP violation), it is desirable to determine experimentally whether CPT theorem is violated. For this one needs to focus attention on some of the consequences of the CPT theorem and see to what extent the consequences are borne out. Particles and antiparticles, according to the CPT theorem, must have the same mass and lifetime, and magnetic moments which are the same in magnitude but opposite in sign. Fractional difference in masses of π^+ and π^- is found to be less than 10^{-3}, for proton and antiproton it is less than 8×10^{-3}, and for K^+ and K^- it is less than 10^{-3}. The best test of mass difference comes from the study of neutral K decays: $(M_{\bar{K}^0} - M_{K^0})/M_{K^0} \leq 10^{-18}$. Lifetime equalities for particles and antiparticles have also been tested, and for pions, muons, and charged kaons, the fractional difference is found to be less than 10^{-3} for all of them. Magnetic moments for particles and antiparticles have also been compared; in the case of muons, the fractional difference is less than 3×10^{-9}. For electrons and positrons a very high precision test of the equality of the g-factors is provided by the development of the ion trap technique by Dehmelt [135] with the result $g(e^-)/g(e^+) = 1 + (0.5 \pm 2.1) \times 10^{-12}$. Thus, it might be stated that CPT invariance is generally valid.

Crossing Symmetry

The amplitudes for different processes are obtained from the matix elements of the scattering matrix which relates physical particle states and, hence, is only defined in the physical region, where all the momenta and all the scattering angles are real. In quantum field theory, however, the nonphysical region also has significance. Suppose we have a matrix element $T(p_3, p_4; p_1, p_2)$, which gives the physical amplitude for some process in which particles 1 and 2 with four-momenta p_1, p_2, respectively, are incident on one another, and particles 3 and 4 with four-momenta p_3, p_4, respectively, leave the interaction region. In extending this matrix element to the nonphysical region, we suppose that there exists an analytic continuation in some variables, which takes us to the nonphysical region. An explicit expression for the matrix element T can be obtained to any order in perturbation theory, and its analytic continuation to the nonphysical region can be checked.

To understand what is involved, we consider a process involving the interaction of two particles, leading to a final state involving two particles also. Let us introduce the Mandelstamm variables, $s = (p_1 + p_2)^2$, $t = (p_1 - p_3)^2$, and $u = (p_1 - p_4)^2$. The four-momentum conservation gives $p_1 + p_2 = p_3 + p_4$, where $p_i^2 = m_i^2$, m_i being the mass of the ith particle, $i = 1, 2, 3, 4$. The Mandelstamm variables are Lorentz invariant scalars and satisfy $s + t + u = \sum_{i=1}^{4} m_i^2$, so that only two of the three variables are independent. Thus the variables on which the matrix T depends can be chosen to be any two of these three variables.

The channel in which the particles undergo the reaction $1+2 \to 3+4$ is called the s-channel. Here the physical region for s is $s \geq (m_1 + m_2)^2$. In this channel t and u are physical four-momentum transfer variables. We can define a crossed channel reaction by crossing particle 3 to the left-hand side and replacing it by its antiparticle $\bar{3}$, and crossing particle 2 to the right-hand side and replacing it by its antiparticle $\bar{2}$: $1 + \bar{3} \to \bar{2} + 4$. (It could also be achieved as: $\bar{4} + 2 \to \bar{1} + 3$ by crossing particles 4 and 1.) The channel obtained by this procedure is called the t-channel. Because the momenta of the antiparticles are opposite in sign to the momenta of the particles, in the t-channel the physical region corresponds to $t \geq (m_1 + m_3)^2$, and $s = (p_1 - p_2)^2$ and $u = (p_1 - p_4)^2$ become the physical four-momentum transfer variables. The other crossed channel we can get is called the u-channel in which $1+\bar{4} \to \bar{2}+3$, the physical region corresponds to $u \geq (m_1 + m_4)^2$, and $s = (p_1 - p_2)^2$ and $t = (p_1 - p_3)^2$ are physical momentum transfer variables. Thus, as an example, consider the s-channel reaction $\pi^+ + p \to \pi^+ + p$. The crossed t-channel reaction is $\pi^+ + \pi^- \to p + \bar{p}$. The crossed u-channel reaction is $\pi^- + p \to \pi^- + p$ where, instead of crossing the protons, we have crossed the pions. In perturbation theory, the s- and u-channel amplitudes are simply related by the substitution rule; p_2 and p_1 in the s-channel become $-p_2$ and $-p_1$, respectively, in the u-channel. Similarly one can give the substitution rule to get the t-channel amplitude from the s-channel amplitude. The relation of the amplitudes for crossed processes is the statement of *crossing symmetry*. Thus in our example, the amplitudes for π^+p scattering, π^-p scattering, and $p+\bar{p} \rightarrow \pi^+ + \pi^-$ are related by crossing symmetry. The nonphysical region in one channel becomes the physical region in the crossed channel.

Current Algebra

This is an approach to dealing with dynamical effects of strong interactions on various weak processes, which involve hadrons, and was pursued vigorously in the 1960's. This approach enabled one to obtain relations between different weak processes suffered by hadrons. It was also able to give relations between weak and strong interactions of hadrons. As it was known that the axial vector currents are only partially conserved (a concept called *Partial Conservation of Axial Currents (PCAC)*), it was combined with some other techniques developed by Adler, and others to obtain these relations [136]. We include here a brief description of the current algebras which have been applied to weak interaction problems.

A simple form of current algebra which was first used by Adler [137] and Weissberger [138] to derive a relation which goes under their names will be described to illustrate the procedure. We consider the $\Delta Y = 0$ (hypercharge non-changing) weak vector $J^{i\mu(V)}(x)$ and axial vector $J^{i\mu(A)}(x)$ currents, with $i = 1, 2, 3$, and consider introducing equal time commutation relations between them as shown below. The hypothesis of the conservation of the vector current (CVC) leads to the identification of $J^{1\mu(V)}(x) \pm iJ^{2\mu(V)}(x)$ with the charge changing isospin currents and $J^{3\mu(V)}(x)$ with the isovector part of the electromagnetic current. When electromagnetic interactions are absent, all three components of the vector current, $J^{i\mu(V)}$ $i = 1, 2$, or 3, are conserved separately. We can define generators of isospin rotations (isospin vector charges) by

$$I^i(t) = \int d^3x J^{i0(V)}(x), \qquad i = 1, 2, 3.$$

These generators satisfy the equal time commutation relations

$$[I^i(t), I^j(t)] = i\epsilon_{ijk}I^k(t), \qquad i, j, k = 1, 2, 3,$$

where ϵ_{ijk} is totally antisymmetric in its indices and $\epsilon_{123} = 1$. In the absence of electromagnetism these generators are time independent. When electromagnetism is present, the components with $i = 1$ and 2 will not be conserved, will become time dependent, and will not commute with the electromagnetic part of the Hamiltonian. The basic idea of the approach of current algebra is to propose that the equal time commutation relations hold exactly even when electromagnetism is present. One can extend these considerations to the isospin axial vector charges also. Define

$$I^{i,A}(t) = \int d^3x J^{i0(A)}(x), \quad i = 1, 2, 3.$$

Because these axial vector current components are not conserved, these quantities are time dependent, but they are isospin vectors. They will satisfy the equal time commutation relations,

$$[I^i(t), I^{j,A}(t)] = i\epsilon_{ijk} I^{k,A}(t), \quad i, j, k = 1, 2, 3.$$

To close the algebra, we assume that

$$[I^{i,A}(t), I^{j,A}(t)] = i\epsilon_{ijk} I^{k,A}(t), \quad i, j, k = 1, 2, 3.$$

Just as with the vector isospin currents we assume that these relations for the axial vector isospin currents are valid even when electromagnetism is present. The above three sets of commutation relations between, vector-vector charges, vector-axial vector charges, and axial vector-axial vector charges form the $SU_2 \times SU_2$ algebra of charges. Using these, different applications have been considered, most of which are in the nature of sum-rules.

The one involving axial vector-axial vector commutation relations was the one which was used to derive an expression for calculating the axial vector coupling constant g_A for the proton. The starting point for this work is the matrix element between two proton states of equal momentum, of the commutation relation between an axial charge and an axial current. This is written in the form of a dispersion integral, over unknown matrix elements of the divergence of the axial vector current, which involves not only single particle intermediate states but also more complicated states. After integrating over the internal degrees of freedom in the intermediate states, one makes the approximation of keeping only the contribution to the matrix element from the pion pole. Without going into the details, we quote the final result derived by Adler [137] and Weissberger [138]. This relation is essentially in the form of a sum-rule:

$$1 = |g_A|^2 \left[1 + \frac{2M_p^2}{g^2} \frac{1}{\pi} \int_{m_\pi}^{\infty} dE_\pi^L \frac{p_\pi^L}{(E_\pi^L)^2} (\sigma_{\pi^- P}^{tot}(E_\pi^L) - \sigma_{\pi^+ P}^{tot}(E_\pi^L)) \right],$$

where the integral involves the difference of laboratory total cross sections for negative pions on protons and positive pions on protons at a laboratory energy E_π^L for the pion. The integral is over all pion laboratory energies starting at the mass of the pion m_π and going to infinity. p_π^L is the corresponding laboratory momentum for the pion, M_p is the mass of the proton, and g is the pion nucleon coupling constant. In the integral, the $\Delta(1236)$ resonance gives a rather large contribution, due to which the right-hand side becomes less than 1 and, hence, g_A^2 is somewhat larger than 1. The evalutation of the integral can be performed by using pion-proton scattering data obtained from experiments, and finally one obtains $|g_A| \simeq 1.16$. This compares reasonably well with the experimental value of 1.24.

Current algebras have been extended to $SU_3 \times SU_3$ currents and many interesting results involving the multiplet of particles related by SU_3 symmetry have been derived. For details reference must be made to original literature.

Dalitz Plot

Many reactions or decays of particles lead to a final state involving just three particles. A plot was invented by Dalitz [139] to investigate whether there are any correlations among the product particles in three particle final states and has come to be called the *Dalitz plot*. The energy distribution of the produced particles is a product of the square of the matrix element for the reaction (or decay) and the phase space factor for the three particles. If the matrix element is a constant, the distribution is directly determined by the phase space factor. If the observed distributions differ markedly from the distribution determined by phase space factors alone, the difference must be attributed to the non-constancy of the square of the matrix element. Knowledge of regions where the matrix element is large or small can be used to learn about the kinds of interactions among the particles which give rise to that behavior of the matrix elements. The plot that Dalitz invented incorporates, in a clever way, the constraints imposed by energy and momentum conservations and allows one to draw conclusions about the square of the matrix element and, hence, about the possible correlations among the produced particles.

The phase space factor represents the number of quantum states available for a particle of momentum $\vec{p}_i$ between $\vec{p}_i$ and $\vec{p}_i + d\vec{p}_i$ and is given by $\frac{V d^3 \vec{p}_i}{(2\pi)^3}$, where V is the normalization volume. We consider three spinless particles in the final state, so the phase space factor is $\prod_{i=1}^{3} \frac{V d^3 \vec{p}_i}{(2\pi)^3}$. Momentum conservation is expressed by a delta function in the three momenta. In a frame in which the total momentum is zero (usually called the center of mass system), momentum conservation gives $\vec{p}_3 = -(\vec{p}_1 + \vec{p}_2)$ resulting from the integration over $\vec{p}_3$. The total energy of the final particles is $E_f = E_1 + E_2 + E_3$, where E_i is the energy of particle i, with $i = 1, 2, 3$. The matrix element for the process is expressed in a Lorentz invariant form, and the phase space expression can be made relativistically invariant if each of the volume factors is supplied its Lorentz contraction factor m/E, where m is the mass and E the energy of the particle.

It is then not difficult to show that the number of quantum states available per unit final energy interval in E_f, $\frac{dN}{dE_f}$ = const $\cdot dE_1 dE_2$. This comes about as follows. The number of states, after performing the integration over $\vec{p}_3$, is proportional to $p_1^2 dp_1 d\Omega_1 p_2^2 dp_2 d\Omega_2$, where Ω's represent the appropriate solid angles. If the initial state is unpolarized, the angular integrations can be performed by changing variables to the polar and azimuthal angles, θ_{12} and ϕ_{12}, between the particles 1 and 2, and integrating over the solid angle of 1, which gives 4π. The integration over ϕ_{12} leads to a further factor of 2π. Taking the Lorentz contraction factors mentioned above, we have for the number of states the expression,

$$dN = \text{constant} \cdot \frac{p_1^2 dp_1 p_2^2 dp_2 d(\cos\theta_{12})}{E_1 E_2 E_3},$$

where in E_3 we must remember that the momentum $\vec{p}_3 = -(\vec{p}_1 + \vec{p}_2)$. Using the energy-momentum relation for the particles, $E^2 = p^2 + m^2$, we can rewrite the expression for the number of states as

$$dN = \text{constant} \cdot \frac{E_1 dE_1 E_2 dE_2 E_3 dE_3}{E_1 E_2 E_3},$$

where the last factor $E_3 dE_3$ in the numerator comes from $p_1 p_2 d\cos\theta_{12} = E_3 dE_3$. The final total energy is $E_f = E_1 + E_2 + E_3$, and for fixed values of E_1 and E_2, $dE_f = dE_3$, so

$$\frac{dN}{dE_f} = \text{constant} \cdot dE_1 dE_2.$$

This says that the phase space for three bodies, represented by the density of points per unit area in the E_1, E_2 plane, is uniformly distributed.

Schematically, the Dalitz plot is obtained if one plots the number of events with energies E_1 and E_2 (the third energy is fixed because it is $E_f - E_1 - E_2$), plotted in the E_1, E_2 plane and is shown in Figure 4.9. The event rate is given by the phase space multiplied by square of the matrix element $|M(E_1, E_2)|^2$. Any deviation from uniformity of the density of points indicates the effect due to the non-constant behavior of the matrix element.

To illustrate how the Dalitz plot is made and how it is used in the determination of spin and parity of particles undergoing decays, let us consider the three pion decay mode of the charged K meson. In this case, all the three final particles have the same mass. The Q value

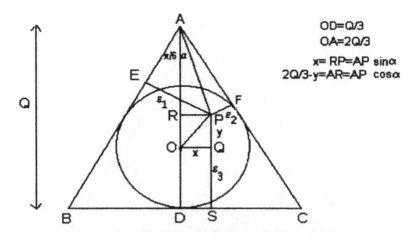

Figure 4.9: Construction of the Dalitz plot. (Adapted from *Introduction to High Energy Physics* by Donald H. Perkins. Copyright © 1982 by Addison-Wesley Publishing Company, Inc. Reprinted by permission of Addison Wesley Longman Publishers, Inc.)

for the reaction is 75 MeV, and the pions in the final state are close to being nonrelativistic. Their kinetic energies are $\epsilon_i = E_i - m_\pi, i = 1, 2, 3$. The Dalitz plot is obtained by drawing an equilateral triangle of height Q, and plotting points inside the triangle with values $\epsilon_1, \epsilon_2, \epsilon_3$ measured perpendicularly from the three sides of the triangle as shown in Figure 4.9

From the property of the triangle, it follows that $Q = \epsilon_1 + \epsilon_2 + \epsilon_3$. Further, the points are restricted to a region in which every $|\cos\theta| \leq 1$, where θ is the angle between any two particles, θ_{12} or θ_{23} or θ_{31}. The boundary values $\cos\theta_{12} = \pm 1$ correspond, for example, to particles 1 and 2 moving parallel or antiparallel to one another, and similarly for the others.

In the case of three nonrelativistic particles, this boundary turns out to be the circle inscribed within the triangle as shown, and all the points will lie inside the inscribed circle. If the origin O is taken as the center of the inscribed circle and the plotted point has the coordinates x, y as shown, then the following relations can be deduced from the figure: $\epsilon_1 = AP\sin 30 + \alpha$, $\epsilon_2 = AP\sin 30 - \alpha$, and $\epsilon_3 = y + (Q/3)$. It is also seen from the figure that $AR = (2Q/3) - y$, $RP = x$, and $AP = \sqrt{AR^2 + RP^2}$. Note that $AP\cos\alpha = AR$ and $AP\sin\alpha = RP$, so that ϵ_1, ϵ_2, and ϵ_3 become, $\epsilon_1 = (Q/3) - (y/2) + (3^{1/2}/2)x$, $\epsilon_2 = (Q/3) - (y/2) - (3^{1/2}/2)x$, and $\epsilon_3 = y + (Q/3)$. From these we can

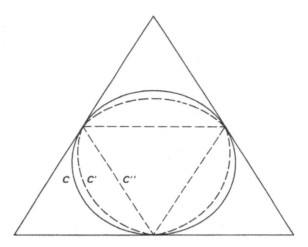

Figure 4.10: Boundary shapes in Dalitz plot for relativistic final particles. (From *Introduction to High Energy Physics* by Donald H. Perkins.
Copyright © 1982 by Addison-Wesley Publishing Company, Inc. Reprinted by permission of Addison Wesley Longman Publishers, Inc.)

deduce the x, y coordinates, $x = \frac{\epsilon_1 - \epsilon_2}{\sqrt{3}}, y = \frac{2Q}{3} - (\epsilon_1 + \epsilon_2)$ belonging to
the values ϵ_1 and ϵ_2.

As the final state particles become relativistic, the boundary changes
from the inscribed circle C to other shapes, C′, C″, etc., as shown in
Figure 4.10, with the inscribed inverted triangle becoming the shape in
the extreme relativistic limit.

Now we turn our attention to the expected distribution of points in
the Dalitz plot for different assumed values for the spin of the kaon.
In the decay of positive kaons into three charged pions, two of them
will be π^+, while one will be π^-. Let pions 1 and 2 be the positive
pions and 3 the negative one. In this case the distribution of points
must be symmetrical in the two triangles BAD and CAD in Figure 4.9.
We now use angular momentum conservation in the decay. toward this
end, we form the total angular momentum J of the final three pion
state by the vector sum of the orbital angular momentum of the positive
pions, $\vec{L}_+$, about their center of mass, and the angular momentum of the
remaining negative pion, $\vec{L}_-$, about the center of mass of the positive
pions. Thus $\vec{J} = \vec{L}_+ + \vec{L}_-$, which implies that the magnitude of J
satisfies $|L_+ - L_-| \leq J \leq |L_+ + L_-|$. Bose statistics for the two like
pions imply L_+ can take on only even values 0,2,.... Remembering that
the pion has intrinsic odd parity, the possible values of J and parity

L_-	$L_+ = 0$	$L_+ = 2$
0	0^-	2^-
1	1^+	$1^+, 2^+, 3^+$
2	2^-	$0^-, 1^-, 2^-, 3^-, 4^-$

Table 4.1: Values of J and parity for assumed values of L_+, L_-.

for different values of L_+ and L_- can be worked out and are shown in Table 4.1 We see that, if $J \neq 0$, then one of L_+ or L_- is not zero. The experimental data obtained by Orear et al. [140], and reproduced here in Figure 4.11, for the decay of charged kaons, shows how one can obtain information about the spin of the kaon.

Consider $L_+ \geq 2$, then we expect the matrix element to vanish when the two positive pions are at rest because of the angular momentum barrier effects. This corresponds to the point when ϵ_3 of the negative pion is at its maximum value. Thus there will be a depletion of events at the top of the inscribed circle, in the region labeled A in Figure 4.11 on the following page. If there is no noticeable depletion around the region A, we will conclude that $L_+ = 0$. Similar arguments can be made for $L_- \geq 1$, in which case, the region of the inscribed circle around C of Figure 4.11, corresponding to the negative pion at rest, will show a depletion of points. The data [140] show no deviation from uniformity in the entire figure, hence one must conclude that the spin-parity assignment for the kaon is 0^-.

The kaon, however, is also observed to decay into a two pion mode, $K^+ \rightarrow \pi^+ + \pi^0$, in which case we would conclude that the parity has to be even, if it has zero spin. Measurements of mass and lifetimes of the particles which decay into the 3 pion and 2 pion modes revealed that they were equal within experimental errors.

Historically, the three pion decay mode was called the τ-mode and the two pion decay mode, the θ-mode. That these decay modes represent the decay modes of a single particle presented a puzzle, called the τ-θ puzzle, and was widely discussed in the mid-1950's. The resolution of the puzzle demands that either there are two different particles accidentally having the same mass and lifetimes or something is wrong in the deduction of the parities assuming parity conservation in the decay process. It is the latter possibility, which prompted T. D. Lee and C. N. Yang to ask what evidence there was for parity conservation in weak interactions in general [141]. The detailed examination of this question led to the

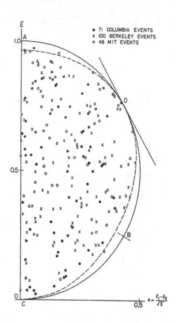

Figure 4.11: Dalitz plot for 3 pion decay of the Kaon. (Figure from J. Orear, G. Harris, S. Taylor, *Physical Review* 102, 1676, 1956. Copyright 1956 by the American Physical Society. Reproduced with permission from J. Orear and the APS.)

overthrow of parity conservation in weak interactions and one cannot determine the parity of the kaon from its decay modes.

Dark Matter, Machos, Wimps

The name *dark matter* has been given to matter in the universe about whose existence we learn only through its gravitational effects. There is also *luminous matter* in the universe about whose existence we learn through the fact that it emits or absorbs radiation.

The amount of a given type of matter is given as a ratio of the density ρ of that type of matter to the critical density, where the critical density ρ_c is defined in big bang cosmology as $\rho_c = [3H^2/(8\pi G)]$. Here H is the Hubble parameter, and G is the Newtonian gravitational constant. The ratio (ρ/ρ_c) is denoted by the symbol Ω. The present value of the Hubble parameter has been deduced to be $H = 100\ h_0$ km s^{-1} Mpc^{-1} (Mpc = megaparsec = 3.26×10^6 light years) with h_0 lying in the range between 0.4 and 1. The value of Ω for luminous matter, denoted by Ω_l, is estimated to be of the order 0.01 or less.

There are several methods by which the amount of dark matter present can be determined. These include (1) observations of rotation curves of spiral galaxies, (2) gravitational lensing effects by single galaxies and clusters of galaxies, (3) estimates of cluster masses by using the virial theorem, and (4) estimates by studying the peculiar velocity measurements of galaxies distributed within distances of the order of a few hundred Mpc.

Of these, method (1) presents the strongest evidence that more than visible matter is involved. What is observed are the velocities v of hydrogen clouds in circular motion around a galaxy center by measuring the Doppler shifts as a function of r, the distance from the center. Simple considerations suggest that $v^2 \simeq GM_l/r$, where M_l is the visible mass of the galaxy mostly concentrated at its center. Such observations have been done on many spiral galaxies. The measurements show that the velocities, instead of varying inversely with r, are actually independent of r for large r. The velocity has a flat behavior as a function of r. Such a velocity curve as a function of r is called *flat rotation (velocity) curve*. This gives an estimate of $\Omega \simeq 0.1$ which is about ten times that for luminous matter. The other methods, (2), (3), and (4), give even larger values of Ω, but these methods involve additional assumptions about the formation of the galaxies.

From all these observations, the value of Ω has been estimated, and it is found to lie in the range 0.1 to 2. These values, which are considerably in excess of $\Omega_l = 0.01$ of luminous matter, indicate that there is a lot of mass that has not been accounted for. In accounting for the discrepancy, one invokes the existence of *dark matter* which is not luminous, as mentioned above.

What is the nature of this dark matter? Is it baryonic or non-baryonic? The composition is called baryonic if it is made up of ordinary matter, such as protons, neutrons, and electrons. This form of matter readily emits and absorbs radiation and hence would be luminous on the whole. Exceptions are matter in the form of white dwarfs, neutron stars, black holes, and Jupiter-like large planets having masses of the order of a tenth of the mass of the Sun and not luminous because of no nuclear reactions in their cores. Massive objects of the last type, which are not self-luminous are called *machos* and may be found in the halos around galaxies. The gravitational lensing methods can be used to search for such objects. Such methods indicate that a significant part of the mass in the halo of our galaxy could be composed of machos. However, stringent limits on the total amount of baryon content of the universe have been deduced from observations of the abundance of deuterium, helium, and lithium synthesized by nuclear reactions right after the big bang. The value of Ω_{baryonic} derived from these limits is considerably less than the range mentioned above.

Thus a significant contribution of non-baryonic dark matter is indicated. There are additional arguments for non-baryonic contribution. Without non-baryonic contribution to seed galaxy formation, it turns out to be difficult to obtain sufficiently small scale fluctuations in the cosmic microwave background radiation. Also, inflationary models of the universe require $\Omega = 1$.

Non-baryonic dark matter has been further classified as *hot* or *cold*. This classification has to do with whether the dark matter particles were relativistic or nonrelativistic at the time in the universe when there was sufficient matter to form a galaxy. If neutrinos have a small mass, they will be relativistic and could constitute hot dark matter. Supersymmetric extensions of the standard model have neutralinos which are mixtures of the neutral higgsinos, z-inos, and photinos. The neutralinos may have a mass in the range of tens to hundreds of GeV and will have weak interactions with other matter. Such weakly interacting massive particles are called *wimps* and may contribute to cold dark matter. The lightest supersymmetric particle (LSP) may be stable and contribute to dark matter. Axions, introduced in extensions of the standard model to avoid CP violation in strong interactions, could also be part of cold dark matter.

There are at present a number of experimental efforts to find dark matter. These efforts have not resulted in any discoveries so far but have served to set limits on the mass of the dark matter candidates.

Deep Inelastic Scattering with Polarized Particles

In our discussions of Bjorken scaling, we already introduced the subject of deep inelastic scattering of electrons on protons. (For details of kinematics, etc., please refer to our discussions under "Bjorken Scaling".) In our previous discussions we considered the deep inelastic scattering of unpolarized electrons on targets of unpolarized protons. If we perform deep inelastic scattering experiments involving polarized incident particles and targets, we can gain knowledge of the spin distribution inside the proton from measurements of the polarized structure functions.

Such experiments have indeed been carried out by the EMC collaboration [142]. This collaboration measured the spin asymmetry in the deep inelastic scattering of longitudinally polarized muons on longitudinally polarized protons over the range $0.01 < x < 0.7$ of the Bjorken scaling variable $x = Q^2/(2M\nu)$, where $q^2(=-Q^2)$ is the square of the four-momentum transfer and ν is the energy transferred by the incident particle in the laboratory. The spin asymmetry is a measure of $\frac{d\sigma(L,R)-d\sigma(R,R)}{d\sigma(L,R)+d\sigma(R,R)}$, where the arguments L, R of the $d\sigma$'s refer to the longitudinal polarizations of the electron and proton, respectively. Using the asymmetry data, they derived the spin-dependent structure function $g_1^p(x)$ for the proton. This function can be integrated over the variable x and produce a quantity called I_1^p.

The *Bjorken sum-rule* [143] involves integrating $g_1^p(x) - g_1^n(x)$ over x, where $g_1^n(x)$ is the neutron spin dependent structure function, and it gives

$$\int_0^1 (g_1^p(x) - g_1^n(x))dx = \frac{1}{6}\left|\frac{g_A}{g_V}\right|(1 - \frac{\alpha_s}{\pi}),$$

where g_A and g_V are the axial vector and vector coupling constants determined from neutron beta decay. This sum-rule includes QCD effects. It is derived using light cone algebra, relates the quark-parton model with weak currents, and is hence on a good theoretical footing. The numerical value of the right-hand side of this sum-rule is 0.191 ± 0.002, for an $\alpha_s = 0.27 \pm 0.02$.

Separate sum-rules for the proton and the neutron have been derived by Ellis and Jaffe, called the *Ellis-Jaffe sum-rule* [144], using SU_3 current algebra and assuming an unpolarized strange quark sea. These sum-rules for the proton and the neutron are

$$\int_0^1 dx g_1^{p,n}(x) = \frac{1}{12}\left|\frac{g_A}{g_V}\right|\left[+(-)1 + \frac{5}{3}\left[\frac{3(F/D)-1}{(F/D)+1}\right]\right],$$

where in the first term of the square bracket, the plus sign is for the proton, and the minus sign is for the neutron, and F/D is the SU_3 ratio of F to D type couplings. Using the values, $(F/D) = 0.632 \pm 0.024$ and $(g_A/g_V) = 1.254 \pm 0.006$, one can work out the right-hand sides for the proton and the neutron. For the proton, this is the above mentioned integral I_1^p and has the theoretical value 0.189 ± 0.005. Using the experimental data from asymmetry measurements, which give $g_1^p(x)$, I_1^p is found by integrating $g_1^p(x)$ over x. The numerical value thus obtained is $0.114 \pm 0.012 \pm 0.026$, which is distinctly smaller than the theoretical value for the right-hand side.

If one assumes that the Bjorken sum-rule is valid, the small value measured from experiment for the quantity I_1^p, leads one to conclude that quark spins contribute very little to the proton spin (about 10%), and most of the contribution to the proton spin must come from the polarization of the gluons and/or orbital angular momentum of the quarks. (This is referred to as the *proton spin crisis*.) This is in marked contrast to the successful results of the constituent quark model and the parton model where the spin of the nucleon is assumed to be carried by the quark-partons. It also means that the result of integrating the neutron spin-dependent structure function $g_1^n(x)$ over x must give a significant negative value for the integral, a result which can be checked when the spin dependent structure function for the neutron is also measured experimentally.

$\Delta(1232)^{++}$

This is the first baryonic resonance discovered in collisions of positive pions on protons. The work was done at the Chicago cyclotron with pion beams of kinetic energy in the range from 80 MeV to 150 MeV and subsequently confirmed with pions of much higher energy at the Cosmotron in Brookhaven National Laboratory. A dramatic peak in the cross section was observed at a total center of mass energy of 1232 MeV. The scattering amplitude for this process is analyzed in terms of isospin and spin of the combined pion-nucleon system, and the phase shift for the isospin 3/2, spin 3/2 state goes through 90° at a total center of mass energy of the pion nucleon system of 1232 MeV corresponding to the peak in the cross section.

$\Delta(1232)^0$

This charge neutral resonant state was first indicated in experiments done by Anderson et al. [145] in collisions of negative pions on protons. They measured the cross section for negative pions in the energy range 80 MeV to 230 MeV. The cross section for this process also exhibits a peak at a total center of mass energy of 1232 MeV.

Deuteron

This is an isotope of hydrogen of mass 2, ^2H. It is the simplest nucleus next to the nucleus of hydrogen—the proton. The chemical element is *deuterium*. The nucleus of this atom has a proton and a neutron and plays a role in nuclear physics similar to that of the hydrogen atom in atomic physics.

Historically, evidence for its existence [146] came from efforts to understand the discrepancy between the atomic weight as determined chemically and as determined by mass spectrographic methods. If it was assumed that a hydrogen isotope of mass 2 was present to the extent of 1 part in 4,500 of hydrogen of mass 1, then the discrepancy between the two values could be removed. The actual discovery of the heavy isotope came from analyzing atomic spectra of hydrogen in a discharge tube [147]. Samples of hydrogen for the discharge tube were obtained by concentrating the isotope of mass 2 by evaporating large quantities of liquid hydrogen. The discharge was run in such a way as to make prominent the atomic spectrum over the molecular spectrum. In the Balmer lines they saw faint lines on the short wavelength side of the $^1\mathrm{H}_\beta$, $^1\mathrm{H}_\gamma$, $^1\mathrm{H}_\delta$ shifted by 1 to 2 Å. The wavelength of the faint lines agreed with those calculated for isotope of mass two. They obtained an idea of the relative abundance of $^2\mathrm{H}$ by comparing the times required to record the faint lines relative to the strong ones. In this way they estimated the relative abundance to be about 1 part in 4,000 in natural hydrogen and those in the concentrated samples to be about five times greater.

The studies of atomic energy level structure are carried out by using light to excite and ionize the atoms. Analogously, the deuteron structure can be studied by exciting and disintegrating ("ionizing") it with light of suitable wavelength. From the mass measurements, it was estimated that the binding energy of the deuteron was small—of the order of 2 MeV—which in this case would require gamma rays. Using *Th C"* gamma rays of energy 2.62 MeV, Chadwick and Goldhaber [148] achieved the disintegration of the deuteron: $\gamma + d \rightarrow p + n$. By studying the details of this reaction they were able to come up with a first good measurement of the mass of the neutron. Since those first measurements, many more accurate measurements of the deuteron binding energy are available.

Very shortly after the work of Chadwick and Goldhaber [148], the nuclear spin of deuterium and the magnetic moment of the deuteron were determined. For nuclear spin determination, the method used depends on the fact that a measurement of intensities of alternate lines in the molecular spectra of diatomic molecules with identical nuclei should show a variation. This comes about because interchanging identical nuclei affects the sign of the wave function of the molecule as shown just below.

The wave function of the molecule is expressible as the product $\psi = \psi_{elec}\psi_{vib}\psi_{rot}\psi_{nucl.spin}$, where ψ_{elec} is the electronic wave function of the molecule, ψ_{vib} is the wave function of the vibrational state, ψ_{rot} is the wave function of the rotational state, and $\psi_{nucl.spin}$ is the wave function referring to the nuclear spin. Under interchange of nuclei, the electronic wave function does not change sign for the ground state of most molecules, and the vibrational wave function is likewise symmetric. The rotational wave function of the molecule with rotational quantum number j acquires a factor $(-1)^j$ when the two nuclei are interchanged.

Turning to the behavior of the wave function for nuclear spin, a nucleus with spin I has a degeneracy $(2I + 1)$ due to the projection M, which can take any of the values, $I, I - 1, \cdots, -I$. For two identical nuclei, there are $(2I + 1)^2$ wave functions of the form $\psi_{M_1}(1)\psi_{M_2}(2)$, where the arguments of the functions refer to the coordinates of the nuclei. For identical nuclei, one must use combinations of these functions which are symmetric and antisymmetric under the interchange of the nuclei. The product function with $M_1 = M_2$ leads to symmetric wave functions. There are $(2I + 1)$ of these. The remaining ones, $(2I + 1)^2 - (2I + 1) = 2I(2I + 1)$, have $M_1 \neq M_2$. Every such product function must be replaced by symmetric and antisymmetric functions: $(1/\sqrt{2})[\psi_{M_1}(1)\psi_{M_2}(2) \pm \psi_{M_2}(1)\psi_{M_1}(2)]$, (upper sign, symmetric, and lower sign, antisymmetric). Thus half of the $2I(2I + 1)$ product functions with $M_1 \neq M_2$ are symmetric and the other half are antisymmetric with respect to the interchange of the nuclei. The total number of symmetric functions is $(2I + 1) + I(2I + 1) = (I + 1)(2I + 1)$, and there are $I(2I + 1)$ antisymmetric functions. For nuclei obeying Bose statistics, symmetric nuclear spin functions must go with even rotational wave function (even j), and antisymmetric nuclear spin functions must go with odd rotational wave functions (odd j). In this way the statistical weight associated with the nuclear spin influences the intensity of transitions in the rotational spectrum of the molecule between states of different j. Because the ratio of the symmetric to antisymmetric nuclear spin functions is $(I + 1)/I$, we expect that, for nuclei obeying Bose statistics, the intensity of even rotational lines will be $(I + 1)/I$ times that of the odd rotational lines in the rotational spectrum. For nuclei obeying Fermi statistics, the opposite will be the case. This method was used by Murphy and Johnston [149] and they found intensity ratios in two separate measurements to be 1.95 ± 0.06 and 2.02 ± 0.04, leading to the result 1 for the nuclear spin of deuteron.

Historically, the first determination of the magnetic moment of the deuteron was carried out by Esterman and Stern [150] using a Stern-Gerlach apparatus. They obtained a value between 0.5 and 1 nuclear magneton for the magnetic moment of the deuteron. Taking the value of 2.5 nuclear magnetons for the magnetic moment of the proton as determined earlier and using additivity of the magnetic moments of the proton and the neutron in the deuteron, they suggested that the neutron would have a magnetic moment between -1.5 and -2 nuclear magnetons. These values were considerably improved by I. I. Rabi and collaborators [151] by developing the molecular beam magnetic resonance method. The values obtained in this work were 2.785 ± 0.02 nuclear magnetons for the proton and 0.855 ± 0.006 nuclear magnetons for the deuteron. These values have been further improved over the years with the development of nuclear magnetic resonance techniques. Modern values for the proton and deuteron are $2.79284739 \pm 0.00000006$ and 0.85741 nuclear magnetons, respectively.

Dirac Equation

This equation arose as a result of the quest to obtain a relativistic equation for an electron which contains time and spatial derivatives only to first order. Dirac invented this remarkable equation in 1928 [152]. This equation has been found to describe particles of spin $1/2$ very well. The Dirac equation for a free particle of mass m has the form

$$(-i\vec{\alpha} \cdot \vec{\nabla} + \beta m)\psi(\vec{x}, t) = i\frac{\partial}{\partial t}\psi(\vec{x}, t),$$

where the set of quantities, $\alpha^k, (k = 1, 2, 3)$, and β are four-by-four square matrices satisfying the anticommutation relations,

$$\{\alpha^k, \alpha^l\} \equiv \alpha^k \alpha^l + \alpha^l \alpha^k = 2\delta^{kl}, \ (k, l = 1, 2, 3),$$

$$\{\alpha^k, \beta\} = 0, \ (k = 1, 2, 3),$$

$$\beta^2 = 1,$$

and the function $\psi(\vec{x}, t)$ is a one-column matrix containing four components. (Here δ^{kl} is the Kronecker delta symbol, equal to 1 when $k = l$ and zero when $k \neq l$.) The anticommutation relations are imposed in order that the Klein-Gordon equation

$$(\frac{\partial^2}{\partial t^2} - \vec{\nabla}^2 + m^2)\psi(\vec{x}, t) = 0$$

also hold for each of the four components of the wave function, which guarantees that the energy and momentum for a free particle satisfy the relativistic equation $E^2 = p^2 + m^2$. The matrix function $\psi(\vec{x}, t)$ stands to the right of the four-by-four matrices. The Hermitian adjoint matrix function, $\psi^\dagger(\vec{x}, t)$, is a one-row matrix containing four elements (complex conjugates of those in ψ) and will occur to the left of the four-by-four matrices. The products of the matrices with ψ or $\psi^\dagger$ will follow the multiplication rule for matrices.

A canonical form for the matrices $\alpha^k, k = 1, 2, 3$, and β is

$$\alpha^k = \left(\begin{array}{c|c} 0 & \sigma^k \\ \hline \sigma^k & 0 \end{array}\right) \text{ and } \beta = \left(\begin{array}{c|c} I & 0 \\ \hline 0 & -I \end{array}\right).$$

Here the elements 0 and I are 2×2 zero matrix and unit matrix, respectively, and the σ^k are 2 x 2 Pauli matrices,

$$\sigma^1 = \begin{pmatrix} 0 & 1 \\ 1 & 0 \end{pmatrix}, \sigma^2 = \begin{pmatrix} 0 & -i \\ i & 0 \end{pmatrix}, \sigma^3 = \begin{pmatrix} 1 & 0 \\ 0 & -1 \end{pmatrix}, I = \begin{pmatrix} 1 & 0 \\ 0 & 1 \end{pmatrix}.$$

We may rewrite the Dirac equation in a more compact form. For this we introduce the γ^μ, ($\mu = 0, 1, 2, 3$) matrices according to the definitions, $\gamma^0 = \beta$, $\gamma^k = \beta \alpha^k (k = 1, 2, 3)$. Multiplying the equation by γ^0 from the left, we have

$$(i\gamma^\mu \partial_\mu - m)\psi(x) = 0.$$

The anticommutation relations for the γ matrices follow from those of the α and β matrices,

$$\{\gamma^\mu, \gamma^\nu\} = 2g^{\mu\nu},$$

with $g^{00} = 1$, $g^{kl} = -\delta^{kl}$, $(k, l = 1, 2, 3)$. From the definitions, we can see that the following Hermiticity properties follow

$$\gamma^{0\dagger} = \gamma^0, \ \gamma^{k\dagger} = -\gamma^k.$$

These can be combined into the single relation:

$$\gamma^0 \gamma^{\mu\dagger} \gamma^0 = \gamma^\mu.$$

The function defined by $\bar{\psi}(x) = \psi^\dagger(x)\gamma^0$ is easily seen to satisfy the equation

$$i\partial_\mu \bar{\psi}(x)\gamma^\mu + m\bar{\psi}(x) = 0.$$

The Dirac equation for ψ and its conjugate $\bar{\psi}$ given above can be used to show that the quantity $\bar{\psi}(x)\gamma^\mu\psi(x) = j^\mu(x)$ satisfies the conservation law $\partial_\mu j^\mu(x) = 0$. The zero component of $j^\mu(x)$ is $j^0(x) = \bar{\psi}(x)\gamma^0\psi(x) = \psi^\dagger(x)\psi(x) = |\psi(x)|^2$ (or, written more explicitly in terms of the four components, $\sum_{k=1}^4 |\psi_k(x)|^2$) and is clearly a positive quantity. This is just as in the Schrodinger theory, $\psi(x)$, resembling the wave function with four components and the spatial integral of $j^0(x)$, is independent of time.

In the early stages of the development of this theory by Dirac, the equation was considered as suitable for describing the motion of a single relativistic particle. However, there are problems with this single particle interpretation. Fourier analysis of $\psi(x^0, \vec{x})$ in terms of plane wave states $e^{-i(p^0 x^0 - \vec{p} \cdot \vec{x})}$ will contain integral over p^0 from $-\infty$ to $+\infty$. In the single particle picture, $p^0 = \sqrt{\vec{p}^2 + m^2}$ is the energy of the particle of momentum $\vec{p}$, and both positive and negative energies occur in the superposition. In a relativistic theory there is no reason to exclude negative energy states because the particle may change its energy by arbitrarily large amounts (through some interaction) and go into a negative energy state. In such a case there is nothing to prevent the particle from going into lower and lower energy states and ending up at negative infinity in energy, which is clearly physically absurd.

Faced with this problem, Dirac proposed the idea of the "hole" theory. Because the theory is for electrons, which obey the Pauli exclusion principle, he proposed that all the states of the negative energy sea be filled with electrons. Pauli principle will prevent electrons making transitions to negative energy states because they are all occupied. If one defines the ground state of the theory to be one in which all the positive energy states are unoccupied by any particles while all the negative energy states are occupied, only measurable quantities are differences from the ground state values. The infinite energy and the infinite charge possessed by the negative energy sea are thus not observable. A vacancy in the negative energy sea will clearly be observable and manifest as a particle of the same mass as the electron but of opposite sign of charge. In this way, Dirac was led to propose the idea of an antiparticle to every particle that is described by his equation. This was a revolutionary idea at the time but was remarkably soon proven to be correct with the discovery of the positron in cosmic rays by C. D. Anderson in 1933.

It was soon realized that although the idea of the hole theory solved the problem for the electrons, because of Pauli exclusion principle, the Klein-Gordon equation considered as a relativistic equation for a single spinless particle also suffers from the negative energy problem. Here there is no exclusion principle and hole theory will not help. Consequently, some other method had to be found to solve the negative energy problems in relativistic theories. A solution was found by going away from the idea that the relativistic equations describe single particle behavior. Instead, it was proposed that the functions which satisfy the relativistic equations are not wave functions of single particles, but represent field functions. These field functions when quantized give rise to quanta of the field which manifest as particles. Thus one has a quantum field theory whose field equations are the Klein-Gordon equation or the Dirac equation. The Fourier decompositions of the fields involving positive and negative values of p^0 are referred to as positive and negative frequency modes and do not have any direct connection with positive and negative energies of particles. In quantum field theory, the field functions become operators, the positive frequency parts of which destroy particles and negative frequency parts create antiparticles, all of positive energy. Correspondingly, the Hermitian conjugate field functions become operators which create particles and destroy antiparticles. Relativistic quantum field theories which yield particles of spin 0, 1/2, and 1 have been successfully constructed (see section under "Quantum Field Theory" for more details). ,

For further reference, it is useful to write the Dirac equation for the electron of charge e, $(e < 0)$ in an electromagnetic field described by the four-vector potential $A_\mu(x)$:

$$(i\gamma^\mu \partial_\mu - e\gamma^\mu A_\mu(x) - m)\psi(x) = 0.$$

Negative energy solutions of this equation will be associated with the antiparticle to the electron, namely, the positron.

The correspondence between negative energy solutions of the Dirac equation and the positron solutions can be formalized in terms of an operation known as *charge conjugation*. Let us describe the positron by the charge conjugate wave function ψ^c. It will satisfy an equation, similar to the one above for the electron, except that, $e \to -e$ in it,

$$(i\gamma^\mu \partial_\mu + e\gamma^\mu A_\mu(x) - m)\psi^c(x) = 0.$$

Now we explore the relation between ψ^c and ψ. It is clear that in going from the equation for ψ to that for ψ^c, one has to have the same sign for the derivative term and the eA term. This suggests that we take the complex conjugate of the equation for the ψ, and we get (factoring an overall negative sign)

$$[(i\partial_\mu + eA_\mu)\gamma^{\mu*} + m]\psi^* = 0.$$

What we now need to do is to find a nonsingular matrix $C\gamma^0$, such that

$$(C\gamma^0)\gamma^{\mu*}(C\gamma^0)^{-1} = -\gamma^\mu,$$

then we can cast the equation for ψ^* in the form

$$[(i\partial_\mu + eA_\mu)\gamma^\mu - m](C\gamma^0\psi^*) = 0.$$

Now we can identify $C\gamma^0\psi^* = C\bar{\psi}^T = \psi^c$ (T represents transpose), and we have the desired connection. Because $\gamma^0\gamma^{\mu*}\gamma^0 = \gamma^{\mu T}$, we have to find a matrix C, such that

$$C^{-1}\gamma^\mu C = -\gamma^{\mu T}.$$

It is easy to verify that $C = i\gamma^2\gamma^0$. Thus we have $\psi^c = C\gamma^0\psi^* = i\gamma^2\psi^*$. This is the desired relation between the electron wave function and the charge conjugate wave function, which is the wave function for the antiparticle, the positron.

Dispersion Relations

An approach popularly known as *dispersion relations* for treating scattering processes involving strongly interacting particles was vigorously pursued in the 1960's. This method provides relationships between matrix elements for processes based on very general principles such as Lorentz invariance, causality, and the unitarity of the scattering matrix (or completeness of the set of all physical states). The detailed knowledge of an underlying field theory was abandoned in this method, since at that time, one neither knew what the fundamental fields were nor had any knowledge as to what the Lagrangian might be. Even if one knew these, strong interactions could not be treated using perturbation theory, and no method other than perturbative solution of field theories was known. The *dispersion method* draws its inspiration from the field of optics which treats the propagation of light through a dielectric medium.

In treating the propagation of light through a medium, one introduces the concept of the refractive index, n. The refractive index is in general a complex quantity with real and imaginary parts representing the refractive and absorptive parts. Dispersion relations in optics are relations between the real and imaginary parts of the refractive index. It can be shown that these relations follow from the principle of causality; namely, no disturbance can emanate from a scatterer before the incident wave has reached it. The propagation through a medium occurs as a result of the incident wave accelerating the electrons bound in atoms with a set of natural frequencies $\omega_k, k = 1, 2, 3, \ldots$ and subsequent re-radiation by these accelerated electrons. If the electrons with these natural frequencies are described by simple damped harmonic oscillators labeled by k, one can show that the refractive index is related to the amplitude for the forward scattering of light by the atoms, with an amplitude $f(\omega)$, where

$$f(\omega) = \sum_k a_k \frac{\omega^2}{\omega_k^2 - \omega^2 - i\gamma_k \omega},$$

where $\gamma_k, k = 1, 2, 3, \ldots$ are damping constants and $a_k, k = 1, 2, 3, \ldots$ are related to oscillator strengths which determine how effectively a given oscillator k contributes to the scattering amplitude. The important point one notices from this expression for the forward scattering amplitude is that, considered as a function of the complex variable ω, it has poles given by

$$\omega = -(1/2)i\gamma_k \pm (\omega_k^2 - (1/4)\gamma_k^2)^{1/2}.$$

As the term due to the damping constants is usually small compared to the term due to ω_k^2, the square root is real and the poles are only in the lower half ω-plane (Im $(\omega) < 0$). (We will show later that the location of the poles of the forward scattering amplitude being only in the lower half ω-plane is directly a consequence of causality.) The function $f(\omega)$ does not fall off for large ω, so let us consider the function $f_1(\omega) = f(\omega)/\omega^2$, which, because it has poles only for Im $(\omega) < 0$, can be represented by the Cauchy integral formula

$$f_1(\omega) = \frac{1}{2\pi i} \int_C \frac{f_1(\omega')}{\omega' - \omega} d\omega',$$

where the contour C consists of the real axis $(-\infty < \omega' < +\infty)$ and the infinite semicircle in the upper half plane. Because the function $f_1(\omega)$ falls off like ω^{-2} for large ω, the contribution from the infinite semicircle is zero, and we have

$$f_1(\omega) = \frac{1}{2\pi i} \int_{-\infty}^{+\infty} \frac{f_1(\omega')}{\omega' - \omega} d\omega'.$$

Let us approach the real axis for ω from the upper half plane, then $\omega \to \omega + i\epsilon$ for small ϵ, and we have

$$f_1(\omega) = \lim_{\epsilon \to 0} \int_{-\infty}^{+\infty} \frac{f_1(\omega')}{\omega' - \omega - i\epsilon} d\omega'.$$

Using the fact that $[1/(x - i\epsilon)] = P(1/x) + i\pi\delta(x)$, where P stands for the Cauchy principal value, we immediately get

$$f(\omega) = \frac{1}{\pi i} P \int_{-\infty}^{+\infty} \frac{f_1(\omega')}{\omega' - \omega} d\omega'.$$

Taking the real part of both sides we have

$$\text{Re} f_1(\omega) = \frac{1}{\pi} P \int_{-\infty}^{+\infty} d\omega' \frac{\text{Im} f_1(\omega')}{\omega' - \omega}.$$

Similarly taking the imaginary parts of both sides, we get

$$\text{Im} f_1(\omega) = -\frac{1}{\pi} \int_{-\infty}^{+\infty} d\omega' \frac{\text{Re} f_1(\omega')}{\omega' - \omega}.$$

We can put these in a usable form if we notice that $f_1(\omega)$ satisfies the properties Re $f_1(\omega) = $ Re $f_1(-\omega)$ and Im $f_1(\omega) = -$Im $f_1(-\omega)$, we can rewrite the above as an integral over positive ω only and get

$$\text{Re} f_1(\omega) = \frac{1}{\pi} P \int_0^\infty 2\omega' \frac{\text{Im} f_1(\omega')}{\omega'^2 - \omega^2} d\omega'$$

and

$$\text{Im} f_1(\omega) = -\frac{1}{\pi} P \int_0^\infty 2\omega \frac{\text{Re} f_1(\omega')}{\omega'^2 - \omega^2}.$$

These pair of relations relating real and imaginary parts of the forward scattering amplitude are called *dispersion relations*. They were first derived for optics by Kramers [153] and Kronig [154].

The way these relations may be used is as follows. Using the fact that unitarity relates the imaginary part of the forward scattering amplitude to the total cross section

$$\text{Im} f(\omega) = \frac{\omega}{4\pi} \sigma_{tot}(\omega),$$

one can use the measured total cross sections at various energies to evaluate the right-hand side of the first relation. This gives us the real part of the forward scattering amplitude. Consistency demands that if we use this on the right-hand side of the second relation, we should get back the imaginary part which we used in the first relation.

We now derive the connection between causality and the fact that $f(\omega)$ has poles only for $\text{Im}\,\omega < 0$. Suppose we have an incident light wave given by $A(x,t)$. Fourier analyzing it, we can write

$$A(x,t) = \frac{1}{(2\pi)^{1/2}} \int_{-\infty}^{+\infty} d\omega\, A(\omega) \exp i\omega(x - t),$$

If $A(\omega)$ has no poles in the region $\text{Im}\,(\omega) > 0$, then $A(0,t)$ can be evaluated by residue theorem by completing the contour for $t < 0$ in the upper half ω plane. Because there are no poles for $\text{Im}\,(\omega) > 0$, we get $A(0,t) = 0$ for $t < 0$. If an atom was located at $x = 0$, this says that the incident wave does not reach the atom before time $t = 0$. Let the incident wave reach an atom located at $x = 0$. This atom will produce a scattered wave in the forward direction at a location x at time t given by

$$S(x,t) = \frac{1}{(2\pi)^{1/2}x} \int_{-\infty}^{+\infty} d\omega\, f(\omega) A(\omega) \exp i\omega(x - t),$$

where $f(\omega)$ is the forward scattering amplitude. Causality demands that the scattered wave should not be present for times $t < x$. We have already seen that $A(\omega)$ has no poles in the region $\text{Im}\,(\omega) > 0$. Thus if $f(\omega)$ also has no poles in the region $\text{Im}\,(\omega) > 0$, $S(x,t)$ will be zero for $t < x$ and demand of causality will be satisfied.

These ideas from optics have found extensions in applications to the scattering of strongly interacting particles such as pion-nucleon scattering. All such activities started with the work of Gell-Mann, Goldberger, and Thirring [155]. They considered the case of scattering of photons by a proton. They pointed out that the form of the scattering amplitude permits analytic continuation to complex values of the photon energy. They further showed that the forward scattering amplitude for a photon on a proton is proportional to the integral

$$\int d^4x e^{i(k \cdot x)} \theta(x_0) A_{\mu,\mu}(x_0, \vec{x}),$$

where, $k(\omega, \vec{k})$ is the four-momentum of the photon, and the function $A_{\mu,\mu}(x_0, \vec{x})$, which is labeled by the photon polarization vectors with the index μ, depends on the matrix element of the commutator of two electromagnetic currents with index μ each, between proton states with momentum p. The arguments of the currents are separated from one another by the four vector $(x_0, \vec{x})$. In our discussion of causality (see section under "Causality Condition—Quantum Field Theory"), we point out that a way of stating this condition is that the commutator vanishes for space-like separations. Thus $A_{\mu,\mu}(x_0, \vec{x}) = 0$ for $x_0 < |\vec{x}|$. The θ function with argument x_0 means that the integral over x_0 extends over positive x_0 in a region with $x_0 > |\vec{x}|$.

The exponential in the integrand above can be written $e^{i\omega(x_0 - \hat{k} \cdot \vec{x})}$, where ω is now the photon energy. Since the integration is restricted to the region $x_0 - \hat{k} \cdot \vec{x} > 0$, we see that the form of the integral permits us to let the photon energy become complex with a positive imaginary part. This analytic continuation of the expression in the photon energy to the upper half of the complex ω plane allows us to write a Cauchy integral representation for the forward scattering amplitude where the contour consists of the real axis and the infinite semicircle in the upper half ω plane. Calling the integral $f(\omega)$, a parallel is established between the $f(\omega)$ in the optics case above and $f(\omega)$ here. Notice that $f(-\omega)$ is given by the same integral with $-\theta(-x_0)$ in it, which is just $f^*(\omega)$. This helps us to do the same manipulations for ω in the region $-\infty < \omega < 0$ as in the optics case.

In this case, we could derive dispersion relations only for $f_1(\omega) = [f(\omega)/\omega^2]$ and not for $f_1(\omega)$ due to its slow fall off as $\omega \to \infty$. In the present case, the behavior of $f(\omega)$ for $\omega \to \infty$ depends on how $A_{\mu,\mu}(x_0, \vec{x})$ behaves when $x_0^2 - \vec{x}^2 \to 0$. This behavior may be very singular. If this singularity is no worse than the derivative of a delta function of finite order, then $f(\omega)$ is bounded by a polynomial in ω for large ω. If this polynomial is of order n, dispersion relations can be derived for $f(\omega)/\omega^{2n}$, and it has the form

$$\text{Re}\, f(\omega) = \sum_{m=1}^{n} C_m (\omega^2)^{m-1} + \frac{\omega^2}{\pi} P \int_0^\infty d\omega'^2 \frac{\text{Im}\, f(\omega')}{\omega'^{2n}(\omega'^2 - \omega^2)}.$$

The derivation of dispersion relations for scattering of massive particles, such as the case of pions on protons, is not a straight forward extension of the photon-proton case. This is because, for the massive particle, the exponential which appears in the integral becomes $\exp(i\omega x_0 - i\sqrt{\omega^2 - m^2}\,\hat{k} \cdot \vec{x})$. It is not possible now to continue analytically to complex ω without resorting to complicated methods. Thus, proofs of dispersion relations for pion-nucleon scattering or other scattering processes are much more complicated and we do not go into them here.

DONUT Experiment
The acronym here stands for Direct Observation of the NU Tau. This is an experiment to be performed at Fermilab in which tau leptons produced directly by ν_τ are to be detected. The ν_τ's may have arisen from the oscillation of another neutrino flavor.

Drell-Yan Mechanism
The first observation of muon pairs with high invariant mass in hadron-hadron collisions occurred in an experiment performed at the Brookhaven proton synchrotron in 1970 [156]. The reaction studied the collision of protons in the momentum range of 22–29 GeV with a uranium target, $p + U \to \mu^+ + \mu^- + X$, and observed pairs of muons in the final state having a certain mass range. (This was a prelude to the experiments which discovered the J/ψ and the Υ; for details, see sections under "J/ψ meson" and "Bottomonium".)

Away from resonances, the production of massive muon pairs in the continuum, is through a mechanism called the *Drell-Yan* process [157]. Here a quark (or an antiquark) from the incident particle annihilates with an antiquark (or a quark) from the target and produces a virtual photon, which materializes into a lepton pair $q + \bar{q} \to \mu^+ + \mu^-$. Study of such processes yields information about the quarks, quark distribution functions in the hadrons, etc.

In a study of collisions of π^+ and π^- with ^{12}C, an isoscalar nuclear target (that is, one with equal numbers of protons and neutrons), the cross sections have been measured for $\sigma(\pi^\mp + {}^{12}\text{C} \to \mu^+ + \mu^- + X)$ and the ratio $\sigma(\pi^-\text{C})$ to $\sigma(\pi^+\text{C})$ for the production of muon pairs determined. Experimentally, the ratio of the total cross sections for these reactions, $(\sigma(\pi^-\text{C})/\sigma(\pi^+\text{C}))$, is found to be equal to 4 in the region away from any resonances giving rise to the dilepton system. This is what one expects in the Drell-Yan picture. In the case of $\pi^- = (\bar{u}d)$, the u-antiquark from the π^- annihilates with a u quark from the carbon nucleus, the cross section is proportional to $18e_u^2$, where $e_u = (2/3)|e|$ is the charge of the u quark, and there are 18 u quarks in the carbon nucleus. On the other hand, for the case of $\pi^+ = (u\bar{d})$, the d-antiquark from the π^+ annihilates with a d quark from the carbon nucleus, and there are 18 of these also, leading to a cross section proportional to $18e_d^2$, where $e_d = (-1/3)|e|$ is the charge carried by the d-quark. The ratio is clearly equal to $(18(4/9)e^2)/(18(1/9)e^2) = 4$. Measurements of differential cross sections have been used to give information about the quark distribution functions in the hadrons.

Eightfold Way
In the late 1950's, with the development of high energy accelerators, many particles and resonant states were discovered. There were a number of attempts to develop a system of classification of the particles which would give some hints as to any underlying symmetry that was responsible for these particle states. The most successful attempt was that due to Gell-Mann [158] and Ne'eman [159]. They recognized that in the spectrum of particle states, the known particles and resonances could be accommodated in multiplets with same spin J and parity P containing 1, or 8, or 10, or 27 members. These numbers of particles in the multiplets suggested investigation of symmetry groups larger than the isospin group which is associated with the special unitary group in two dimensions SU_2 and which is suggested by the charge independence of nuclear forces. Gell-Mann and Ne'eman investigated the simplest generalization to charge independence, namely, special unitary group in three dimensions SU_3. (For details, see under "SU_3—Model for Hadron Structure").

In SU_3, there are eight generators of the group which form a Lie algebra, just like the three components of the isospin $I_i, i = 1, 2, 3$ which are the generators of SU_2. In the case of isospin, it is known that I_3 and $\sum_i I_i^2$ can be simultaneously diagonalized. The eigenvalues i_3 of I_3 can be any one of the $(2I+1)$ values in the range $-I, -I+1, -I+2, \ldots, +I$, and that of the $\sum_i I_i^2$ is $I(I+1)$, so the states are labeled by giving I and I_3. In the case of SU_3, if the generators are $F_i, i = 1, \ldots, 8$, two of these generators are simultaneously diagonalizable along with $\sum_i^8 F_i^2$. It is customary to choose the two diagonalizable operators as $F_3 = I_3$, called the *third component of the isospin*, and $\frac{2}{\sqrt{3}}F_8 = Y$, called the *hypercharge*. Calling the eigenvalues of these two operators i_3 and y, the states can be labeled by these values. Just as the states in the case of SU_2 can be graphically represented on an i_3 line (abscissa) with the values ranging from $-I, -I+1, \ldots, I-1, +I$, the graphical representation of the states for SU_3 requires a plane, with i_3 as abscissa and y as ordinate, every state being represented by a point (i_3, y) in

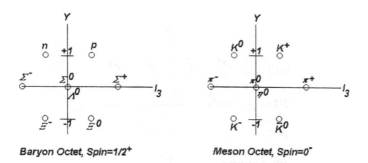

Figure 4.12: Octet representation of SU_3.

this plane. The electric charge carried by a particle Q can be written in terms of these eigenvalues as $Q = |e|(i_3 + \frac{y}{2})$.

One finds that the baryons $p, n, \Sigma^{\pm}, \Sigma^0, \Lambda^0, \Xi^-, \Xi^0$ with $J^P = (1/2)^+$ fit nicely into the eight-dimensional irreducible representation of the SU_3 group. The (i_3, y) values for these particles are: $p(1/2, +1), n(-1/2, +1), \Sigma^+(1, 0), \Sigma^0(0, 0), \Sigma^-(-1, 0), \Xi^0(1/2, -1), \Xi^-(-1/2, -1),$ and $\Lambda^0(0, 0)$. These states group into isotopic spin multiplets for different values of y, (p, n) and $\Xi^{-,0}$ forming isospin doublets with $I = 1/2, \Sigma^{+,0,-}$ isospin triplet with $I = 1$, and Λ^0 an isospin singlet with $I = 0$. These states are degenerate in mass in the limit of exact symmetry. The shape in the (i_3, y) plane of the states occupied by particles in the octet representation resembles a hexagon (see Figure 4.12).

The fact that the masses are not quite the same for the different y values is an indication that the SU_3 symmetry is only approximate. With suitable symmetry breaking put in, one can generate the actual masses of these baryonic states. Gell-Mann [158] and Okubo [160] proposed a mass formula

$$M = M_1 + M_2 y + M_3[I(I+1) - \frac{Y^2}{4}],$$

where M_1, M_2, M_3 are constants in one multiplet. A relation that follows from this formula is

$$\frac{M_\Xi + M_N}{2} = \frac{3M_\Lambda + M_\Sigma}{4},$$

where the subscripts to the masses identify the particle in the multiplet. The equality of the two sides of this expression is extremely well satisfied when the experimentally measured masses are put in.

In the meson sector, the pions and the kaons, along with the η^0, fit into a multiplet of eight particles with $J^P = 0^-$. Here the i_3, y

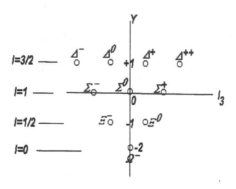

Baryon Decuplet, Spin=3/2⁺

Figure 4.13: Decuplet states of SU_3.

assignments are as follows: K^+, K^0 and $\bar{K}^0, K^-$ are $I = 1/2$ isospin doublets with $y = +1$ and $y = -1$, respectively; $\pi^{+,0,-}$ is an isospin triplet with $I = 1$ and $y = 0$; and η^0 an isospin singlet with $I = 0$ and $y=0$. The mass formula in this sector is satisfied better when squares of masses rather than when first powers of masses are involved. It can be written as

$$M^2 = M_0^2 + M_1^2[I(I+1) - \frac{Y^2}{4}],$$

where M_0^2, M_1^2 are constants. A relation that is derivable from this mass formula is

$$M_K^2 = \frac{3M_\eta^2 + M_\pi^2}{4}.$$

The experimental masses substituted here lead to satisfaction of the equality very well, which is another success for the idea of SU_3 symmetry.

Particles belonging to the 10-dimensional irreducible representation have also been found. The Δ resonances are members of this decuplet. The pattern of states in the (i_3, y) plane leads to a triangular shape, unlike the hexagonal shape for the octet representation (see Figure 4.13). Here we have a total of 10 particles. Four particles, $\Delta^{*++,*+,*0,*-}$, form an isospin quartet with $I = 3/2$ and $y = 1$; three particles: $\Sigma^{*+,*0,*-}$, an isospin triplet with $I = 1$ and $y = 0$; two particles: $\Xi^{*0,*-}$, an isospin doublet with $I = 1/2$ and $y = -1$; and finally, Ω^{*-}, an isospin singlet with $I = 0$ and $y = -2$. In this triangular representation, there is a linear relation between I and Y of the form $I = (1/2)Y + \text{constant}$.

Thus the mass formula in the decuplet representation can be reduced to the form $M = M_0 + M_1 Y$. The states with different y values are spaced linearly. Knowing the spacing between the Δ and the Σ^* states, one can predict what the masses of the Ξ^* and Ω^- states should be. Search for baryonic resonances with these masses were crowned with the spectacular discovery of these particles in experiments and the idea of SU_3 as the underlying group structure for the baryons and mesons was firmly established.

Electromagnetic Form Factors of the Nucleon

In scattering experiments involving targets of strongly interacting particles such as the proton or the neutron (together called the *nucleon*), one studies the matrix element of the electromagnetic current operator $j^\mu(x)$ between the nucleon state N and any state n to which connection is established by the electromagnetic current $(\Psi_n, j^\mu(0)\Psi_N)$ (for explanation of details, see reference [161]). In electron proton scattering, for example, the scattering amplitude in the one photon exchange approximation can be shown to be proportional to

$$\bar{u}_e(p')\gamma^\mu u_e(p) \frac{1}{(p'-p)^2 + i\epsilon} (\Psi_{P'}, j_\mu(0)\Psi_P),$$

where the factor before the fraction $1/((p'-p)^2 + i\epsilon))$ comes from the electron electromagnetic current, and the last factor is that from the proton electromagnetic current. The electron is a lepton and does not participate in any strong interactions. Hence its electromagnetic current is that of a structureless point particle whose matrix element between plane wave states expressed as a Fourier transform is the first factor above. The proton on the other hand is not a point particle and exists for a certain fraction of time dissociated into a $(n, \pi^+), (p, \pi^-, \pi^+), \dots$, etc. These pions themselves also interact with electromagnetic fields. Thus the current matrix element of the proton must include a sum over all these complicated mesonic and other particles into which the proton can dissociate, and it looks as if it might be a formidable job to get an expression for it. However, based on very general requirements such as Lorentz invariance and Hermiticity, one can write down a very general form for the matrix element of the proton current operator $(\Psi_{P'} j^\mu(0)\Psi_P)$. Clearly its structure will have the form $\bar{u}(P')G^\mu(P', P)u(P)$, where the $u(P), u(P')$ are the Dirac spinors for the initial and the final proton, and the quantity $G^\mu(P', P)$ represents the effect of summing over all the mesonic and other intermediate particles. When the effect of the mesons, etc., is neglected, the matrix element of the proton current must resemble the structureless electron current, so $G^\mu(P', P) \to \gamma^\mu$. Hence in general,

$G^\mu(P', P)$ can only depend on γ^μ, P'^μ, and P^μ. Of these, instead of P'^μ and P^μ, we may use the combinations, $(P'^\mu + P^\mu)$ and $q^\mu = (P'^\mu - P^\mu)$. Then $G^\mu(P', P)$ is of the form

$$G^\mu(P', P) = \gamma^\mu A(P^2, P'^2, P \cdot P') + (P'^\mu + P^\mu)B(P^2, P'^2, P \cdot P')$$
$$+ (P'^\mu - P^\mu)C(P^2, P'^2, P \cdot P'),$$

where A, B, C are Lorentz invariant functions of $P^2, P'^2, P \cdot P'$, and are indicated as such in the above expression. For elastic scattering, $P^2 = P'^2 = M^2$ where M is the mass of the proton. The scalar product $P \cdot P'$ can be expressed in terms of $q^2 \equiv (P' - P)^2$, so that $2(P \cdot P') = 2M^2 - q^2$. Thus the functions A, B, C are invariant functions of the square of the four-momentum transfer $q^\mu = (P'^\mu - P^\mu)$ and constants such as M^2. It is customary to suppress the dependence on constants such as M^2, so A, B, C are functions of q^2.

The expression for G^μ can be further simplified using the fact that the electromagnetic current satisfies the conservation law $\partial_\mu j^\mu(x) = 0$. This translates in momentum space to $q_\mu(\bar{u}(P')G^\mu(P', P)u(P)) = 0$. When this procedure is carried out we see that the terms involving A and B vanish (on using Dirac equations for $u(P), \bar{u}(P')$ and $P^2 = P'^2 = M^2$), whereas the term involving C gets the coefficient q^2 which is in general not zero. To enforce the conservation of the current we thus have to choose the function $C = 0$, whereas the functions A, B can be arbitrary. Taking current conservation into account, the form of the current is

$$\bar{u}(P')G^\mu(P', P)u(P) = \bar{u}(P')[\gamma^\mu A(q^2) + (P'^\mu + P^\mu)B(q^2)]u(P).$$

This is usually written in a slightly different form, taking account of an identity called the Gordon identity,

$$\bar{u}(P')\gamma^\mu u(P) = \frac{1}{2M}\bar{u}(P')[P'^\mu + P^\mu + i\sigma^{\mu\nu}q_\nu]u(P),$$

where $\sigma^{\mu\nu} = \frac{i}{2}[\gamma^\mu\gamma^\nu - \gamma^\nu\gamma^\mu]$. This identity allows us to replace the $P'^\mu + P^\mu$ term by the γ^μ term and the $\sigma^{\mu\nu}q_\nu$ term and write,

$$\bar{u}(P')G^\mu(P', P)u(P) = \bar{u}(P')[\gamma^\mu F_1(q^2) + \frac{i\sigma^{\mu\nu}q_\nu}{2M}F_2(q^2)]u(P),$$

where we have put $A(q^2) + 2MB(q^2) = F_1(q^2)$, and $2MB(q^2) = -F_2(q^2)$. The result for $G^\mu(P', P)$ is finally

$$G^\mu(P', P) = \gamma^\mu F_1(q^2) + \frac{i\sigma^{\mu\nu}q_\nu}{2M}F_2(q^2).$$

Multiplying this by the charge e, it contains two unknown functions, $F_1(q^2)$ and $F_2(q^2)$ of q^2, normalized so that $eF_1(0) = e$ and $eF_2(0) = (\kappa/(2M))$, which are called *Dirac charge form factor* and *Pauli anomalous magnetic moment form factor*, respectively. The quantity κ is called the anomalous magnetic moment of the proton because it adds to the Dirac magnetic moment $(e/(2M))$. These are elastic form factors, since here we treated only elastic eP scattering. The cross section for eP scattering can now be calculated using the expression for the proton current matrix element given above and the result in the laboratory frame is

$$\frac{d\sigma}{d\Omega} = \frac{\alpha^2}{4E^2 \sin^4 \theta/2} \frac{\cos^2 \theta/2}{1 + (2E/M)\sin^2 \theta/2}$$
$$\left[F_1^2(q^2) - \frac{q^2}{4M^2}(4M^2 F_2^2(q^2) + 2(F_1(q^2) + 2MF_2(q^2))^2 \tan^2 \theta/2) \right],$$

where E is the electron energy and θ the electron scattering angle in the laboratory. This result is sometimes expressed in terms of two other form factors defined as

$$G_E(q^2) = F_1(q^2) + \frac{q^2}{4M^2} F_2(q^2),$$

$$G_M(q^2) = F_1(q^2) + F_2(q^2).$$

We cannot at present calculate any of these form factors reliably from a theory. Our knowledge of these comes from experimental measurements on eP elastic scattering. From a fit to the experimental data, we find that the q^2 dependence of $G_E(q^2)$ is described well by $\frac{1}{(1-q^2/(0.71GeV)^2)^2}$, where q^2 is measured in GeV^2.

Electron
The first member of the lepton family was discovered in 1898. See section under "Cathode Rays—Discovery of the Electron".

Electron—Anomalous Magnetic Moment, $g - 2$
The first theoretical calculation of the anomalous magnetic moment of the electron was done by Schwinger [162]. In a series of papers, he developed systematic methods for handling divergences which arise in higher orders of perturbative quantum electrodynamics and absorbing them into quantities such as the mass and charge of the electron—a procedure called *renormalization*. After this process is carried out in a careful, consistent, and Lorentz invariant way, many physically measurable quantities obtain finite corrections which modify their values obtained in the lowest order. Such corrections are called *radiative corrections* as

they arise from higher order processes in which more and more virtual (and/or real) quanta are included.

The electron, which in lowest order of interaction with the electromagnetic field is described by the Dirac equation and has a gyromagnetic ratio $g = 2$ associated with the spin. The radiative corrections modify this result which is expressed in terms of the difference $g - 2$. The radiative correction to the magnetic interaction energy was calculated by Schwinger as an application of his method, and he showed that there is an additional contribution to the spin magnetic moment of magnitude $[(g - 2)/2] = \frac{\alpha}{2\pi} \approx 0.0011614$ where α is the fine structure constant (1/137). (See section under "Electromagnetic Form Factors of the Nucleon"; $F_2(0) = 0$ in lowest order and gets a correction $\alpha/(2\pi)$ to first order in α).

The first measurement of $g-2$ for the electron was done by Foley and Kusch [163] who obtained the value 0.00229 ± 0.00008 agreeing with α/π from theory. The most accurate measurement of $g - 2$ for the electron and the positron has been carried out by Van Dyck et al. [135] (see section under "*CPT* Invariance").

Electron—Charge Measurement

The first precision measurement of the electric charge carried by the electron was done by Millikan [164] by an ingenious method. His method is now well known and will be described only briefly. He produced a fine spray of oil drops in a region between two metal plates. These drops were so fine that they fell very slowly under gravity. Besides gravity, these droplets suffered a viscous drag due to the air in which they were falling and, hence, soon attained a terminal velocity. The magnitude of the viscous force is given by Stokes' law, $6\pi\eta a v$, where η is the coefficient of viscosity, a the radius of the drop, and v the velocity of the drop. By observing a droplet with a microscope, one could time its motion between specified locations in the field of view and, hence, get a measurement of its velocity. Then he imposed an electric field E between the metal plates. It turns out that the droplets acquire charge due to friction with the air, or for some other reason. If a droplet has a charge q, it will be subject to a force qE due to the electric field, in addition to the gravitational force and the viscous force. Thus, it would have a changed velocity, up or down, depending on the sign of the electric field and the sign of the charge. If the fundamental unit of charge is e, a drop carrying n units will have a charge $q = ne$. By measuring the altered velocity in the electric field, he could determine the charge q. Using this technique, he observed many droplets over long periods of time and measured the charges they carried. He then observed that the charges carried by the drops had a set of discrete values, and these values were always an integer multiple of a fundamental charge. He identified that fundamental charge as, e, the unit of charge. He extracted this number from his observations on many droplets. He obtained a value which, when expressed in SI units, is $|e| = 1.60 \times 10^{-19}$ Coulomb. Since then the electron charge has been measured by many improved methods and is one of highly precisely determined numbers, the most recent value being $1.60217733 \times 10^{-19}$ Coulomb, with an uncertainty of 0.3 part per million.

Electron Cooling

In the development of accelerators to provide high energy particles, protons or antiprotons, the intensity of particles produced in the beam, is a very important consideration. Methods to increase the intensity to as high a value as possible are highly desirable. A method to achieve this was put forward by Budker [165]. The method suggested exploits the effect of the sharp rise in cross section for interactions of electrons with heavy particles such as the proton, or positrons with antiprotons, at small relative velocities. This results in damping out synchrotron and betatron oscillations in the proton or the antiproton beam. The method, called *electron cooling*, has been shown to be capable of compressing and accumulating proton and antiproton bunches in the beam and, hence, increasing the intensity per bunch.

To achieve electron cooling experimentally, one injects a beam of "cool electrons" into a storage ring of heavy particles, say protons, in a straight section of the storage ring. When the velocity of the electrons and that of the heavy particles coincide in magnitude and direction, an effective friction is introduced by the electrons on the heavy particles, which causes the phase-space volume of the heavy particles to decrease. This means the volume in which the heavy particles are contained is considerably reduced and the spread of the energy in the bunch in the beam is also reduced. The first experimental studies of using electron cooling techniques on a beam of 35–80 MeV protons was carried out by Budker et al. [166], and since then it has been successfully used in many heavy particle storage rings. At a proton energy of 65 MeV and electron current of 100 mA, they reported measurements of the betatron oscillation damping time and the equilibrium proton beam dimensions. The latter quantity was found to be ≤ 0.8 mm and angular spread $\leq 4 \times 10^{-5}$. The momentum spread of the proton beam as measured by $\delta p/p$ was $\leq 1 \times 10^{-5}$.

Electron-Nucleus Scattering

The first theoretical paper directing attention to the possibility of determining charge distributions in atomic nuclei by electron scattering was carried out by Rose [167]. Considering the nucleus as a point charge but treating the electron relativistically with the Dirac equation, Mott had earlier derived an expression for the scattering cross section for the electron, which is referred to as the *Mott scattering formula* [168]. Rose pointed out that the finite size of the nucleus will give rise to sizable deviations from the Mott scattering formula, when the change in the wavelength of the electrons is of the order of the nuclear diameter. Thus, the deviation is expected for large scattering angles of the electron. For a spherically symmetric charge distribution in the nucleus, the Fourier transform of the observed angular distribution gives the nuclear charge distribution directly. For high energy electrons, one needs to consider also other competing processes, such as nuclear excitations, atomic excitations and ionization, and bremsstrahlung, and separate out their effects.

The first studies of electron scattering on nuclei were carried out in detail using the Stanford linear accelerator for electrons by a team headed by Hofstadter [169] These studies gave the first measurement of the electromagnetic radius of the proton. Results for helium nucleus are also presented in this early paper. Following on this work, electron scattering studies on a number of nuclei were done, and their electromagnetic structure was determined.

Electron-Positron Pair Production

While the discovery of the positron was made by Anderson [170] in cosmic rays, confirming one of the predictions of the Dirac equation that antiparticles to electrons must exist, Blackett and Occhialini developed cloud chamber techniques in a magnetic field [171] and showed that electron-positron showers are also generated in cosmic rays. These observations, not only confirmed the earlier findings of Anderson [170] but also led to detailed studies of the properties of cosmic radiation. A method was developed by them so that the high energy cosmic ray particles, trigger the cloud chamber and take their own photograph in the chamber. The result of the passage of the cosmic ray particles is to create a shower of particles in the cloud chamber, whose charges could be analyzed by the curvatures of the tracks in the magnetic field. They analyzed the nature of the particles in the showers and the complex tracks they left in the cloud chamber. The points of origin of electrons and positrons in the showers were observed. It was shown that they arise, as a result of collision processes involving a cosmic ray particle and a particle in the cloud chamber. The frequency of occurrence of the showers was measured by them and shown to be in conformity with other work related to the occurrence of bursts of ionization in cosmic rays. This work further confirmed the correctness of Dirac theory for the electron and paved the way for the construction of a theory of cascade showers based on the interaction of electromagnetic radiation (photons) with electrons and positrons.

Electron-Proton Elastic Scattering
See the section under "Electromagnetic Form Factors of the Nucleon".

Electron—Relativistic wave equation
See section under "Dirac Equation" for this item.

Electron—Spin
Much of the early information on the properties of the electron came from a study of atomic spectra. The introduction of spin as an intrinsic property possessed by the electron emerged from these efforts. Experimental work on atomic spectroscopy revealed a number of regularities in the spectra of atoms. The work of Bohr on the spectrum of the hydrogen atom was produced in an effort to understand how these spectra arose. It was found that not only hydrogen but also the alkali atoms had spectra which presented similar regularities. The levels in these atoms fell into closely grouped multiplets and the spacings between the levels of the multiplets followed simple relationships.

The study of Zeeman effect (that is, the study of spectra of atoms placed in a magnetic field) showed that the different levels of a multiplet in these atoms split in different ways due to the magnetic field. To understand these observations, Landé and others pointed to the effect of several different angular momenta in an atom (those of the different electrons), which coupled to form a resultant. The resultant was quantized as in the Bohr theory of the hydrogen atom. In the case of just two angular momenta, for example, the relative orientation of these vectors could be parallel, antiparallel, or some other orientation in between. Then the resultant angular momentum can take all integral quantized values between the minimum and the maximum values (in units of $\hbar$, which we have taken as 1).

It was assumed that the various levels of a multiplet could somehow arise from the different relative orientations of the angular momentum vectors as follows. If, to each such angular momentum vector, one could associate a small magnet with a tiny magnetic moment and the magnets interacted exerting torques on one another, the energy of interaction would then depend upon the relative orientation of the magnets. This would explain the separation in energy of different members of a multiplet. If the energy is proportional to the cosine of the angle between the magnetic moments, a simple law is derived for the spacings of the levels in a multiplet. This law was called the *Landé interval rule*, and the spacings derived from the experimental values followed this rule to a very high degree. In a magnetic field, J would have components along the magnetic field (taken as the quantization axis) ranging from $J, (J - 1), \ldots, -J$, leading to $(2J + 1)$ possibilities in all. Each of these would lead to a different energy level in the magnetic field, and just by counting them one could determine J. In this way, the vector sum J of the angular momenta in the atom could be determined by Zeeman effect.

Applying these considerations to the levels in an alkali atom, as there is only one outer electron around a core, one would not expect to see a multiplet structure in the levels; yet experiments do show a multiplet structure. Experiments in fact showed that the levels of alkali atoms were doublets with the quantum numbers J turning out to be not integers (as expected in Bohr's theory) but half odd integers. For example, only an s state was found to be single with one value $J = 1/2$, while a p state turned out to be double with $J = 1/2$ and $3/2$, and a d state double with $J = 3/2$ and $5/2$, etc. This information on the J values was extremely puzzling and the whole situation was very confused. The resolution of the problem came in two parts. Pauli [172] made the suggestion that an electron was to be labeled by a new two-valued quantum number σ in addition to the known labels (n, l, m_l) corresponding to the principal quantum number n, orbital angular momentum l, and the projection m_l of orbital angular momentum on the z-axis; this would obviously lead to a doubling of the levels. Uhlenbeck and Goudsmit [173] went further and suggested that this new degree of freedom be associated with an intrinsic property of the electron, the ability to carry an intrinsic angular momentum. They suggested that the value of $1/2$ for J in an atomic s state has nothing to do with the atomic core around which the outer electron moved but that it possessed an intrinsic angular momentum $(1/2)$ leading to two possible orientations in a magnetic field, $m_J = \pm 1/2$. (The intrinsic angular momentum carried by the electron is essentially quantum mechanical in nature and cannot be thought of in classical terms.) This intrinsic angular momentum, which could be called the *spin*, combines with the orbital angular momentum to produce the total angular momentum J which is what is observed. They also proposed that for spin, the ratio of the magnetic moment to spin (called the *spin gyromagnetic ratio g*) was two times that for the orbital gyromagnetic ratio. With these suggestions, most of the Zeeman splittings in alkali spectra could all be understood and the notion of an elementary particle possessing an intrinsic degree of freedom called the *spin* was born.

Electroweak Synthesis
As early as 1964, in a paper written by Salam and Ward [174], the requirements for a theory which unifies electromagnetism and weak interactions were enunciated. They drew inspiration from the fact that both electromagnetism and weak interactions act on leptons as well as hadrons universally, and both are vector in character. Universality and the vector character are features of a gauge theory and so they started looking for a gauge theory of weak interactions despite the profound differences between electromagnetism and weak interactions. If the strength of the weak force was determined by the fine structure constant, they showed that the mass for the gauge mediator needs to have a value approximately 137 times the mass of the proton and a group structure for the weak interaction, which is a combination of vector and axial vector couplings.

Independent of the efforts by Salam and Ward, Weinberg was also working on the idea of unifying electromagnetism and weak interactions and his preliminary work was published in 1967 [175]. He too was motivated to look for a gauge multiplet, which would include the mediators of the weak interactions along with the photon. The impediment to this unification was, of course, the huge difference between the masses of the gauge mediator for the weak interaction and the photon, and the size of their couplings. He suggested that a way to bridge these differences might arise from the fact that the symmetries relating electromagnetic and weak interactions, which are exact symmetries of the Lagrangian, might be broken by the vacuum. Such symmetry breaking had been examined by Goldstone earlier [176], who showed that it would give rise to massless bosons which are now called *Goldstone bosons*. Since such massless bosons were not observed in nature, this symmetry breaking mechanism was not taken seriously at that time. Weinberg proposed a model in which the symmetry between electromagnetic and weak interactions was spontaneously broken, but massless Goldstone bosons were avoided by introducing photon and intermediate boson fields as the gauge fields [177]. He wrote down a Lagrangian for leptons interacting with such gauge fields and ended with a conjecture that such a field theory with spontaneously broken symmetry might be renormalizable.

Another aspect of this synthesis was addressed by Glashow even as early as 1961 [178]. He examined a theory in which the weak interactions are mediated by vector bosons. He showed that in a theory with an isotopic spin triplet of leptons coupled to a triplet of vector bosons, there are no partial symmetries. To establish such partial symmetries one would be obliged to introduce additional leptons or additional intermediate bosons. Because there was no evidence for additional leptons, he suggested introducing at least four vector boson fields including the photon. This was the first introduction of the idea of a neutral intermediate vector boson, in addition to the charged intermediate vector bosons, to mediate weak interactions. He showed that such a theory contains partially conserved quantities which are the leptonic analogues of strangeness and isotopic spin.

One of the common features of the unification proposal made by Salam, Weinberg, and Glashow is the introduction of the neutral intermediate vector boson as a mediator of weak interactions. Thus in addition to weak decay processes which involve a charge change (charged current weak interactions mediated by charged gauge bosons $W^{\pm}$), there should also be processes in which weak interactions participate without the change in charge of the participating particles (neutral current weak interactions mediated by neutral gauge boson Z^0). Such a neutral current effect was discovered in 1978, lending support to the idea of unification as pursued by Salam, Weinberg, and Glashow. The renormalization of the spontaneously broken gauge theory was proven by G. 't Hooft in 1971 [179]. The weak interaction mediators, the $W^{\pm}$ and Z^0, were also first found in 1983–84 by the UA1 and the UA2 collaborations working at the CERN $\bar{p}p$ collider [180,181]. (See further under "Standard Electroweak Model".)

Energy Quanta—Photon

At the beginning of the 20th century, there were intense investigations on the nature of radiation emitted by a hot body, called *black body radiation*. In experimental measurements of the energy density of the black body radiation, when examined as a function of the frequency of emitted radiation, it was found to have a maximum (which shifted to higher frequencies as the temperature of the emitting body was increased), and then it gradually fell off to zero.

The only theoretical derivation of this energy spectrum available at that time was due to Rayleigh and Jeans; the crucial ingredient for this theory was the classical notion that the energy of radiation could take on all possible continuous values. This Rayleigh-Jeans formula for the energy density showed a behavior proportional to the square of the frequency for all frequencies with no maximum, clearly disagreeing with experimental results at high frequencies, and further leading to the absurd result that the integral of the energy density over all frequencies would be infinite. These were indications that there was something seriously wrong with the assumptions which were the basis of the Rayleigh-Jeans derivation.

Planck studied this problem anew and started examining the consequences of assuming that the energy carried by electromagnetic radiation may be in the form of discrete packets or quanta. Planck assumed that the energy E associated with radiation of frequency ν is given by $E = h\nu$, where h is a constant (having the dimension of energy times time) introduced by Planck (hence called *Planck's constant*). The quantum of radiation has come to be called the *photon*. Using this definition of energy, Planck derived the expression for the energy density of black body radiation which bears his name. It gave an energy spectrum for black body radiation which agreed with what was found from experimental studies.

According to the special theory of relativity, no massive particle can travel with the velocity of light. The photon, the quantum of radiation, does travel with the velocity of light c, hence, its rest mass must be zero. This implies that the photon has a momentum $p = E/c = h\nu/c$ associated with it. (In terms of units $\hbar = c = 1$, we would have, introducing the angular frequency $\omega = 2\pi\nu$, $p = E = \omega$.)

The firm establishment of the idea of the light quantum as being correct came from Einstein's explanation [182] of the photoelectric effect using the ideas of Planck. The photoelectric effect was discovered by Hertz in 1887. The initial observations of Hertz were refined by the later and much more precise work of Millikan [183] and others, who confirmed all Hertz's findings. They found that when a freshly cleaned metal surface was irradiated with light, the surface emitted electrons. No emission of positive ions was found. It was found that there is a threshold frequency of the irradiating light ν_0 (which depends on the metal), and for frequencies less than this threshold frequency, there is no electron emission. The number of photoelectrons emitted, as measured by the photoelectric current, was found to be proportional to the intensity of the light and independent of the frequency of the light (as long as it is greater than ν_0). The energy of the photoelectrons was found to be independent of the intensity of the light but was found to vary linearly with the frequency of the light.

It is easy to see that these observations cannot be explained by using Maxwell's classical electromagnetic wave theory for the nature of light. According to the classical theory, the energy carried by an electromagnetic wave is proportional to the intensity, which is the square of the amplitude of the wave. On this basis, we would expect the energy of the photoelectrons to vary with the intensity of the source of the waves, and it should have no frequency dependence—exactly contrary to what was observed. Einstein's explanation of this effect used the idea that the incident light consists of photons of energy $h\nu$, and that these quanta release electrons from the metal by colliding with them. However, to have an electron leave the metal, a minimum amount of energy has to be supplied to it, characteristic of the metal, called the *work function W* of the metal. This minimum energy determines the threshold frequency $\nu_0 = W/h$. Further, for the same frequency, a high intensity light source puts out a much higher number of photons than a low intensity one. Thus one would expect that the higher intensity source would lead to emission of more photoelectrons than would a lower intensity one. This is in accord with what is observed. The kinetic energy that the electron acquires is determined by conservation of energy: kinetic energy of the electron $= h\nu - h\nu_0$. This varies linearly with the frequency as is observed experimentally. With this explanation of the photoelectric effect, the idea of photons, as light quanta, was firmly established.

The fact that photon arises from quantizing the electromagnetic field, which is a vector field described by electric and magnetic field vectors, suggests that the photon should carry intrinsic angular momentum or spin. Early experimental proof that the photon has spin 1 came from the work of Raman and Bhagavantam [184]. (See further under "Photon—Mass, Spin, Statistics".) Thus the photon is an electrically neutral, massless particle with spin 1.

Eta meson

A particle that is predicted to exist when SU_3 symmetry is extended to hadrons is the η meson (see section under "Eightfold Way"). It was discovered in the reaction $\pi^+ + d \rightarrow p + p + \pi^+ + \pi^- + \pi^0$ by Pevsner et al. [185]. It was found that the number of events as a function of the invariant mass of the three pion system, as measured by

$$M^2 = (E_{\pi^+} + E_{\pi^-} + E_{\pi^0})^2 - (\vec{p}_{\pi^+} + \vec{p}_{\pi^-} + \vec{p}_{\pi^0})^2,$$

showed a sharp peak at an energy of about 550 MeV. This is interpreted as the particle called the η^0, which decays into three pions.

Careful measurements of its properties have been carried out since then. Its mass has been measured to be 547.3 ± 0.12 MeV. A number of other decay modes have also been seen in experiments. Among these, $\eta^0 \rightarrow 2\gamma$ has been seen with a branching ratio of about 39%, and the three pion decay mode has a comparable branching ratio of about 23%. The η^0 meson decays just like π^0 into 2γ, and because the other decay modes have comparable branching ratios, it is concluded that the decay occurs through electromagnetic interactions and not through weak interactions. The observation of the 2γ decay mode of the η^0 suggests that its spin cannot be 1 and its lifetime must be of the order of the lifetime for π^0. Since parity is conserved in electromagnetic interactions, one may also determine the spin and parity of the η^0 from its decays. The Dalitz plot for the three pion decay mode shows a uniform distribution of events [186] thus allowing one to conclude that its parity is odd and its spin is zero. (See section under "Dalitz Plot".)

Searches for any other states with mass close to 550 MeV have found negative results. The mass of the η^0 substituted in the meson mass formula (see section under "Eightfold Way") shows that the relation is well satisfied. Thus, η^0 is concluded to be an isosinglet with $y = 0$ in the SU_3 octet.

Eta′ Meson

First evidence for a mass state around 960 MeV in mass came from a study of the invariant mass distribution of the five pions in the reaction

$$K^- + p \to \Lambda^0 + \pi^+ + \pi^+ + \pi^0 + \pi^- + \pi^-,$$

observed in the hydrogen bubble chamber at the LBL Bevatron [187] and confirmed by Goldberg et al. [188] in the reaction $K^- + p \to \Lambda^0 +$ neutrals (and also the one involving charged pions in the final state) observed in the hydrogen bubble chamber at the Brookhaven proton synchrotron (AGS). The invariant mass distribution showed a peak at a mass of about 960 MeV. A more precise measurement of its mass is 957.78 ± 0.14 MeV. The width of this state is small, and it decays predominantly in the mode $\eta^0 + 2\pi$ with a branching ratio of about 44%. Notable absence is the three pion decay mode. The absence of the three pion decay mode allows one to conclude that the interaction responsible for the decay is not electromagnetic in this case but involves strong interactions. The decay mode to $\eta^0 + 2\pi^0$ has been observed, which allows one to conclude isospin 0 or 2 for the decaying particle. However, isospin 2 can be excluded from the production mechanism, as it involves K^- and p, both of which have isospin 1/2. Thus one must attribute isospin 0 to it. Decay mode leading to $\rho^0 + \gamma$ (including nonresonant $\pi^+ + \pi^- + \gamma$) has also been seen with a branching ratio of about 30%. A Dalitz plot analysis of these decay events as well as of the $\eta^0 + 2\pi$ events allows one to draw conclusions about the spin and parity of the decaying particle. The conclusion from this study for J^P is 0^-. This particle is given the name η'^0.

According to the SU_3 classification of mesons, there is room for a particle which is a singlet under SU_3. There will be two particles both with $I = 0$ and $Y = 0$ of which one (η^0) is in a SU_3 octet and the other

(η'^0) is in a SU_3 singlet. In general there could be mixing between these states when SU_3 symmetry is broken leading to more complicated mass formulae. (This, in fact, happens when we consider the vector mesons with $J = 1$.) As the mass formula in the SU_3 octet is well satisfied with the experimental value of the η^0 put in, it seems that for the pseudoscalar mesons, consideration of $\eta^0 - \eta'^0$ mixing is not necessary.

Exchange forces

The implications of the suggestion that atomic nuclei are made up of protons and neutrons rather than protons and electrons were first examined by Heisenberg [47]. He wrote the Hamiltonian function of the nucleus and made a general discussion of the forces between protons and neutrons in nuclei, called *nuclear forces*. In analogy with the notion of electrical forces between charged particles arising as a result of exchange of photons, he suggested that exchanges of some quanta were responsible for the nuclear forces. These exchange forces must be capable of acting between two neutrons, two protons, or a proton and a neutron. That the specifically nuclear forces (as distinct from the electrical forces between protons) between any pair of these particles were similar was established by examining the stability of nuclei with different numbers of protons and neutrons. Together with the knowledge of the fact that neutron and proton had almost the same mass, it led Heisenberg to propose the isotopic spin formalism for the description of proton and neutron as two states of a particle called the *nucleon*. This was further explicitly elaborated in a work by Cassen and Condon [189]. In analogy with ordinary spin, in the formalism of isotopic spin, proton and neutron are the "up" and "down" states of a doublet of isotopic spin $1/2$. The equality of force between any pair of nuclear particles (protons or neutrons) can be expressed in this formalism as a symmetry under rotations in the I-spin space, which rotates protons into neutrons. The notion of charge independence of nuclear forces is a consequence of the invariance of nuclear interactions under isotopic spin rotation transformations, or I-spin symmetry.

Exclusion principle

It was Pauli who made the statement of the exclusion principle in the context of electrons in atoms [172]. He stated that an atom cannot contain two electrons with identical sets of quantum.

This statement applies to particles other than the electron also. Nature has two kinds of particles, namely particles with spin of $1/2$ like the electron and others with integral spin like the photon. The exclusion principle of Pauli applies to all particles with half odd integral spin. In

quantum mechanics, this means that the wave function of a two particle system obeying the exclusion principle has to be *antisymmetric* with
respect to the exchange of all the variables, such as position, spin, and
isospin, describing the particles.

Fermi Interaction—Weak Interaction

Fermi constructed a field theory of β decay, treating the emission of
electrons and neutrinos from a nucleus much as the emission of photons
from an excited atom (for details see under "Beta Decay—Theory").
The form of the interaction that Fermi proposed, which was successful
in deriving the form of the continuous spectrum of beta decay and the
lifetime, has come to be called *Fermi interaction*. It is characterized by
the dimensionful Fermi coupling G_F.

Many other particles have been found in nature which also undergo
decay processes. A question that naturally arises concerns the nature
of the interaction responsible for the observed decays. This question
was examined for the first time by Lee, Rosenbluth, and Yang [190].
In particular, they looked at muon decay to electron and two neutrinos
and the nuclear capture of muons and tried to describe these with Fermi-
type interactions. In muon decay the Fermi interaction would involve
the product of the currents, one formed from an electron and a neutrino
and the other from the muon and the other neutrino. In muon nuclear
capture, a μ^- may be absorbed by a proton with resulting emission of
a neutron and a neutrino. Here the two currents would be formed, one
from the proton-neutron system and the other from the muon-neutrino
system. In each case there is a phenomenological coupling constant g,
which would have the dimensions of energy times volume as in beta
decay. The value of the g's could be determined by fits to the available
data on muon decays and nuclear capture probability of muons. Lee,
Rosenbluth, and Yang found values, $g = 3 \times 10^{-48}$ erg cm^3, and $g =
2 \times 10^{-49}$ erg cm^3, respectively, from these processes. They noted that
these numbers are strikingly near each other and also near the value
found from nuclear beta decays. The equality of these interactions led
them to suggest that the interaction responsible for these decays may be
universal. Thus nuclear beta decays and the decays of various elementary
particles are governed by a universal interaction which is called *weak
interaction* in general. Since the early days much progress has been made
in understanding weak interactions. These interactions are transmitted
through an intermediate field, the quanta of which are the W and Z
bosons.

Fermions

When we consider a system of two *identical* particles, the behavior of the wave function of the two particle system under the exchange of two particles is decided by a deep law of nature. Particles come in two classes—those for which the wave function of a two particle system is symmetric under their interchange and those for which the wave function is antisymmetric under the interchange. The former category of particles with symmetric wave functions are called *bosons* named after their discoverer S. N. Bose, and the latter with antisymmetric wave functions are called *fermions* named after their discoverer E. Fermi [191]. Examples of particles which are bosons are, the photon, the pion, the deuteron, and the alpha particle. Examples of fermions are, the electron, the proton, the neutron, the muon, and the neutrino.

A particle which is a bound state made up with an even number of fermions is a boson. The wave function of a system of many such particles which are bosons, is described by wave functions which are symmetric under the interchange of any two identical bosons. For a many fermion system, the wave function of the system is antisymmetric under the interchange of any two identical fermions. This behavior of identical particles with antisymmetric wave functions under their interchange was also independently discovered by Dirac [192]. The statistics of an assembly of fermions is called *Fermi-Dirac statistics*, while the statistics of an assembly of bosons is called *Bose-Einstein statistics*. The antisymmetry principle for two identical fermions can be shown to lead to Pauli's exclusion principle; namely, no more than one fermion can be in a given quantum mechanical state.

Feynman diagrams

Particle physics experiments involve the scattering of particles and the measurement of cross sections. Using quantum field theory, one tries to calculate the cross section theoretically, in terms of the matrix element M for the scattering process. Unfortunately, even for the simplest of processes, the *exact* expression for M is not calculable. If the interaction responsible for the scattering is weak and can be characterized by a small coupling parameter, an approximation to M can be obtained as a power series in terms of this small parameter, keeping the first few terms of this series.

Feynman [193] devised a beautiful way to calculate the matrix elements of processes involving electrons and photons (the coupling characterized by the fine structure constant $(1/137)$) in the different orders of perturbation theory, organizing these calculations with the aid of visual diagrams, which have now come to be called *Feynman diagrams*. The

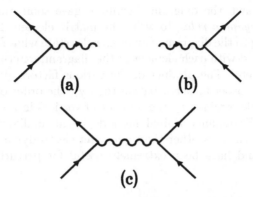

Figure 4.14: Examples of Feynman diagrams in quantum electrodynamics.

diagrams show a flow of electrons and photons with lines belonging to these. The simplest diagram that can be drawn is one for the emission or absorption of a photon by an electron (Figure 4.14). We see that the diagram has straight lines, which represent electrons, and wavy lines which represent photons. Where these lines meet (as in (a) and (b) of the figure), we have what is called a *vertex*. We can join the fundamental vertices in various ways to represent different scattering processes. As an example, if we join the photon lines from the vertices, we get the diagram shown in part (c) of the figure, which can be interpreted as two electrons coming in, exchanging a photon between them at the vertices, and going out as the scattered electrons. Lines coming into a vertex or leaving a vertex are called *external lines*. In the figure in part (c), we have four external lines. The wavy line, which joins the two vertices, is called an *internal line*, and represents the virtual photon exchanged between the two electrons. If we assign four-momentum vectors to each line in the diagram, p_1, p_2 to the incoming electron lines, p'_1, p'_2 to the outgoing electron lines, and q to the internal photon line, four-momentum conservation at either vertex can be represented by $p'_1 = p_1 - q$, $p'_2 = p_2 + q$. Eliminating q between these two, we get, the overall four-momentum conservation for the scattering process: $p_1 + p_2 = p'_1 + p'_2$. To specify the electrons further, we may include a specification of the spin states s_1, s_2, s'_1, s'_2 of the electrons. Figure 4.14 represents the scattering of two electrons with momenta p_1, p_2 and spins s_1, s_2 to momenta p'_1, p'_2 and spins s'_1, s'_2.

Having drawn the diagram, Feynman gave some rules, which are now called *Feynman rules*, to write the matrix element for the process corresponding to the diagram. Feynman rules state what factor in M has to be associated with each element of the diagram, external line, vertex, internal line, etc. The product of the various factors gives M (in this case to second order in perturbation theory), the order of perturbation theory being determined by the number of vertices in the diagram.

Following Feynman's original methods, Feynman diagrams and Feynman rules for theories other than quantum electrodynamics have been constructed and have been extremely useful for perturbative calculations.

Flavor-Changing Neutral Currents

Weak decay of hadrons into leptons occurs by the emission of the charged W boson or the neutral Z boson by one of the quarks in the hadron, and subsequent decay of the W or the Z into the outgoing leptons. Weak decays also lead to hadrons in the final state, which arise from the W or the Z decaying into quarks. Let us focus attention on the neutral Z bosons which transmit neutral current weak interactions. At the quark level, we can have, d-quark to d-quark transitions, s-quark to s-quark transitions, as well as d-quark to s-quark transitions (and vice versa). Experimental observations show that the neutral Z do not mediate strangeness changing transitions. Prior to the introduction of the idea of the charm quark, with only three quarks, u, d and s, it was not possible to understand the absence of strangeness changing neutral currents. With Cabibbo mixing among the d and the s quarks, $d' = d \cos \theta + s \sin \theta$ pairs with the u to form a left-handed doublet. The orthogonal combination $s' = s \cos \theta - d \sin \theta$ has no other quark with which to form a doublet, and so has to be treated as a singlet. In this situation the neutral Z mediates $d' \to d'$ (doublet) transitions and $s' \to s'$ (singlet) transitions with different strengths. Adding these, give $d \to d$ and $s \to s$ transitions, and in addition give $d \leftrightarrow s$ transitions. The latter are the strangeness-changing neutral currents, which would conflict with the lack of such effects in experiments. The situation was resolved with the introduction of the idea of the charm quark, which could pair with the orthogonal combination s' into a second left-handed doublet. In this case it is easy to see that there are no $d \leftrightarrow s$ transitions in the neutral weak current. Thus with two families of quarks and Cabibbo mixing, there are no strangeness-changing neutral currents.

More generally, suppose the quarks are arranged in three families, (u, d), (c, s), and (t, b), and the d, s, b quarks undergo mixing through a unitary matrix to d', s', b' respectively. A general feature of such mixing is that neutral weak currents involving a flavor-change are absent. This is in accord with the absence in experiments of flavor-changing neutral current effects in general. (See further discussion under "GIM Mechanism".)

Fock space

This is a concept that was introduced by Fock [194] and is a method for describing states of a quantum system with an arbitrary number of particles, including the possibility of an infinite number of particles. When one deals with a physical system possessing an infinite number of degrees of freedom, the concept of a field proves to be useful. The field is quantized to obtain quanta of the field, which are the particles. For simplicity we will consider only the case of a scalar field. In the procedure for quantizing the field, the field functions and their time derivatives, which are functions of space and time, become operators. Fourier expansions of these operators are carried out. The expansion coefficients are operators, called *creation* and *annihilation operators*, which obey certain commutation relations. If the creation and annihilation operators for a particular mode of wave number k are $a^\dagger(k)$ and $a(k)$, respectively, expressions for the Hamiltonian H of the system (the total energy) (we are ignoring complications due to zero point energy for the present) and for the number N of particles, can be written in terms of these operators as

$$H = \sum_k E_k a^\dagger(k)a(k), \qquad N = \sum_k a^\dagger(k)a(k).$$

Here E_k is the energy associated with the mode k. If $a^\dagger(k)a(k) = N_k$ is interpreted as the number of quanta in the mode k, the Hamiltonian, which represents the total energy, is obviously equal to the sum, over all modes, of the energy per quantum in mode k multiplied by the number in mode k, and the total number of particles N is the sum, over all modes k, of the number of particles in mode k. If we want the ground state of the system to have zero energy, we introduce a ground state or the vacuum state vector $|0\rangle$, which is such that $a(k)|0\rangle = 0$. In such a state the energy is zero and the number of quanta is zero. One can create a one particle state by $a^\dagger(k)|0\rangle$. Using the commutation relations of the creation and annihilation operators, it is possible to show that this state has the eigenvalue 1 in mode k for the number operator and that the energy has the eigenvalue E_k in this state. Thus we have created a one particle state by operating on the vacuum state vector once with a creation operator. Using this procedure, a two particle state will be a state obtained by $a^\dagger(k_1)a^\dagger(k_2)|0\rangle$. One can extend this procedure for constructing state vectors with more particles. In this manner one can construct the full Hilbert space of states generated by the state vectors such as

$$|n\rangle = a^\dagger(k_n)a^\dagger(k_{n-1})\cdots a^\dagger(k_1)|0\rangle,$$

for various values of n. A general state of the system is a linear superposition of states formed with such state vectors. This space is called the *Fock space*. In making this simple introduction to Fock space we have sidestepped a number of subtleties which have to be taken care of, relating to questions of normalization of the states, identity of the particles, etc. Clearly, the concept of Fock space is very useful for the description of a state with an arbitrary number of particles.

Form Factors
See under "Electromagnetic Form Factors of the Nucleon".

Forward-Backward Asymmetry
The forward-backward asymmetry in $e^+e^- \rightarrow l^+l^-$ is defined as

$$A_{FB} \equiv \frac{\sigma_F - \sigma_B}{\sigma_F + \sigma_B},$$

where $l^\pm$ are leptons, and σ_F and σ_B are the cross sections for the l^- to travel forward and backward, respectively, with respect to the incident electron direction. High precision measurements of this quantity at the position of the Z mass are available from LEP and SLC.

The theoretical expression for the forward-backward asymmetry depends upon the axial-vector and vector couplings, and the values of Z mass and width. The experimental data have to be adjusted for straight QED corrections as well as electroweak corrections. One must remove the standard QED corrections to verify the size of the electroweak corrections. The quality of the measurements at LEP and SLC have made this precision comparison between theory and experiment possible. It is found that theory and experiment agree very well thus establishing the correctness of the Standard Model.

Froissart Bound
Clearly, the knowledge of the behavior of total cross sections for hadronic interactions at very high energies is of great importance in the study of high energy collisions of elementary particles. Froissart [195] proved a very important theorem on the large energy behavior of hadronic cross sections. The behavior is usually expressed in terms of a variable s (called a *Mandelstam variable*), which is the square of the total energy of the colliding particles in a frame in which the total three-momentum of the colliding particles is zero (usually referred to as the center of

mass frame). In dealing with pion-nucleon scattering, for example, Mandelstam [196] introduced a representation for the scattering amplitude, which exhibits its analytic properties as a function of the energy and momentum transfer variables. This representation, called the *Mandelstam representation*, has been very useful in the study of pion-nucleon scattering using dispersion relations (see section under "Dispersion Relations") and unitarity. Froissart proved that a reaction amplitude involving two scalar particles, satisfying Mandelstam representation, can grow at most like $Cs \ln^2 s$, as a function of s, where C is a constant. This implies that the total cross section can grow at most like $\ln^2 s$. Intuitively one would expect a bound on the cross section at high energies based on the fact that the strong interactions of hadronic particles are of finite range, and the derivation by Froissart reflects this expectation. Experimental data on hadronic interactions at high energies are in qualitative agreement with the predictions of the Froissart bound.

G-Parity
In our discussion on charge conjugation, we noticed that the charge conjugation parity quantum number can only be defined for uncharged particles. In the case of charged particles, for example, charged pions, a quantum number called the *G-parity*, closely related to charge conjugation, can be introduced. It can be used to derive selection rules for states consisting of a system of charged pions or for a nucleon-antinucleon system with total baryon number zero [197]. The operation G consists of a combination of the operation of charge conjugation and a rotation in isospin space. Specifically, $G = CR$ where C represents the charge conjugation operation and R represents a rotation in isospin space. We choose the operation R to be a rotation by 180 degrees around the y axis in isospin space. If this operation is applied to a state with a z-component of isospin, I_3, it results in changing the value of I_3 to $-I_3$. Thus, for example, if we consider a π^+ meson, its I_3 has the value $+1$ before the rotation, which changes to -1 under the rotation; the charge conjugation operation changes the I_3 value back from -1 to $+1$. This makes it plausible that π^+ may be an eigenstate of G. To derive its eigenvalue, let us look at the particular state with $I = 1, I_3 = 0$. The wave function for this state behaves like the wave function of orbital angular momentum I and z-component zero, $Y_I^0(\theta, \phi)$. The rotation operation of π about the y axis makes $\theta \rightarrow \pi - \theta, \phi \rightarrow \pi - \phi$. Under these transformations of θ and ϕ, $Y_I^0 \rightarrow (-1)^I Y_I^0$. Thus the isospin wave function $\psi(I, 0) \rightarrow (-1)^I \psi(I, 0)$. Because the strong interactions are invariant under rotations in isospin space (a fact which is based on the observed independence of strong interactions on the electric charge

of the interacting particles), this behavior of the isospin wave function must hold also for the states with $I_3 \neq 0$. Thus G operating on a π^+, or π^-, or π^0 state, will give us

$$
\begin{aligned}
G|\pi^{(+,-,0)}\rangle &= C(-1)^1|\pi^{(+,-,0)}\rangle \\
&= (-1)C|\pi^{(+,-,0)}\rangle.
\end{aligned}
$$

We know what the eigenvalue of C in the $|\pi^0\rangle$ state is; it is $+1$ (see section under "Charge Conjugation Operation"). This leads to $G|\pi^0\rangle = (-1)|\pi^0\rangle$. Thus the eigenvalue of G, called the G-parity, is clearly -1 for the neutral pion. For the charged pions, the effect of operating with C may introduce an arbitrary phase. It is convenient, however, to define the phases in such a way that all the charged and the neutral members have the same G-parity, namely (-1). This amounts to defining the phase factor in the C operation on the charged states as leading to (-1). With this choice, the charged as well as the neutral pions have G-parity (-1). From the definition of G-parity, it is also clear why we must have baryon number zero for the system. Under the operation C, the baryon number changes sign, and this would not allow one to define an eigenvalue for G for a system with baryon number nonzero. One can show that the G-parity for a state containing n pions is $(-1)^n$. It can also be shown that the operator G commutes with the isospin operator.

We may consider an application of the concept of G-parity to derive certain selection rules in nucleon-antinucleon annihilation. Consider, for example, the annihilation of antiprotons on neutrons at *rest*. The reaction involves the emission of pions in the final state. We will show that the annihilation cannot give rise to three pions in the final state due to conservation of G-parity, in the reaction

$$
\begin{aligned}
\bar{p} + n \;\; &\rightarrow \;\; \pi^- + \pi^0 \\
&\nrightarrow \;\; 3\pi.
\end{aligned}
$$

To derive this, we first note that the effect of the C operator on a nucleon-antinucleon system with total spin s and orbital angular momentum l is to give a factor of $(-1)^{l+s}$ (refer to section under "Charge Conjugation Operation" and "Positronium"). Thus the G-parity of such a nucleon-antinucleon state is given by $G = (-1)^{l+s+I}$, where I is the total isospin of the nucleon-antinucleon system. The isospin assignments (I, I_3) for the proton and neutron are $(1/2, +1/2)$ and $(1/2, -1/2)$, respectively, while for the antiproton and antineutron they are $(1/2, -1/2)$ and $(1/2, +1/2)$, respectively. The total I_3 value for the antiproton-neutron system is (-1), so it cannot belong to $I = 0$ and must belong to $I = 1$. Thus the G-parity for the antiproton-neutron system is $(-1)^{l+s+1}$. For the two pion annihilation, the G-parity must be $(-1)^2 = +1$, and therefore $(l + s)$ must be odd if G-parity is conserved. For 3-pion annihilation, the G-parity is $(-1)^3 = -1$, and $(l + s)$ must be even.

Consider now the 2-pion annihilation from a singlet spin state $(s = 0)$ of the $(\bar{p}, n)$ system; the G-parity is $(-1)^{l+1}$ and l has to be *odd*. For low values of l, the state on the left-hand side is $^1P_1$, which has a spatial even parity $(+1)$ (remembering antiparticles have opposite intrinsic parity to that of the particle). On the right-hand side the two pions in orbital angular momentum state $l = 1$ have a spatial parity (-1). Thus parity conservation will forbid the $^1P_1$ antiproton-neutron state to go into 2 pions. ($^1S_0$ is not possible because $l = 0$ is even.) Thus for the singlet spin state, no annihilation is possible for $l = 0$ or $l = 1$ states.

Now consider the triplet spin state of the $(\bar{p}, n)$ system. Since $l+s$ has to be odd, this excludes $^3P_{0,1,2}$ states, and only $^3S_1$ state can annihilate into 2 pions. For antiprotons annihilating at rest, the antiproton is preferentially captured from S-states, and therefore, only annihilation from $^3S_1$ state need be considered. By G-parity conservation, this state cannot decay into 3 pions.

Another applications of G-parity selection rules is in the decay of vector meson resonances, ρ (770 MeV), ω (782 MeV), ϕ (1020 MeV), f (1270 MeV). These states decay into pions, and the multiplicity of the decay pions can be derived by the application of $G = (-1)^n$ rule for the n pion state. Thus, the ρ and f mesons, which decay into 2 pions $(G = +1)$, cannot decay into 3 pions, and the ω and ϕ, which decay into 3 pions $(G = -1)$, cannot decay into 2 pions. The η (547 MeV) and η' mesons (both pseudoscalar mesons) whose dominant pion decay modes are 3 pion and 5 pion modes respectively, however, are not strong interaction decays. They decay due to electromagnetic interactions. This is because of the existence of the 2γ decay mode for these mesons, which implies that $C = +1$. Further, since $I = 0$ for these, the G-parity must be $+1$. Strong decay via the 2 pion decay mode is forbidden by parity conservation. Thus η and η' states decay by electromagnetic interactions, which violate G-parity.

$g - 2$ Factor
(See section under "Electron—Anomalous Magnetic Moment, $g - 2$")

GALLEX Experiment
This is a gallium radiochemical solar neutrino detector currently in operation in the Gran Sasso laboratories in Italy. The capture reaction of solar neutrinos in gallium-71 leads to the production of germanium-71,

which are separated by radiochemical means and counted. This reaction measures the most abundant neutrino flux from the Sun, that from the *pp* reaction. The target is 30 tons of gallium in the form of an aqueous solution of gallium trichloride. Shielding for the experiment is provided by the rock overburden to the extent of 3300 meters of water depth equivalent. The efficiency of the radiochemical method has been tested by calibrating with a chromium neutrino source. The expected rate of production of ^{71}Ge induced by solar neutrinos is about one atom per day. Proportional counters are used to detect the K and L x-rays emitted by ^{71}Ge during its decay. The experiment finds a deficiency in the solar neutrino flux nearing a factor of two.

Gamow-Teller Selection Rules
(See section under "Beta Decay—Theory", referring to Gamow-Teller transitions)

Gauge Theories
The impetus for studying gauge theories came from the realization that the principle of gauge invariance can lead to strong restrictions as to the form of interactions between elementary particles. To illustrate this statement, we consider first, in brief, classical electromagnetism, where the idea of gauge invariance originated.

The interactions of charged particles with electromagnetic fields are governed by the laws of electromagentism as embodied in the famous Maxwell's equations. As is well known, these equations describe precisely how the electric and magnetic fields vary in space and time when the sources of these fields, which are charges and currents, are given. In a system of units convenient for studies in theoretical particle physics, we may write Maxwell's equations for the electric ($\vec{E}$) and magnetic ($\vec{B}$) fields in the form

$$\vec{\nabla} \cdot \vec{E} = \rho$$
$$\vec{\nabla} \times \vec{E} = -\frac{\partial \vec{B}}{\partial t}$$
$$\vec{\nabla} \cdot \vec{B} = 0$$
$$\vec{\nabla} \times \vec{B} = \vec{j} + \frac{\partial \vec{E}}{\partial t}$$

where all the quantities, $\vec{E}, \vec{B}, \rho, \vec{j}$, are in general functions of space and time, and ρ and $\vec{j}$ are charge and current densities. An immediate consequence of the fourth equation is conservation of the electromagnetic current:

$$\vec{\nabla} \cdot \vec{j} + \frac{\partial \rho}{\partial t} = 0,$$

which follows from the mathematical identity that the application of divergence operator to a curl operator acting on a vector leads to zero. This conservation law states that if the amount of charge changes in a certain volume V, it must be accounted for by a flow of charge (that is, a current) through the surface bounding the volume V, no matter how small the volume V. Since the volume can be made as small as we like, this conservation law of charge must hold at every space time point, that is, it must hold *locally*. In treating electromagnetic problems in classical (or in quantum) mechanics, it is found convenient to introduce a scalar potential $V(x)$ and a vector potential $\vec{A}(x)$, instead of the electric and magnetic field vectors, through the definitions

$$\vec{B} = \vec{\nabla} \times \vec{A}; \qquad \vec{E} = -\vec{\nabla}V - \frac{\partial \vec{A}}{\partial t}.$$

With these definitions two of the Maxwell's equations are automatically satisfied. The other two can be rewritten in a different form. But before we write those down, we note that the potentials belonging to given $\vec{E}$, $\vec{B}$ are not uniquely defined.

Let us consider transformations $\vec{A} \to \vec{A}' = \vec{A} + \vec{\nabla}\chi$ and $V' = V - \frac{\partial \chi}{\partial t}$, called *gauge transformations*. It is easily seen that the $\vec{A}'$ and V' lead to the same values for $\vec{E}$ and $\vec{B}$. Thus, $\vec{A}'$ and V' can serve as vector and scalar potentials as well as $\vec{A}$ and V.

To incorporate special relativity, we need to treat space and time on the same footing, so we introduce four vector notation. Let $A^\mu(x)$, $\mu = 0, 1, 2, 3$ be a four vector potential with $A^0(x) = V(x)$, $A^i(x)$, $i = 1, 2, 3$ be three components of $\vec{A}(x)$. (We use the metric $g^{00} = -g^{11} = -g^{22} = -g^{33} = 1$). The gauge transformation may then be written as $A^\mu \to A'^\mu = A^\mu - \partial^\mu \chi$, where $\partial^\mu = (\partial^0, -\vec{\nabla})$. If we define the antisymmetric tensor, $F^{\mu\nu} = \partial^\mu A^\nu - \partial^\nu A^\mu$, we see that the components with one of μ or ν equal to zero, coincide with the definition for the electric field vector components, while those components with μ, ν taking on spatial values, $1, 2, 3$, give the magnetic field components. Thus the definition of $F^{\mu\nu}$ incorporates two of the Maxwell's equations. The other two can be compactly rewritten as $\partial_\mu F^{\mu\nu} = j^\nu$ with $j^\nu = (\rho, \vec{j})$. The current conservation equation takes the simple form $\partial_\mu j^\mu = 0$. Thus all the Maxwell's equations can be written as,

$$
\begin{aligned}
F^{\mu\nu} &= \partial^\mu A^\nu - \partial^\nu A^\mu, \\
\partial_\mu F^{\mu\nu} &= j^\nu.
\end{aligned}
$$

The latter equation, in terms of the potentials, takes the form,

$$\Box A^{\nu} - \partial^{\nu}(\partial_{\mu} A^{\mu}) = j^{\nu},$$

where the operator $\Box = \partial^{\mu}\partial_{\mu}$. As a consequence of the last equation, we immediately see that $\partial_{\nu} j^{\nu} = 0$ is automatically satisfied. Thus *local gauge invariance* of Maxwell's equations and *local charge-current conservation* are intimately connected. We further note that, by choosing a suitable function $\chi(x)$, we can make the potential A^{μ} satisfy, $\partial_{\mu} A^{\mu} = 0$, in which case, the last equation takes the simple form

$$\Box A^{\nu} = j^{\nu}.$$

The choice of gauge such that $\partial_{\mu} A^{\mu} = 0$ is called the *Lorentz gauge*. In the Lorentz gauge the equation for the four vector potential takes the simplest form.

Consider the problem of a spin $(1/2)$ particle of mass m and charge q interacting with an electromagnetic field. The Lagrangian for the fields without interaction is

$$\mathcal{L} = -\frac{1}{4}F^{\mu\nu}F_{\mu\nu} + \bar{\psi}(i\gamma^{\mu}\partial_{\mu} - m)\psi.$$

The free particle of spin $(1/2)$ is described by the Dirac equation,

$$(i\gamma^{\mu}\partial_{\mu} - m)\psi = 0.$$

Now we would like to know how the interaction with the electromagnetic field is to be introduced. Here a principle known as the gauge principle comes in handy. The demand that the Dirac equation above be invariant under *local* phase transformations of the wave function ψ of the form, $\psi \to \psi' = e^{iq\chi(x)}\psi$, is the gauge principle. Using this, we transform ψ to ψ' and work out the equation satisfied by ψ'. It is

$$e^{iq\chi(x)}[i\gamma^{\mu}(\partial_{\mu} + iq\partial_{\mu}\chi) - m]\psi' = 0.$$

Clearly the equation is not invariant under this transformation because of the appearance of the $\partial_{\mu}\chi$ term which is nonzero when χ is a function of space and time. This, however, suggests a way to construct an equation which would be invariant under such a transformation. Suppose, instead of the free Dirac equation, we start with the equation

$$[i\gamma^{\mu}(\partial_{\mu} - iqA_{\mu}) - m]\psi = 0,$$

where $A_{\mu}(x)$ is the four vector potential of the electromagnetic field. Now if we demand that the equation be invariant under the phase transformation of the ψ, the equation satisfied by ψ' is

$$[i\gamma^{\mu}(\partial_{\mu} - iq(A_{\mu} - \partial_{\mu}\chi)) - m]\psi' = 0.$$

This equation may be rewritten as

$$[i\gamma^\mu(\partial_\mu - iqA'_\mu) - m]\psi' = 0,$$

where $A'_\mu = A_\mu - \partial_\mu\chi$. We now see that if we absorb the extra term that was introduced by the phase transformation of the ψ as a gauge transformation of the potential $A_\mu(x)$, we have invariance of the equation under the phase transformation. The physics described by the equation for ψ with the original $A_\mu(x)$ is the same as that described by the equation for ψ' with the gauge transformed potential $A'_\mu(x)$.

This simple example tells us that, to get an equation which is invariant under phase transformations of the wave function, we replace the ordinary derivative ∂_μ by $D_\mu = \partial_\mu - iqA_\mu$, called the *covariant derivative*. The local phase transformations of the ψ can be absorbed into local gauge transformations of the A_μ. The gauge principle has dictated the form of the interaction between the charged particle and the electromagnetic field and is, hence, a very powerful principle.

The function $\chi(\vec{x}, t)$ can be viewed as the generator of local gauge transformations of the electromagnetic potentials on the one hand, and on the other, as generator of local phase transformations of the form $e^{iq\chi(\vec{x},t)}$ on the wave function. These phase transformations form a group, called the *group of U_1 transformations*. Since the elements of this group commute with one another, it forms an Abelian group. Thus, electromagnetic theory is known as a U_1 *gauge theory*, or an *Abelian gauge theory*.

One can envisage more complicated gauge transformations. If the particle wave function has two components, as it well might if it has isotopic spin $(1/2)$ for example, the generator of SU_2 phase transformations can be written in the more complicated form, $e^{iq\sum_{i=1}^3 T_i \cdot \chi_i(\vec{x},t)}$, where the T_i are $(1/2)$ times the Pauli matrices $\tau_i, i = 1, 2, 3$. The gauge potentials $W_{i\mu}$ would also be more complicated objects being four vectors in ordinary space-time (index μ) and also have 3 components (index i) in the space of isospin. The covariant derivative now would be $D_\mu = \partial_\mu - iq\sum_{i=1}^3 T_i \cdot W_{i\mu}$. The gauge transformation of $W_{i\mu}$ is more complicated in this case because the Pauli matrices do not commute. It is

$$W_{i\mu}(x) \to W_{i\mu}(x) - \partial_\mu\chi_i(x) - q\sum_{j,k} e_{ijk}\chi_j(x)W_{k\mu}(x),$$

where e_{ijk} is the totally antisymmetric tensor with $e_{123} = 1$.

Such transformations are called *non-Abelian gauge transformations* and the corresponding gauge potentials lead to non-Abelian gauge fields. Such non-Abelian gauge fields are called *Yang-Mills fields* [198]. One can define field strengths $W_i^{\mu\nu} = D^\mu W_i^\nu - D^\nu W_i^\mu$, and the Lagrangian in this case can be written as

$$\mathcal{L} = -\frac{1}{4} \sum_{i=1}^{3} W_i^{\mu\nu} W_{i\mu\nu} + \bar{\psi}(i\gamma^\mu D_\mu - m)\psi.$$

It is possible to go further: if we write the generator of phase transformations in the form $e^{iq \sum_{i=1}^{8}(\lambda_i/2)\chi_i(\vec{x},t)}$, where λ_i are the Gell-Mann matrices, we have SU_3 phase transformations. If we attach an index i to the wave function which can take three possible values called *color* (red, blue, or green), then we have color SU_3 symmetry and we get local color charge conservation. Gauge invariance will need a set of eight non-Abelian gauge potentials which undergo gauge transformations. When quantized, these lead to eight colored gluons, and we have the theory called *Quantum Chromodynamics (QCD)*.

An important point must be noted here. The vector bosons, which one has to introduce according to the gauge principle, have to be massless. Clearly a mass term in the Lagrangian would be of the form $M^2 \sum_i W_i^\mu W_{i\mu}$; it is not invariant under gauge transformations of the W_i^μ.

Gauge Theories—Massive Vector Bosons

The idea of a gauge principle has also been extended to describe the weak interactions of leptons and quarks. In this case, however, the theory is much more subtle, because the gauge invariance turns out to be hidden, unlike the cases of electromagnetism or Yang-Mills fields or QCD, where it is manifest. The story of the explicit construction of a theory of weak interactions based on the gauge principle, where the gauge symmetry is hidden, is a fascinating one and we owe it to the work of Glashow, Salam, and Weinberg (see also the section under "Electroweak Synthesis" for the references to their original work). The most crucial contribution they made was to incorporate a mechanism for the generation of masses for the vector bosons, and still retain many of the features of the gauge principle so that theory is renormalizable. We present below, in brief, some of the general features that such a construction must possess.

In our discussion on the gauge principle, we introduced the SU_2 gauge model. In that model one has to introduce an isospin triplet of non-Abelian gauge fields $W_i^\mu, i = 1, 2, 3$ which undergo gauge transformations. Let us recall that the weak force has been established to be of the $V - A$ form and involves only left-handed fermion currents (see discussions in section under "Beta Decay—Theory"). It is thus clear that, if the weak force is mediated by the exchange of quanta, the exchange quanta have to arise from vector fields. The fact that the effective low energy weak interaction theory involves only left-handed fermion currents is not difficult to build into the gauge theory. In view of these facts, it is natural to attempt to construct a theory of weak interactions also based upon a gauge principle. The SU_2 Yang-Mills gauge model provides such a possibility. Linear combinations of the isotriplet vector fields can be formed,

$$W_\mu^\pm = \frac{1}{\sqrt{2}}(W_{1\mu} \mp W_{2\mu}), \qquad W_\mu^0 = W_\mu^3.$$

The corresponding field operators can lead to annihilation (and creation) of the charged W bosons which has the potential to mediate charged-current weak interactions. A feature of such a gauge approach is that neutral-current reactions of the same strength as the charged-current reactions, mediated by the neutral W_μ^0 partner of the charged W's, must also occur. Since electromagnetism is also based upon a gauge theory, this brings in the hope of unifying electromagnetism and weak interactions into one gauge theory.

The most serious impediment to this program of unification is the fact that weak interactions are of extremely short range and hence must be mediated by only massive vector bosons, while the W's provided by the gauge theory are massless. The mechanism for generating masses for the gauge bosons and the fermions is called the *Higgs mechanism.* The mechanism involves the spontaneous breaking of gauge symmetry. (See further under "Higgs-Kibble Mechanism—Non-Abelian Gauge Theories" and under "Standard Electroweak Model".)

Gauge Theories—Renormalizability

We have mentioned earlier that Quantum Electrodynamics (QED) is based on a gauge field theory. The gauge invariance of the theory guarantees that the mass of the quanta of the field, the photons, is zero. Feynman [193], Schwinger [199], and Tomonaga [200] created the covariant theory of quantum electrodynamics. In constructing this theory, demand is made that the Lagrangian function for the system should be invariant when a local phase transformation of the electron field is carried out. This necessitates that a gauge field be introduced, which is altered in a suitable way, so that the Lagrangian is invariant. In QED the phase factors are elements of the U_1 symmetry group.

The solution of the gauge theory has been possible in perturbation theory which relies on the smallness of the coupling between the charged particles and the photons. Physical quantities are calculated in this theory to various orders of perturbation theory, that is, to given powers of the coupling constant of the theory. Gauge invariance has to be respected at all stages of the calculations. In high orders of perturbation theory, intermediate states occur in which particles can go around in loops and the momenta associated with particles in the loop can take an infinite range of values. This leads to infinities in the theory. A limiting procedure respecting all the symmetries of the theory, including gauge invariance, has to be devised to handle the infinities without introducing any ambiguties. Such limiting procedures are called *regularization*. A theory is said to be *renormalizable*, if these regularized infinite contributions can all be consistently eliminated and absorbed into redefinitions (or renormalizations) of a finite number of properties of the particles, such as mass, charge, renormalization of the wave functions.

Gauge theories possess the vital property of being renormalizable. A renormalizable theory involves only a finite number of such parameters and these are not at present calculable *ab initio*. What is done is to put in the experimentally measured values for these quantities. Once these are put in, the theory can be used to calculate various physical quantities which can then be compared with experimental measurements of those quantities. The agreement between theoretical and experimental values of the anomalous magnetic moments of the electron and the muon, and the Lamb shifts in hydrogen and helium to high precision, is the cornerstone of QED.

One can construct other gauge theories by taking other symmetry groups to which the phase factors can belong and, hence, introduce other gauge fields with more complicated gauge transformations. Examples of such gauge theories has been given under the section on "Gauge Theories". All such theories are renormalizable. G. 't Hooft established that theories, in which spontaneous breaking of gauge symmetry occurs, such as the electroweak theory of Glashow, Salam, and Weinberg, are also renormalizable [201].

Gell-Mann, Nishijima Formula

In the period around 1953, besides the neutron, proton, and pions, a number of new particles were found in experiments. A set of new mesons, now called the *Kaons*, charged and neutral, were found. Also hadrons, heavier than the proton, which decay into protons and pions were found. These were the Λ^0, $\Sigma^{+,-,0}$, $\Xi^{-,0}$. Gell-Mann [202] and independently Nishijima [203] made an effort to classify these particles by extending the notion of charge independence to them. Historically this is the first formula attempting to classify all the known particles at that time and discover any underlying symmetry. The Gell-Mann, Nishijima formula relates the charge of a particle (in units of $|e|$) in terms of the isotopic

Particle	B	I_3	S	Q
P	1	$(1/2)$	0	1
N	1	$-(1/2)$	0	0
Λ^0	1	0	-1	0
Σ^+	1	1	-1	1
Σ^0	1	0	-1	0
Σ^-	1	-1	-1	-1
Ξ^0	1	$(1/2)$	-1	0
Ξ^-	1	$-(1/2)$	-2	-1
π^+	0	1	0	1
π^0	0	0	0	0
π^-	0	-1	0	-1
K^+	0	$(1/2)$	1	1
K^0	0	$-(1/2)$	1	0
$\bar{K}^0$	0	$(1/2)$	-1	0
K^-	0	$-(1/2)$	-1	-1

Table 4.2: Assignment of quantum numbers in Gell-Mann, Nishijima formula.

spin, strangeness and baryon number. It is

$$Q = \frac{1}{2}B + \frac{1}{2}S + I_3,$$

where I_3 is the third component of the isotopic spin, B the baryon number, and S the strangeness. The assignment of these quantum numbers for the particles is given in Table 4.2. Hadrons (strongly interacting particles) heavier than the proton and which decay into protons and mesons are called *Baryons* and have a baryon number B associated with them. Mesons are assigned baryon number zero. The quantum number called *strangeness* was introduced to explain the copious associated production of certain particles and their slow decay violating strangeness. The discovery of more particles and resonances, and the realization that these particles could be grouped in multiplets of eight and ten, led to the assignment of the particles to irreducible representations of the group SU_3. (See section under "Eightfold Way".) The sum $B + S$ has come to be called the *hypercharge Y*.

GIM mechanism
In an important work, the idea of introducing symmetry between leptons and quarks in weak interactions of hadrons was explored by Glashow,

Iliopoulos, and Maiani (GIM) [204]. The motivation for introducing this symmetry was as follows. It was known at this time that the lepton family consisted of the electron and its neutrino, and the muon and its neutrino. On the quark side, only three quarks were introduced, namely, the u, d, and s quarks, which results in an asymmetry in the number of leptons and quarks. To rectify this, Glashow et al. [204] proposed the introduction of a fourth quark, now called the *charm quark*. They proposed a model of weak interactions in which the $V - A$ currents constructed out of the four basic quark fields interact with charged vector boson fields (now known to be the $W^{\pm}$) and the neutral vector boson field (now known to be the Z^0). They showed that such a theory leads to no violations of strong interaction symmetries. They extended the theory to the case of leptons and quarks having Yang-Mills symmetry. (The idea of introducing a fourth quark, which they called *charm quark* had already been explored in a paper by Bjorken and Glashow in 1964 [102]; see details under "Charm Particles—Charmonium".)

The mechanism proposed in reference [204] provides for the absence of *flavor-changing neutral currents*, in accord with experimental findings. Considering just two families of quarks, (u, d) and (c, s), the d and s quarks undergo Cabibbo mixing and become $d' = d \cos \theta + s \sin \theta$, and $s' = -d \sin \theta + s \cos \theta$, respectively. This mixing of the families is necessary to account for $\Delta S = \pm 1$ charged current decays. The 2 x 2 mixing matrix is unitary, which guarantees that the weak interactions are universal (see "Cabibbo Angle"). The charged current $V - A$ structure involves the sum of $\bar{u}\gamma^{\mu}(1 - \gamma_5)d'$ and $\bar{c}\gamma^{\mu}(1 - \gamma_5)s'$ over the two families. The neutral $V - A$ current involves the sums:

$$\bar{u}\gamma^{\mu}(1 - \gamma_5)u + \bar{c}\gamma^{\mu}(1 - \gamma_5)c$$

and

$$\bar{d}'\gamma^{\mu}(1 - \gamma_5)d' + \bar{s}'\gamma^{\mu}(1 - \gamma_5)s'.$$

In the latter sum we see that the cross term involving the d and s quarks has a zero coefficient. Thus, with Cabibbo mixing among the d and s quarks, the *neutral current* has no flavor-changing elements in it.

The GIM mechanism has also been extended to the case of three families of quarks, involving mixing among d, s, and b quarks. It leads to the general absence of flavor-changing neutral currents in the weak decays of hadrons.

Gluino

A hypothetical particle of spin 1/2, which is the supersymmetric partner of the spin 1 gluon. Because the gluons form a color octet, the gluino likewise must be a color octet. (See under "Supersymmetry".)

Gluon

In the section on gauge theories, we have seen that if we demand invariance under local phase transformations which are elements of SU_3 symmetry, we get non-Abelian gauge fields. The SU_3 symmetry is associated with color, which can take on one of three possible values, red, blue, and green. The non-Abelian gauge fields carry color and mediate interaction between particles which carry color. The structure of the color fields is similar to that of the electromagnetic field, so they are called *chromo-electric fields* and *chromo-magnetic fields*. Quantization of such a gauge theory gives rise to quanta of the color electric and magnetic fields which are called *gluons*. The fermions of the theory which carry color charges are called *quarks*. Each flavor of quark, u or d or s, etc., comes in three colors. The strong interactions between colored quarks are mediated by the exchange of colored gluons in analogy with the interaction between electrically charged particles being mediated by the exchange of photons. Details of the dynamics of quarks and gluons have been constructed in analogy with QED and the theory is called *Quantum Chromodynamics (QCD)*. The fact that gluons carry a color charge, unlike the photon which is electrically neutral, leads to the possibility of interactions among gluons. This interaction is of the same strength as the interaction between a quark and a gluon, and has to be included at a fundamental level. When the full calculations are carried out, it is found that the color interaction between quarks becomes weaker at short distances and stronger at long distances. Thus at high energies, which corresponds to probing short distance scales, the interactions become weak. One can use perturbation theory in this regime. Correspondingly, the theory involves strong coupling at long distance scales or at low energies, and perturbation theory is not valid. One of the features that is conjectured to come out of such non-perturbative region is the confinement of color; that is, free quarks or free gluons are not observable. They are permanently confined in hadrons. Hadrons are viewed as bound states of quarks, and require for their solution, non-perturbative methods because of the strong coupling. Despite these limitations, perturbative QCD calculations are applicable in a number of phenomena involving hadrons and impressive successes do exist which put perturbative QCD on a firm footing.

Gluons like photons are massless quanta and carry a spin of 1. Direct observation of gluons is from the production of gluon jets in e^+e^- annihilations. As mentioned above, because of color confinement, one cannot see free quarks or gluons in electron-positron annihilations. The primary quarks and gluons which are produced in the annihilation rapidly materialize into colorless hadrons, and it is the jets of the hadrons which are seen in experiments.

First evidence of quark jets in e^+e^- annihilations came from the work of Hanson et al. [31]. Working at center of mass energies of 6.2 and 7.4 GeV, they found the hadrons produced in the reaction had a two jet structure. They determined the jet axis angular distribution integrated over azimuthal angles. It was found to be proportional to $(1 + a\cos^2\theta)$ with $a = 0.78 \pm 0.12$. The angular distribution is consistent with that expected from two quarks being initially produced in the annihilation process, $e^+e^- \to q\bar{q}$.

First evidence for the gluon jet came from e^+e^- annihilation into a three jet structure of hadrons [205]. The process that they observed was $e^+e^- \to q\bar{q}g$, where g is a hard gluon emitted in a non-collinear direction with respect to the quarks. The data were completely consistent with predictions from QCD for the emission of non-collinear hard gluon from either the quark or the antiquark arising from the annihilation process. Such three jet e^+e^- annihilation events afford the possibility to determine the spin of the gluon. The first experimental determination of the spin of the gluon using three jet events was carried out in the TASSO detector at PETRA [206]. The experiment analyzed the angular correlations between the three jet axes from the experiment. When compared with theoretical expectations based on QCD, they found that the data were in agreement with theory when they assume vector (spin 1) gluons, while scalar (spin 0) gluons were disfavored by about four standard deviations. Confirmation of these results came from the independent work of Berger et al., of PLUTO collaboration [207]. This collaboration also found that the scalar gluon was strongly disfavored by their data.

Gluon Distribution in the Proton

Deep inelastic scattering of electrons on protons has revealed the quark structure of the proton. The measured structure functions give information on the momentum distributions of the quarks in the proton. Since in the parton model, the quarks carry a fraction x of the proton's momentum, one would expect the momentum sum-rule $\sum_i \int_0^1 dx q_i(x) x = 1$ to be valid, if the proton contained nothing but quarks, where $q_i(x)$ is the quark distribution function for a quark of flavor i. When the sum-rule was evaluated using experimental data on the quark distribution functions extracted from deep inelastic scattering, it did not give the value 1 but only half. This established that there were other partons, electrically neutral, which contribute to the momentum sum-rule. This neutral component is the gluon associated with the quanta of the chromdynamic fields. It carries the remaining fraction of the momentum of the proton. It is invisible in the low to moderate q^2 deep inelastic electron proton scattering experiments.

The gluon component in the proton can manifest through the process of photon-gluon fusion leading to the production of the J/ψ in the deep inelastic scattering experiment. The virtual photon may dissociate into a $c\bar{c}$ (charm-anticharm) pair and interact with two gluons from the proton, eventually forming the J/ψ, a color singlet bound state. When the process of J/ψ production is calculated, it will depend upon the gluon distribution in the proton. Thus the measurement of the J/ψ production gives information on the gluon distribution in the proton. The process of charm meson production, in which the produced $c\bar{c}$ do not form J/ψ, but where each c and $\bar{c}$ pick up $\bar{d}$ and d quarks, respectively, from the vacuum and form D mesons, can also give information on the gluon distribution in the proton.

Goldberger-Treiman Relation

This relates the axial vector coupling constant in β decay with the pion decay constant and the pion-nucleon coupling constant. It was first derived by Goldberger and Treiman using dispersion relation techniques [208]. We have seen that the vector current in β decay is conserved.

What can we say about the conservation of the axial vector current? That it cannot be exactly conserved is seen from the following argument. The exact conservation implies a symmetry called *chiral symmetry*. Through the *Goldstone theorem*, spontaneous breaking of chiral symmetry leads to spin 0 massless particles corresponding to each of the four axial vector currents. These four axial vector currents are one isosinglet current $J^{\mu(A)}$ and three isovector currents $J^{\mu i(A)}, i = 1, 2, 3$ (superscript A for axial vector). Nature does not seem to have four massless pseudoscalars which participate in strong interactions but it does have an isotriplet of pions which have a very low mass. Let us treat these as the Goldstone bosons of spontaneous chiral symmetry breaking and write the matrix element of the $J^{\mu i(A)}$ between the vacuum and a one pion state as

$$\langle 0|J^{\mu i(A)}(x)|\pi^j(p)\rangle = -iq^\mu F_\pi \delta_{ij} e^{-iq\cdot x}.$$

Here i, j stand for isospin indices and F_π is a constant having the dimensions of mass and called the *pion decay constant*. Its value can be determined by using the above matrix element to calculate the rate for pion decay and fit it to the experimental value. The value of F_π comes out to be 95 MeV. If we take the four-divergence of the above current, we should get zero for a conserved current. On the right-hand side we get q^2 which is m_π^2; that is, the pion must be massless (Goldstone boson).

Using pions as the Goldstone bosons of chiral symmetry breaking, we can derive implications for hadronic matrix elements involved in certain processes [98]. Let us consider the β decay of the neutron, and in particular, the matrix element of the axial vector isospin current of the nucleon N. The most general form it can have involves three form factors:

$$\langle p|J^{\mu(A)}(q)|n\rangle \;\; = \;\; \bar{u}_p \left[\gamma^\mu \gamma^5 F_1^A(q^2) + q^\mu \gamma^5 F_2^A(q^2) \right.$$
$$+ \;\; \left. i\sigma^{\mu\nu} q_\nu F_3^A(q^2) \right] u_n.$$

Unlike the case of the conserved vector current, the value of F_1^A at zero momentum transfer is renormalized and we want to know what it is. Let us write $F_1^A(0) = g_A$. If we take the divergence of the above axial current, it amounts to multiplying the matrix element with q_μ, and conservation of the axial current constrains the form factors to satisfy

$$\bar{u}_p(p')[-2m_N F_1^A(q^2) + q^2 F_2^A(q^2)]\gamma^5 u_n(p) = 0.$$

Thus

$$g_A = \lim_{q^2 \to 0} \frac{q^2}{2m_N} F_2^A(q^2).$$

From this it is clear that g_A will be zero unless $F_2^A(q^2)$ develops a pole at $q^2 = 0$. Now we must remember that we do have a massless particle, the pion, which can play a role here. The π^- emitted by the neutron could disappear into the the vacuum. This could lead to a pole at $q^2 = 0$. To evaluate this contribution we note that the pion-nucleon interaction Lagrangian can be written as

$$\mathcal{L}_{\pi NN} = g_{\pi NN} \pi^i \bar{N} \gamma^5 2T^i N,$$

where $2T^i = \tau_i$, $i = 1, 2, 3$ are the Pauli isospin matrices, and $g_{\pi NN}$ is the pion-nucleon coupling constant. From this the matrix element for the neutron to emit a π^- (and change into a proton), and for the pion to propagate with a four-momentum q and disappear into the vacuum is

$$g_{\pi NN} 2\bar{u}_p \gamma^5 u_n \frac{i}{q^2} (-iq^\mu F_\pi),$$

where the last factor $(-iq^\mu F_\pi)$ represents the pion disappearing into the vacuum (pion decay). The coefficient of $q^\mu \bar{u}_p \gamma^5 u_n$ is to be identified with $F_2^A(q^2)$ occurring in the matrix element of the axial vector current between nucleon states given above. From this we can read off the contribution to $F_2^A(q^2)$; it is

$$F_2^A(q^2) = \frac{2}{q^2} F_\pi g_{\pi NN}.$$

Using this we get for g_A the relation,

$$g_A = \frac{F_\pi}{m_N} g_{\pi NN}.$$

This is the Goldberger-Treiman relation. Putting in the value of $g_{\pi NN} = 13.3$, obtained from nucleon-nucleon scattering at low energies, and $F_\pi = 95$ MeV, we find for g_A the result 1.3. This compares favourably with the experimental determination of the ratio $g_A/g_V \simeq 1.25$. The agreement is to within 5%. The use of the idea of pion as the Goldstone boson has other successes in describing the behavior of pion-pion and pion-nucleon scattering amplitudes at low energies.

Goldstone Theorem

This theorem relates to the appearance of massless spinless particles as a result of the spontaneous breaking of some continuous symmetry [188]. We present a derivation of this theorem [98]. Let us consider a field theory involving a number of fields, $\phi^i, i = 1, 2, \ldots, N$. Let the Lagrangian be

$$\mathcal{L} = (\text{kinetic energy terms} - V(\phi),$$

where the kinetic energy terms are those which involve derivatives of the fields, and $V(\phi)$ stands for the potential energy terms. Let the minimum of V as a function the ϕ's occur at some nonzero value of the fields ϕ_0^i so that the first derivative of the V vanishes at $\phi^i = \phi_0^i$. Expanding the V about this minimum, we get

$$V(\phi) = V(\phi_0) + \frac{1}{2}(\phi^i - \phi_0^i)(\phi^j - \phi_0^j)\frac{\partial^2 V}{\partial \phi^i \partial \phi^j}\bigg|_{\phi_0} + \cdots$$

The coefficient of the quadratic terms in the expansion, which involve the second derivative of the potential evaluated at the minimum ϕ_0, is a symmetric matrix. Its eigenvalues give the square of the masses of the fields, which have to be positive because the minimum of the potential is positive. If we can show that any continuous symmetry of the Lagrangian which is not respected by ϕ_0 gives eigenvalue zero for the mass matrix, then we would have established Goldstone's theorem. To do this let us consider an infinitesimal transformation of the ϕ^i,

$$\phi^i \rightarrow \phi^i + \epsilon \delta^i(\phi),$$

where ϵ is an infinitesimal parameter and $\delta^i(\phi)$ is some function of the fields. If we consider the special case of constant fields, the kinetic energy terms do not give any contribution, then the invariance of the

Lagrangian implies that the potential energy terms have to be invariant under the above transformation. Thus we must have

$$V(\phi^i) = V(\phi^i + \epsilon\delta^i(\phi)).$$

This is satisfied if

$$\delta^i(\phi)\frac{\partial V(\phi)}{\partial \phi^i} = 0.$$

Let us differentiate this expression with respect to ϕ^j and set $\phi = \phi_0$. Then we get

$$\frac{\partial}{\partial \phi_j}\left[\delta^i(\phi)\frac{\partial V(\phi)}{\partial \phi^i}\right] = \left.\frac{\partial \delta^i}{\partial \phi^j}\right|_{\phi_0} \cdot \left.\frac{\partial V}{\partial \phi^i}\right|_{\phi_0} + \delta^i(\phi_0)\left.\left(\frac{\partial^2 V}{\partial \phi^i \partial \phi^j}\right)\right|_{\phi_0}.$$

Of these the first term already vanishes because we are at the minimum of V, so the second term has to vanish by itself. If the ground state ϕ_0 is invariant, then $\delta^i(\phi_0) = 0$ and the relation is satisfied for any value of the mass matrix term. The $\delta^i(\phi_0) \neq 0$ when the ground state is not invariant under the symmetry transformation, and in that case, this will be associated with eigenvalue zero of the mass matrix. Thus one has a massless boson associated with a spontaneously broken symmetry and the Goldstone theorem is established.

Grand Unified Theories (GUTS)
The idea behind grand unification is that the Standard Model group $SU_3(color) \times SU_2 \times U_1$ is a subgroup of some larger gauge group G with quarks and leptons belonging to the same multiplets of G. It is assumed that the symmetry represented by G is exact above some mass scale M_X, called the *grand unification mass*, and broken below that scale. The gauge couplings g_1, g_2, and g_3 belonging, respectively, to the U_1, SU_2 and $SU_3(color)$ factors are related to the single gauge coupling g_G through known G-dependent factors above the mass scale M_X. Above the mass scale M_X, the renormalization group equations with G as the gauge group determine the evolution of g_G. Below the mass scale M_X, the group G is assumed to be broken spontaneously, and the three couplings evolve according to their individual renormalization group equations. A good test as to whether grand unification works is to take the known values of the couplings at the weak scale, of the order of the W-boson mass, evolve them to the mass scale M_X, and see whether they meet there.

An early attempt at grand unification was that based on the group G=SU_5 [209]. The first generation leptons and quarks were assigned to two different representations, the 5* and the 10, the 5* containing

(d^c, e^-, ν_e) and the 10 containing (e^c, d, u, u^c), where the superscript c represents the conjugate. One of the immediate consequences of putting leptons and quarks in the same representation is that transformations among them are possible, which leads to the prediction of proton decay, through one of its quarks transforming into a lepton by virtual emission or absorption of a heavy gauge boson X. The calculated rate for $p \to e^+ \pi^0$ is in the range $2 \times 10^{29 \pm 1.7}$ years for a heavy gauge boson mass of the order 10^{14} GeV. Experimental measurements for this decay mode place a lower bound of 10^{32} years. Thus SU_5 may not be the unifying grand gauge group. The fact that the assumption of a large unification mass scale, and the evolution of couplings to that scale, predicts a rate for proton decay which is close to the experimental bound makes the idea of GUTS schemes extremely interesting. Unifying groups other than SU_5 have also been explored. Pati and Salam [210], in 1973, proposed a gauge group based on a left-right symmetric $SU_2 \times SU_2 \times U_1$ electroweak interaction. In this theory, parity violation naturally arises as a result of the spontaneous breaking of symmetry, and "lepton number" was identified with a "fourth color". Another group which was actively investigated was SO_{10} [211]. The nice thing found about this group is that $SU_5 \times U_1$ is one of its subgroups. The general problem of embedding $SU_3 \times SU_2 \times U_1$ in a larger grand unifying group was studied by Gell-Mann, Ramond, and Slansky in 1978 [212]. There is much literature on the subject of grand unified theories in the period 1974 to 1982. We do not go into details here; interested readers may wish to refer to the original literature on the subject.

Hadronic Collisions

Hadrons are strongly interacting particles such as the proton, neutron, pion, kaon, and Σ. Accelerators produce beams of these particles to study collision processes involving them. Such collisions involving strongly interacting particles come under the category of hadronic collisions. The size of the cross section for strongly interacting particles is of the order of the square of the radius of a typical hadron (10^{-13}cm) and is typically in the milli-barn range (10^{-27}cm^2). At high energies many inelastic processes occur and the total cross section which takes account of all these processes increases. However, this increase of the total cross section with energy has to respect *unitarity*, which is essentially a statement of the conservation of probability. Theoretically, Froissart derived a unitarity bound for the growth of this cross section, known as the *Froissart bound*. If s is the square of the center of mass energy of the colliding particles, Froissart bound says that the cross section cannot grow faster than $\ln^2 s$. (See more in the section under "Froissart Bound".)

Hadronic Shower
In high energy collisions, the resulting product is a jet of hadrons, called
a *hadronic jet* or a *hadronic shower*. The reaction that is studied involves
a high energy hadron impinging on a nucleus leading to the production
of many hadrons. In the early studies, nuclear emulsions were utilized to
study collisions of cosmic ray particles with the atoms of the emulsion;
the nuclear emulsions served both as the target and detector. The characteristics of secondaries produced in such collisions were studied by a
number of different nuclear emulsion groups [213]. The specific focus in
the experimental studies was on the distribution of the transverse momentum components of the secondaries. Theoretically, statistical and
hydrodynamic models were developed by Fermi and Landau [214] to describe the multiple production of particles at high energies. They found
evidence for limited transverse momentum of secondary particles produced in such showers. Most of the secondary particles (about 80%)
were found to be pions; a few heavy particles were also produced. The
average multiplicity and the average transverse momentum of the pions
and the heavy particles produced were measured. It was found that both
these quantities were practically independent of the primary energy with
the average transverse momenta being limited to around several hundred
MeV/c. The results were in qualitative agreement with the statistical
theories of the multiparticle production at high energies.

With the development of high energy accelerators and colliders, much
more data are available than from cosmic ray studies. An example was
the intersecting storage ring at CERN (ISR; now no longer in operation), which consisted of two counter rotating proton beams each of
30 GeV and intersecting at eight points where the collisions were studied. The center of mass energy is 60 GeV, which would correspond to
an incident proton energy of about 2,000 GeV on a fixed target. Hagedorn [215] formulated and studied in detail a thermodynamical model
of secondary particle production and produced an atlas of momentum
spectra of secondary particle production in proton-proton and proton-
nucleus collisions. He made efforts to take into account the dynamics
of the interactions of the particles. Comparison with experimental data
showed good agreement with the thermodynamical model.

Hadrons—Constituent Quark Model
See discussion under "Constituent Quark Model of Hadrons".

Hadrons—SU_6 Classification
The idea of incorporating the spin of the quarks in an SU_3 symmetric
framework led to the exploration of higher symmetry groups, notably

the SU_6 symmetry. This was first proposed by Gürsey and Radicati in 1964 [216] and also by Sakita [217]. They were motivated by the following considerations. In nuclear physics, Wigner had considered embedding isospin and spin, both of which belong to SU_2, in a larger group SU_4 and classified nuclear states according to the irreducible representations of SU_4. We have seen elsewhere that, on introducing strangeness along with isospin, it is necessary to describe the particles as SU_3 mulitplets. Analogous to Wigner's approach for nuclear states with SU_4, when the spin and SU_3 symmetry are taken together, the enlarged group of interest in particle physics is SU_6. When particle multiplets are classified based on representations of SU_6, some very interesting results are found.

Three quark bound states are of relevance in the baryonic sector. In terms of representations of SU_6, one has to study the decomposition of the product $6 \times 6 \times 6$ into irreducible representations of dimensions, 20, (two) 70's, and 56. Of these, the 56-dimensional representation is completely symmetric in all the quantum numbers and is capable of accommodating the $SU_3(\frac{1}{2})^+$ baryon octet and the $(\frac{3}{2})^+$ baryon decuplet in it. In the meson sector, made up of quark-antiquark bound states, $6 \times 6^*$ can be decomposed into irreducible representations of dimension 35 and a singlet. The 35 states split into a pseudoscalar meson octet and a degenerate vector meson nonet. These are capable of accommodating the (π, K, η) in the octet, and the nonet of vector mesons with negative parity, $(\rho, \omega, K^*, \phi)$. The SU_6 symmetry gives a natural explanation of the degeneracy, of the octet and the singlet states, in the nonet. Mass formulae also have been derived. There is an equal spacing result for vector mesons,

$$M_\phi^2 - M_{K^*}^2 = M_{K^*}^2 - M_\rho^2; \qquad M_\rho^2 = M_\omega^2.$$

Since the mass splittings of the octet of pseudoscalar mesons and the octet of vector mesons are related here, we have also

$$M_{K^*}^2 - M_\rho^2 = M_K^2 - M_\pi^2.$$

One also gets results for baryon magnetic moments which is amazingly good. If we take the proton magnetic moment as the unit, we get the neutron magnetic moment to be $-2/3$, for Λ^0 the value $-1/3$, and for Σ^+ the value $+1$, etc. The ratio of the neutron to the proton magnetic moment is a spectacular success of the SU_6 symmetry.

Hadrons—Weak Interactions

Hadrons are observed to decay through weak interactions. The hadronic currents involved in weak decays in the $V - A$ theory of weak interactions were studied by Cabibbo and he extended the SU_3 symmetry to these currents [95]. (For details, see discussion under "Cabibbo Angle" and under "Hyperons—Decay Selection Rules".)

HERA—The Electron-Proton Collider

This is a high energy electron-proton collider operating at the DESY laboratories in Hamburg, Germany. The collider has 30 GeV electrons colliding against 920 GeV protons (in its upgraded form). There are two detectors which have been collecting data from these collisions, called *H1* and *ZEUS*. Deep inelastic scattering of the electrons on protons give information about the sub structure of the proton in regions of high values of the square of the four-momentum transfer $q^2 = -Q^2$, not reached by any other collider so far. It also probes regions of much smaller values of the Bjorken scaling variable x (see further under "Bjorken Scaling"). The highest Q^2 value reached is 40,000 GeV^2, and the smallest x value reached is 0.000032. At these values of Q^2, the scattering should be analyzed using the full electroweak model, involving not only the virtual photons but also the W bosons and the Z^0 bosons. The deep inelastic scattering cross section has three form factors, F_1, F_2, and F_3, just as is the case for deep inelastic scattering induced by neutrinos. The data obtained by H1 and ZEUS detectors have advanced the knowledge of the deep inelastic structure functions in regions not previously probed. These data have given further valuable information on the quark and gluon distribution functions in regions of very high Q^2 and very low x. The data have provided further tests of QCD. They also have provided information on the structure function for the photon and, through studies of J/ψ production, a very good measure of the gluon distribution function in the region of very small x. They also search for new particles and new currents, and look for deviations from the standard model. For details we refer to some of the papers containing the measurements of the structure functions [218].

Higgs Mechanism—Abelian Gauge Theories

The Lagrangians for the field theories possess certain invariance properties, such as translational, rotational, gauge invariance. Generally, the ground state of the field theory, called the *vacuum state*, shares the symmetry property of the Lagrangian. However, it is known that this is not always the case as is seen from the existence of phenomena such as ferromagnetism, where the ground state of the theory does not possess the rotational symmetry of the Lagrangian or the Hamiltonian. The system of atoms with spins interact with one another described by a Hamiltonian which involves the scalar product of the spin vectors of the particles involved and is, hence, a rotational invariant. When one takes the material to a temperature below the Curie temperature, the system acquires a spontaneous magnetization in some direction, and hence the ground state (or the vacuum state) of the system does not have rotational symmetry. The direction in which the spins align themselves spontaneously is, of course, random. These different directions of alignment correspond to different degenerate vacuum states. There is no energy cost involved in going from one direction of orientation to another because of the degeneracy of the vacuum states. We may say that we have spontaneous breaking of rotational symmetry which has occurred.

An example of a Lagrangian field theory in which the symmetry property of the Lagrangian is not possessed by the vacuum state of the theory involves a complex scalar field ϕ with self-coupling of the field of the form $\lambda\phi^4$. This Lagrangian is

$$\mathcal{L} = \partial_\mu\phi\partial^\mu\phi^* - m^2\phi^*\phi - \lambda(\phi^*\phi)^2.$$

This would be the Lagrangian for the scalar field with a mass term m^2 but for one difference. Here we would like the sign of the m^2 term to be positive or negative. So in this Lagrangian, we do not interpret m as the mass associated with the field; instead, we consider it as a parameter in the potential term: $V(\phi, \phi^*) = m^2\phi^*\phi + \lambda(\phi^*\phi)^2$. The field ϕ has zero mass in our model Lagrangian.

Let us demand that the Lagrangian above be invariant under the local phase transformations of the ϕ,

$$\phi \to \phi' = e^{iq\chi(x)}\phi.$$

This necessitates the introduction of a covariant derivative instead of the ordinary derivative, and a gauge field, $A^\mu(x)$,

$$\partial^\mu \to D^\mu = (\partial^\mu + iqA^\mu),$$

and the Lagrangian which is invariant is

$$\mathcal{L} = (D^\mu\phi)(D_\mu\phi)^* - V(\phi, \phi^*) - \frac{1}{4}F^{\mu\nu}F_{\mu\nu}.$$

Here we have added terms which contain derivatives of the gauge field, $F^{\mu\nu} = \partial^\mu A^\nu - \partial^\nu A^\mu$ so that they can play a dynamical role. This Lagrangian is clearly invariant under the phase transformation of ϕ, because it amounts to a gauge transformation of the A^μ, $A^\mu \to A^{\mu'} = A^\mu + \partial^\mu\chi(x)$.

Now let us pay attention to the ground state of the model. Classically, it is obtained by minimizing the potential function,

$$\frac{\partial V}{\partial\phi} = 0 = m^2\phi^* + 2\lambda(\phi^*\phi)\phi^*.$$

If $m^2 > 0$, this expression is zero only for $\phi = 0$. This is what happens when one interprets the m^2 term as the mass term for the ϕ field. However, if $m^2 < 0$, we can solve for $|\phi|^2 = (\phi^* \phi)$, and the zero occurs at

$$|\phi|^2 = -\frac{m^2}{2\lambda}.$$

Let us call the right-hand side $a^2/2$; this says that the minimum of the potential occurs at $|\phi| = a/\sqrt{2}$. If we treat ϕ as a quantum field, the field ϕ becomes an operator, and the condition for the minimum of the potential holds for the vacuum expectation value of the operator ϕ,

$$|\langle 0|\phi|0\rangle|^2 = a^2/2.$$

We now parametrize the quantum $\phi(x)$ field in a different way. Being a complex field, it can be written as

$$\phi(x) = \rho_1(x) e^{i\theta_1(x)}$$

where ρ_1 and θ_1 are two real fields. For the vacuum expectation value of these fields, it is clear that we must have $|\langle 0|\rho_1(x)|0\rangle| = a/\sqrt{2}$ and $|\langle 0|\theta_1|0\rangle| = 0$. We expand ϕ about the minimum and write $\rho_1 = (a + \rho)/\sqrt{2}$, $\theta_1 = \theta/a$, so that

$$\phi(x) = \frac{1}{\sqrt{2}}(a + \rho(x)) e^{i\theta(x)/a}.$$

Substituting this in the above Lagrangian we may rewrite it as

$$\begin{aligned}
\mathcal{L} &= -\frac{1}{4}F^{\mu\nu}F_{\mu\nu} + (1/2)q^2 a^2 A'^{\mu} A'_{\mu} + (1/2)(\partial^{\mu}\rho)^2 - \lambda a^2 \rho^2 \\
&+ \text{coupling terms,}
\end{aligned}$$

where $A'^{\mu} = A^{\mu} + (1/a)\partial^{\mu}\theta$ is the gauge transformed vector potential. We see from this result that the gauge field with spin 1 and the scalar field ρ have acquired masses qa and $2\lambda a^2$, respectively, and the θ field has disappeared altogether. This mechanism of generation of mass is called the *Higgs mechanism* [219] and our gauge field has acquired a mass. In this model calculation, we chose Abelian gauge symmetry. The Glashow, Weinberg, and Salam work on electroweak unification involves a different choice of gauge symmetries, namely non-Abelian gauge symmetry, and a doublet structure for the scalar fields. But the essential feature of the generation of masses for the gauge bosons is based upon the Higgs mechanism for the scalar fields. It can be so arranged that, apart from the charged W fields and the neutral field Z^0 acquiring masses, the photon can be left massless. Reference may be made to the section under "Standard Electroweak Model" for details of these procedures.

Higgs-Kibble Mechanism—Non-Abelian Gauge Theories

Kibble [220] extended the Higgs mechanism of mass generation for non-Abelian gauge theories, hence it is called the *Higgs-Kibble mechanism*. We give here the essential features of generation of mass in a non-Abelian gauge theory.

Let us consider the SU_2 gauge group with generators denoted by $T_i, i = 1, 2, 3$, satisfying the commutation relations,

$$[T_i, T_j] = ie_{ijk}T_k,$$

where e_{ijk} are the structure constants of the group totally antisymmetric in its indices, with $e_{123} = 1$. We introduce a complex doublet of scalar fields,

$$\phi = \begin{pmatrix} \phi_1 \\ \phi_2 \end{pmatrix},$$

and introduce a potential function $V(\phi)$ which depends on ϕ_i's through combinations which are invariants of the gauge group $V(\phi) = m^2(\phi^\dagger\phi) + \lambda(\phi^\dagger\phi)^2$.

To discuss spontaneous breaking of the gauge symmetry, we choose the potential and parameters in it such that its minimum occurs for nonvanishing expectation value of $\langle 0|\phi|0\rangle = v \neq 0$. Then the local symmetry will be spontaneously broken. If we start with the Lagrangian

$$\mathcal{L} = D^\mu\phi^\dagger D_\mu\phi - V(\phi) - \frac{1}{4}F_i^{\mu\nu}F_{\mu\nu i},$$

where $D_\mu = \partial_\mu + ig\sum_{i=1}^{3} T_iA_{\mu i}$ is the covariant derivative and $F_{\mu\nu i} = \partial_\mu A_{\nu i} - \partial_\nu A_{\mu i} + ge_{ijk}A_{\mu j}A_{\nu k}$ is the field strength tensor, we see that it is invariant under local SU_2 transformations of the ϕ because it simply amounts to a gauge transformation of the gauge potentials as in the Abelian case treated before. We are concerned with spontaneous symmetry breaking. For $m^2 < 0$ the potential has a minimum at $\langle 0|\phi^\dagger\phi|0\rangle = a^2, a = \sqrt{-m^2/(2\lambda)}$. We choose the physical vacuum with the expectation value of ϕ nonzero. Just as in the Abelian case, we rewrite the field $\phi(x)$ as

$$\phi(x) = exp[i\vec{T}\cdot\vec{\xi}(x)]\begin{pmatrix} 0 \\ (v_1(x) + v_2(x))/\sqrt{2} \end{pmatrix},$$

with $\langle 0|\vec{\xi}(x)|0\rangle = 0, \langle 0|v_2(x)|0\rangle = 0, \langle 0|v_1(x)|0 = a$, choosing the gauge called the *unitary gauge*. Let us define a new transformed field $\phi^u(x)$,

$$\phi^u(x) = U(x)\phi(x) = \begin{pmatrix} 0 \\ \frac{1}{\sqrt{2}}(v_1(x) + v_2(x)) \end{pmatrix},$$

where $U(x) = exp[-i\vec{T} \cdot \vec{\xi}(x)]$. In terms of this new field, we can write the covariant derivative as: $UD_\mu \phi = UD_\mu U^{-1} \phi^u = D^u_\mu \phi^u$. Putting in $D_\mu = \partial_\mu + ig\vec{T} \cdot \vec{A}_\mu$, we see that D^u_μ works out to be $D^u_\mu = \partial_\mu + ig\vec{T} \cdot \vec{B}_\mu$, where $\vec{B}_\mu$ is a new gauge potential related to the old potential $\vec{A}_\mu$ by

$$\vec{T} \cdot \vec{B}_\mu = U(x)\vec{T} \cdot \vec{A}_\mu U^{-1}(x) - \frac{i}{g}U(x)\partial_\mu U^{-1}(x).$$

Thus the effect of removing the phase functions $\vec{\xi}(x)$ from ϕ manifests as a gauge transformation on the $\vec{A}_\mu$. With these new gauge potentials, we can define new field strengths, which take the form

$$G_i^{\mu\nu} = \partial^\mu B_i^\nu - \partial_\nu B_i^\mu + ge_{ijk}B_j^\mu B_k^\nu.$$

The Lagrangian in the unitary gauge now becomes

$$\mathcal{L} = (D^u_\mu \phi^u)^\dagger (D^{\mu u}\phi^u) - V(\phi^u) - \frac{1}{4}G_{\mu\nu i}G_i^{\mu\nu}.$$

The gauge boson described by the potentials $B_{\mu i}$ have acquired a mass as can be seen from the quadratic terms in the potentials B contained in $|D^u_\mu \phi^u|^2$,

$$\frac{g^2}{2}(0 \quad a)(\vec{T} \cdot \vec{B}^\mu)(\vec{T} \cdot \vec{B}_\mu)\begin{pmatrix} 0 \\ a \end{pmatrix} = \frac{1}{2}\frac{(ga)^2}{4}\vec{B}^\mu . \vec{B}_\mu,$$

where we have used the vacuum expectation value a for $\langle 0|v_1(x)|0\rangle$. It is clear from this that the gauge field has acquired a mass $\frac{ga}{2}$. This is the way Higgs mechanism works in the case of a non-Abelian SU_2 gauge group. It is possible to extend these procedures for other gauge groups.

Higgs Particle—Searches

The Higgs particle is of central concern to proving the correctness of the standard electroweak model of spontaneous symmetry breaking. Experimental searches for the Higgs particle are a standard procedure when new vistas are opened with higher energy acelerators. There are theoretical arguments as to why the mass of the Higgs particle cannot be small. These are based on the fact that the vacuum state of the theory which leads to the correct value of the mass of the W boson will not be the true ground state of the theory. The value of the mass of the Higgs particle constrained by such a consideration depends upon the top quark mass and the energy up to which the standard model is assumed to be valid. If one assumes that the model is valid up to the Planck scale, then it can be shown that the Higgs mass must be greater than about 130 GeV.

The experiments at LEP have provided the best limits on the mass of the standard model Higgs particle (H). They have looked for the decay of Z to HZ^*, where Z^* is a virtual Z which manifests as one of the decay products, e^+e^-, $\mu^+\mu^-$, $\tau^+\tau^-$, hadrons, etc. It has also been looked for in $e^+e^- \rightarrow ZH$, with the H decaying hadronically (that is, into quark-antiquark pairs) or into tau pairs. At a center of mass energy in LEP of 190 GeV, it should be possible to find Higgs boson of mass up to 95 GeV using $e^+e^- \rightarrow ZH$; however, nothing has been found in this region of energies. The energy of LEP has been pushed up to 208 GeV with which it should be possible to reach a Higgs mass of about 114 or 115 GeV. Since the Higgs boson couples to a fermion in proportion to the mass of the fermion, it has a high probability to decay into $b\bar{b}$. These quarks will hadronize into B mesons. The experiments can tag the B mesons. The problem is that in the standard model, $b\bar{b}$ pairs can be produced even without their coming from the decay of the Higgs particle. These represent a background which has to be removed. If after removing these background B mesons, there is still a signal left over, one could attribute it to the production of the Higgs particle. The experiments have been looking for such a signal above the background. Although elaborate analyses show that there is a hint of a small excess of such events seen in some of the four LEP experiments, there is no clear signal for the Higgs particle. If the Higgs particle has a mass of 115 GeV, unfortunately there is vanishingly small phase space for the production of the Higgs particle at a LEP energy of 208 GeV.

In supersymmetric extensions of the standard model, there are charged Higgs bosons, two neutrals, and one pseudoscalar particle. For a large value of the ratio of the expectation values of the Higgs fields which couple to "up" type quarks and "down" type quarks, the lightest neutral Higgs scalar must be less than 130 GeV in mass. If a neutral Higgs scalar is found in this range, there would also be other accompanying signs of supersymmetry which could be looked for and checked.

If the mass of Higgs particle is higher than can be discovered at LEP, there are chances that it could be discovered at the Fermilab Tevatron through $p\bar{p} \rightarrow HZX$ and $p\bar{p} \rightarrow HWX$. It will occupy center stage when the Large Hadron Collider (LHC) starts operating in 2004. The discovery of the Higgs particle and its properties will point the way to future directions for particle physics.

Hypercharge

The *hypercharge* quantum number Y is a combination of baryon number B and the strangeness number S, $Y = (B + S)$. Mesons have baryon number equal to zero. Baryons, which are defined as strongly interacting particles heavier than the proton and decaying eventually into a proton, are assigned a nonzero baryon number. Conservation law of baryon number prohibits proton decay into any mesons or any other particles with zero baryon number.

When the baryons and the pseudoscalar mesons are classified in terms of hypercharge, one can see a close parallel between the two sets of particles (See discussion under section on "Eightfold Way"). This is seen below.

$$(p, n) \quad (K^+, K^0) \quad \text{have } Y = +1, I = 1/2;$$

$$(\Sigma^+, \Sigma^0, \Sigma^-) \quad (\pi^+, \pi^0, \pi^-) \quad \text{have } Y = 0, I = 1;$$

$$\Lambda^0 \quad \eta^0 \quad \text{have } Y = 0, I = 0;$$

$$(\Xi^0, \Xi^-) \quad (\bar{K}^0, K^-) \quad \text{have } Y = -1, I = 1/2.$$

It is apparent that these particles group themselves into multiplets which are octets when seen in terms of the isospin and hypercharge. This led to the idea of SU_3 symmetry.

Hypernuclei, Hyperfragments

A hypernucleus is one in which a neutron in the nucleus is replaced by a Λ^0. Hypernuclei are unstable and decay emitting nucleons and pions or just nucleons. The mesonic decay modes may be thought of as the decay of the Λ^0 in the nucleus, $\Lambda^0 \to P + \pi^-, N + \pi^0$. First evidence for a hypernucleus was obtained in 1953 by Danysz and Pniewski [221]. They saw an event in a G5 emulsion 600μ thick, which had been exposed to cosmic rays at the very high altitude of 85,000 feet. They saw two stars in the photographic plate. One of these stars was the origin of a heavy nuclear fragment, while the second star appeared to be at the end of the track of the fragment ejected from the first star. From their studies they ruled out the possibility of accidental coincidence responsible for these stars. What they were seeing at the second star was consistent with the explanation based on a delayed disintegration of a heavy nuclear fragment originating at the first star.

Since that first observation, many similar fragments were found in the observations made by many workers. For the detailed interpretation of what they observed, we consider one example of the observations made by Fry et al. [222], which allows one to determine the binding energy of Λ^0 in a nucleus. A negative K meson came to rest in the emulsion and was absorbed by a nucleus of an emulsion atom. The primary reaction that occurred was $K^- + P \to \Lambda^0 + \pi^0$. The Λ^0 stayed in the nucleus and became bound in it. The reaction resulted in the first star, and one of the tracks from the star, which was clearly a heavy particle, was interpreted as a nucleus of ^{9}Be with a Λ^0 bound in it. This heavy nuclear fragment $^9_\Lambda$Be traveled in the emulsion a short distance and decayed into a π^- meson, a proton, and two ^{4}He nuclei (which themselves came from the very shortlived primary ^{8}Be nucleus in the decay) with a total kinetic energy release of 30.92 ± 0.5 MeV. The energy of the negative pion was measured to be 26.6 ± 0.6 MeV.

Using these facts, and the fact that the energy released in the Λ^0 disintegration into P, and π^- is 37.56 MeV, one can determine the binding energy of the Λ^0 in ^{9}Be as follows. Starting from the end products of 2^4He, P, and π^-, we try to reassemble the original nucleus and keep track of the energies involved. First, in forming the Λ^0 from the P and π^-, one needs 37.56 MeV. Assembling the two He nuclei into ^{8}Be, one gains 0.10 MeV. So in getting Λ^0 and ^{8}Be one has to spend 37.46 MeV. The total energy release in the $^9_\Lambda$Be decay is 30.92 MeV, the difference, $(37.46 - 30.92)$ MeV$= 6.54$ MeV, must account for the binding energy of the Λ^0 with ^{8}Be to form $^9_\Lambda$Be. Other hypernuclei in systems of low mass numbers have also been studied.

Evidence also exists for double hypernuclei. First evidence for this came from the work of Danysz et al. [223]. An example is the $^{11}_{\Lambda\Lambda}$Be studied from the interaction of 1.5 GeV K^- mesons in nuclear emulsions. The basic reaction involved is the production of Ξ^- from the K^- interaction which subsequently interacts with a proton in one of the emulsion nuclei according to $\Xi^- + P \to \Lambda + \Lambda$. In this case both the Λ's are bound in the nucleus.

Hyperons—Decay Selection Rules

Hyperons are unstable particles which are heavier than the nucleon possessing nonzero strangeness. Examples are Λ^0 (mass 1115.683 MeV), $\Sigma^{+,0,-}$ (masses 1189.37, 1192.642, and 1197.449 MeV, respectively), and $\Xi^{-,0}$ (masses 1314.9, 1321.32 MeV). The lifetimes of these unstable particles are: $(\Lambda^0, 2.632 \times 10^{-10}$ s), $(\Sigma^+, 0.799 \times 10^{-10}$ s), $(\Sigma^0, 7.4 \times 10^{-20}$ s), $(\Sigma^-, 1.479 \times 10^{-10}$ s), $(\Xi^-, 1.639 \times 10^{-10}$ s), and $(\Xi^0, 2.90 \times 10^{-10}$ s). These lifetimes are incompatible with their copious rates of production in high energy accelerators. The hypothesis of associated production was formulated to explain this (see also the section under "Associated Production"). It was further postulated that in strong interaction processes, strangeness is conserved, while it is violated in weak interaction processes. Thus the production process will lead to a copious production of hyperons along with kaons such that the total strangeness is conserved. The decay of Λ^0 leading to $P + \pi^-$ is a strangeness changing process and proceeds through weak interactions.

The study of weak interaction decays of strange particles has revealed the operation of certain selection rules. In the section on beta decay, it was shown that the weak interaction Hamiltonian density can be written in the form of a scalar product of a $V - A$ (four vector) current with its Hermitian adjoint, multiplied by a coupling constant G_F, called the *Fermi weak interaction constant*. The current contains contributions formed from the nucleon sector and from the lepton sector. To describe the weak decays of all particles in a universal manner, one may extend the weak interaction Hamiltonian density by adding other strongly interacting particles such as hyperons and kaons. A simple way for this extension is to add the $V - A$ currents for all the strongly interacting particles (whether they possess strangeness or not) and form the Hamiltonian density. Certain features observed in the decays of hyperons and kaons lead to some restrictions on the form of the currents in the product. We describe these briefly. Weak decays of strongly interacting particles fall into three classes, hadronic, leptonic and semileptonic. Hadronic decay mode involves only strongly interacting particles among the products of the decay. Leptonic decay modes involve only leptons among the final decay products. Semileptonic decay modes have a mixture of a strongly interacting particle or particles and leptons.

A $V - A$ current of the form $\bar{\psi}_n \gamma^\mu (1 - \gamma_5) \psi_p$, which occurs in beta decay of the neutron, is called a $\Delta S = 0$ *current*. It describes a transformation of a neutron into a proton in which the strangeness does not change, both having strangeness zero. Reading from left to right, the charge Q increases by one unit. This feature is referred to as a current which has $\Delta Q = 1$. (The Hermitian adjoint current will also have $\Delta S = 0$, but will have $\Delta Q = -1$.) All the terms in the current must have $\Delta Q = 1$ to be in accord with charge conservation. A feature that is built into the theory is that the $\Delta S = 0$ part of the weak Hamiltonian be invariant under CP to accommodate features observed in semileptonic decays. It has been found that to accommodate all the features of the decays of strange particles, one adds to the $(V-A)$ $\Delta S = 0$ currents, currents which have $(V - A)$ $|\Delta S| = 1$ also. Transitions in which $|\Delta S| = 2$ do not seem to occur; for example, Ξ^-, which has $S = -2$, has not been observed to decay into $n + \pi^-$ (strangeness 0), even though energy considerations would allow such a decay. It was pointed out by Feynman and Gell-Mann [80] that the absence of $|\Delta S| = 2$ transitions implies that the structure of the currents for which $|\Delta S| = 1$ must have $\Delta S = \Delta Q$. This selection rule seems to hold. Thus for example, $\Sigma^+ \to n + e^+ + \nu$ is forbidden because it would involve the $\Delta Q = 1, \Delta S = -1$ current $\bar{\psi}_n \gamma^\mu (1 - \gamma_5) \psi_{\Sigma^+}$. On the other hand, the decay $\Sigma^- \to n + e^- + \bar{\nu}$ is allowed as it involves the $\Delta Q = \Delta S = 1$ current, $\bar{\psi}_{\Sigma^-} \gamma^\mu (1 - \gamma_5) \psi_n$.

Similar arguments suggest that $K^+ \to \pi^+ + \pi^+ + e^- + \bar{\nu}$ is forbidden, while $K^+ \to \pi^+ + \pi^- + e^+ + \nu$ should be allowed. These are borne out from experimental data.

Another selection rule, called the $\Delta I = 1/2$ *rule*, for the weak Hamiltonian has been formulated in trying to reconcile the rates of the following decays. K^0 is observed to decay to $\pi^+ + \pi^-$ and K^+ to $\pi^+ + \pi^0$. The former decay occurs about five hundred times faster than the second one. The kaon has isospin $1/2$. The allowed isospin states for the final state two pion combinations can be worked out. The two pion $(J = 0)$ state in neutral K can have isospins $I = 0$ as well as $I = 2$, while in the charged K decay, it has only $I = 2$. Thus, the matrix element for the neutral K decay has mixture of $\Delta I = 1/2, 3/2, 5/2$ pieces in it, while for the charged K decay, $\Delta I = 3/2, 5/2$ only. The $\Delta I = 1/2$ piece seems to have an enhancement relative to the $3/2$ and $5/2$ pieces. This is readily accommodated if we demand that the strangeness changing part of the weak interaction transforms as an isospinor. The $\Delta I = 1/2$ rule seems to be successful in explaining many other features; for example, the rate for $\Lambda^0 \to n + \pi^0$ is about $1/3$ of the total $(\Lambda^0 \to n + \pi^0) + (\Lambda^0 \to p + \pi^-)$, which is in accord with experiment.

Cabibbo [95] introduced SU_3 symmetry in the discussion of weak decays of strange particles and was able to show the $\Delta S = \Delta Q$ rule and that the strangeness changing currents obey the $\Delta I = 1/2$ rule. With the introduction of the Cabibbo angle, he was able to accommodate the low decay rates for strange particles and showed that the weak interactions are indeed universal in character.

Inclusive Spectral Distributions

Consider the collision of a high energy particle A with another particle B leading to final state particles $C + X$, where C may be one specific particle, and X may be one particle or a collection of particles. If M_X is the invariant mass of X and X is a single particle, M_X has a fixed value. On the other hand, X may be a collection of particles in various channels, such as $D + E$, $D + E + E', \ldots$, in which case the value of M_X can vary continuously. Sometimes we are interested in the spectral distribution of particles C emerging in the reaction regardless of what other particles X were produced together with C. In this case, we sum over the final state distributions of all particles in channels X and exhibit the distribution of particle C only. Such a spectral distribution for particle C is called an *inclusive distribution*. Processes in which one or more of the particles of X are specifically identified (and measured) lead to what are called *exclusive distributions*.

Inclusive distributions are simple to obtain from a knowledge of the particles A, B, and C, and measurements of their four-momenta. Let p_A, p_B, p_C be the four-momenta of these particles. Four-momentum conservation determines p_X, the total momentum of the collection X, from $p_A + p_B = p_C + p_X$, and we do not need to measure it separately. The four-momentum transfer from A to C is $q = p_A - p_C$, which can also be expressed as $q = p_X - p_B$. We can form the following invariants from these four-vectors,

$$p_A^2 = M_A^2, \ p_B^2 = M_B^2, \ p_C^2 = M_C^2, \ p_X^2 = M_X^2, \ q^2 = (p_A - p_C)^2, \ q \cdot p_B,$$

where M_A, M_B, and M_C are the rest masses of particles A, B, and C, respectively. Since we have $p_X = p_B + q$, we have

$$p_X^2 = (p_B + q)^2 = M_B^2 + q^2 + 2p_B \cdot q.$$

Since $p_X^2 = M_X^2$, one can gain a knowledge of M_X from a knowledge of q, and p_B alone. Let us define

$$x = -\frac{q^2}{2p_B \cdot q} = \frac{q^2}{M_B^2 + q^2 - M_X^2}.$$

Note that $q^2 \le 0$ (spacelike) and, because $M_X \ge M_B$, $0 \le x \le 1$. When $M_X = M_B$, $x = 1$. The kinematics of the process is such that it can be described in terms of three independent invariants, which we choose to be $p_A \cdot p_B$, q^2, and x, regarding M_X as a variable.

Let us illustrate the foregoing with a specific example. Consider the collision of two protons $P1 + P2 \rightarrow \pi^+ + X$ in the laboratory system, where the proton $P1$ is the beam proton and $P2$ the target proton (at rest), and X is a collection of particles which are not measured. We write

$$p_{P1} = (E_{P1}, \vec{p}_{P1}), \ p_{P2} = (M_P, 0), \ p_{\pi^+} = (E_{\pi^+}, \vec{p}_{\pi^+}), \ p_X = (E_X, \vec{P}_X).$$

In this system, $\vec{q} = \vec{p}_X$, and $q_0 = E_X - M_P$ or $E_X = q_0 + M_P$. Also, $p_{P1} \cdot p_{P2} = E_{P1} M_P$, so the three independent variables may be written as

$$E_{P1}, q^2, x.$$

If θ be the angle between $\hat{p}_{\pi^+}$ and $\hat{p}_{P1}$, then $q^2 = M_P^2 + M_{\pi^+}^2 + 2(E_{P1}E_{\pi^+} - p_{P1}p_{\pi^+}\cos\theta)$, $x = -\frac{q^2}{2M_P q^0}$, and $E_{\pi^+} = E_{P1} + M_P - E_X = E_{P1} + \frac{q^2}{2M_P x}$.

Feynman proposed a parton model of hadrons for describing inclusive spectral distributions of hadrons originating in high energy collisions of hadrons [92]. According to this model, the hadron is composed of point-like constituents called *partons*. These partons undergo collisions, and the cross section for the hadron-hadron collision is given by the incoherent sum of the individual parton-parton cross sections summed over the distributions of the partons in the hadron. He showed that the inclusive distributions of hadrons, in this model, scale in the very high energy limit and depend only on the variable x. In the parton model, the variable x can be shown to be simply the fraction of the momentum of the hadron carried by the parton. (See also section under "Bjorken Scaling"). First conclusive evidence for Feynman scale invarince in hadronic inclusive experiments came from the work of Bushnin et al. [224] and of Binon et al. [225] working with the Serpukhov proton synchrotron. Further developments have shown that the parton model has its underpinnings in the field theory of QCD.

Infrared Divergence
An important result of classical electromagnetic theory is that charged particles in accelerated motion radiate electromagnetic energy. This carries over in quantum electrodynamics where, in many nonstationary processes, such as scattering of a charged particle in an external field or the sudden emission of a beta particle from a charged nucleus, the processes are accompanied by the emission of electromagnetic radiation. Such emission is called *bremsstrahlung*. The treatment of emission of radiation by charged particles when they are scattered by the Coulomb field of a nucleus is carried out using time dependent perturbation theory (expanding in powers of $\simeq 1/137$, the fine structure constant). The differential cross section for the radiation of frequency ω, when the charged particle scatters off a nucleus of charge Ze, was calculated by Bethe and Heitler already in 1934. It is called the *Bethe-Heitler cross section* [16]; for a modern treatment see Reference [98]. This cross section has the characteristic feature that when one considers the emission of very low frequency photons ω, in an interval $d\omega$, it behaves like $d\omega/\omega$ as ω tends to zero. The cross section integrated over all frequencies diverges logarithmically as ω tends to zero. This divergence is called *infrared divergence*. The divergence is directly attributable to the fact that the photon has rest mass zero; if the photon had a mass, the lowest frequency it could have would not be zero but its rest mass. To regulate this divergence, one introduces a small photon mass λ; of course, in the final result, the dependence on this mass should disappear.

In the limit of zero frequency, the Bethe-Heitler bremsstrahlung cross section can be shown to be the product of the elastic scattering cross section in the Coulomb field of the nucleus and a factor representing the effect of the emitted radiation, called the *radiative correction* which diverges as the frequency goes to zero (or as the photon mass goes to zero). Since the loss of energy by radiation can also be considered as loss of energy suffered by the charged particle in the scattering process, it is not possible to distinguish between the elastic scattering process, including corrections due to virtual emission and absorption of radiation, and the inelastic process in which real radiation is emitted. Both have to be included to a given order of perturbation theory. When detailed calculations of the virtual corrections to the scattering cross section are made, it turns out that they too are divergent as the photon mass goes to zero. When one adds the two, namely the effects of virtual corrections and of the real emission process, the dependence on the photon mass cancels out, and one is left with a finite result. In a typical experiment, the electron energy can be measured with a finite resolution ΔE. The result of this is that the ultimate correction to order α can be shown to depend on the parameter $(\alpha/\pi)\ln(\Delta E/m)$, where m is the mass of the charged particle which suffers the scattering. If this parameter is large, it is not possible to ignore higher order corrections involving several quanta, including the possibility of emission of an infinite number of quanta. In such a case, the effect of low frequency quanta (or low ΔE) factorize as an exponential which vanishes as $\ln(\Delta E/m) \to 0$.

Already in 1937, Bloch and Nordsieck [226] pointed to the fact that an expansion in powers of $e^2/(4\pi\hbar c)$ does not allow one to take the classical limit $\hbar \to zero$, and so this method does not allow one to recover the classical radiation result for a charged particle. (Here we exhibit the $\hbar, c$ dependence explicitly, as one is interested in the classical limit $\hbar \to 0$.) They developed expansion in terms of another set of parameters, $e^2\omega/(mc^3), \hbar\omega/(mc^2), \hbar\omega/(c\delta p)$, where ω is the frequency of emitted radiation, δp the change in momentum of the particle, and m the mass of the particle. They showed that for frequencies for which these parameters are small, the quantum mechanical result is just the result of rewriting the classical radiation formula. The total probability for a given change in motion of the particle is not altered due to the interaction with radiation. The mean number of quanta emitted is infinite but in such a way that the mean energy radiated is equal in value to what would be emitted in the corresponding classical motion.

Infrared problems exist in all quantum field theories which have quanta with zero rest mass. The problem is considerably more complex in non-Abelian gauge theories and we do not go into them here.

Instantons—'t Hooft-Polyakov Monopole

In some cases, the Euclidean form of the field equations (setting $x_4 = ix_0$ in them) of non-Abelian gauge theories can be solved in the classical regime. Such solutions are called *instantons*, and they lead to solutions with finite energy which are localized both in space and in the Euclidean time variables. In quantum field theory, interpreting these as quantum effects, one gets modifications to the effective Hamiltonian of the quantum field theory. In effect what happens is that, in the non-perturbative regime of a quantum field theory, particles are found which are in addition to the quanta belonging to the fields occurring in the theory. The famous instanton found in four-dimensional non-Abelian gauge theory was shown by 't Hooft to lead to nonconservation of the U_1 axial current in QCD, a fact which leads to an understanding of the light meson spectrum like the pion.

Appearance of extra particle states in theories with spontaneously broken symmetry can be shown to be related to certain topological properties of the vacuum states in such theories. An example is an SU_2 gauge theory with an isovector Higgs field (called the *Georgi-Glashow model*). 't Hooft [227] and Polyakov [228] derived classical solutions for the Higgs field in this theory. It had such a structure that when this gauge theory is interpreted as a unified theory of electromagnetic and weak interactions, they showed that this solution is a magnetic monopole. Also, examples exist which show that the dynamics of theories in the strong coupling regime are related to the particles which are the heavy classical states in the limit of weak coupling.

Integer Spin—Statistics of Light Quanta

The statistics known as Bose-Einstein statistics dealing with particles of integer spin originated with the work of Bose in 1924. He gave a new derivation of Planck's law for radiation. From the historical point of view, it is useful to review Bose's method as it is at the base of quantum statistics. This we do briefly below.

The problem he considered was one of calculating the average energy of a gas of photons contained in a vessel of volume V at a temperature T. He imagined the phase space divided into cells of size h^3 (h=Planck's constant) and considered the distribution of quanta in these cells. Consider a total of N cells, of which N_0 contain 0 quanta, N_1 contain 1 quantum,..., N_j contain j quanta, etc. If M be the maximum number of quanta per cell, then we must have

$$\sum_{j=0}^{M} N_j = N.$$

Focussing attention on the cells only, the number of ways of selecting the N cells such that N_0 contain 0 particles, N_1 contain 1 particle, etc., is

$$W = \frac{N!}{N_0!N_1!\cdots N_M!}.$$

Let us maximize W subject to the condition that the total energy is a constant and the total number of cells is a constant. If E_j is the energy associated with a cell which contains j particles, then the total energy is

$$\sum_{j=0}^{M} E_j N_j = E = constant,$$

and the total number of cells is

$$\sum_{j=0}^{M} N_j = N = constant.$$

In the usual manner we get, for a canonical distribution at a temperature T,

$$N_j = \frac{N \exp\left(-E_j/kT\right)}{\sum_{j=0}^{M} \exp\left(-E_j/kT\right)}.$$

The average energy per cell is

$$\bar{E} = \frac{E}{N} = \frac{\sum_{j=0}^{M} N_j E_j}{\sum_{j=0}^{M} N_j}.$$

Now let us consider the quanta being photons of energy $h\nu$, then $E_j = jh\nu$, then

$$\bar{E} = \frac{\sum_{j=0}^{M} jh\nu \exp\left(-jh\nu/kT\right)}{\sum_{j=0}^{M} \exp\left(-h\nu/kT\right)}.$$

In the limit when the number of photons M per cell is very large, the expression for $\bar{E}$ works to be

$$\bar{E} = \frac{h\nu}{\exp\left(h\nu/kT\right) - 1}.$$

Let us consider the photons to form a perfect gas in a volume V with momenta E/c, where E is the energy of a given photon and c is the velocity of light. The average number of photons with energy in the interval $h\nu$ and $h\nu + d(h\nu)$ and in the volume V is $2 \times \frac{4\pi\nu^2 d\nu}{c^3}$. We have assumed that each photon has two polarization states corresponding to the two transverse polarizations for a massless spin 1 particle. Multiplying this by $\bar{E}$ given above, we get for the energy density $u_\nu d\nu$ of radiation in the frequency range $(\nu, \nu + d\nu)$ Planck's Law

$$u_\nu d\nu = \frac{\frac{8\pi\nu^2 d\nu}{c^3}}{\exp\left(h\nu/kT\right) - 1}.$$

Intermediate Vector Bosons

Electromagnetic interactions of charged particles are known to be mediated by the exchange of photons between them. As an extension of this idea, Yukawa [113] suggested that nuclear forces and β decay could both be explained by meson (boson) exchange [113], if this meson was unstable. This idea of Yukawa, however, proved to be untenable for a number of reasons. The boson for mediating weak interactions has to have spin 1, (hence has to be a vector or axial vector particle) in order to give agreement with data on β decay, etc., while the pions responsible for nuclear forces were spin zero particles and so would not do. Yet the idea of a boson mediating weak interactions was attractive and was not given up. It popped up from time to time, and finally found formal expression in a proposal put forward by Schwinger [229] that the observed weak interactions be mediated by *vector* bosons.

Using this idea of Schwinger, the $(V - A)$ current-current structure of weak interactions can be obtained from a more fundamental interaction between the weak charged current $J^\mu(x)$ and the vector boson field $W_\mu(x)$ by an interaction Hamiltonian density:

$$H = g[J^\mu(x)W_\mu^\dagger(x) + J^{\dagger\mu}(x)W_\mu(x)].$$

The coupling constant g specifies the strength of the (current-W boson) interaction. In second order perturbation theory, this could give rise to an effective current-current interaction if the W particle's propagator has a very big mass M_W.

As an illustration, consider the *beta* decay of the neutron. The weak nucleon current emits a W^- converting a neutron into a proton, the W propagates and is absorbed by the electron-neutrino current, giving rise to the final product, the electron and the electron antineutrino. The matrix element will contain the product g^2, the two currents, and the propagator for the W, $1/(q^2 - M_W^2)$, where q is the four-momentum transfer between the neutron and the proton. If $M_W^2 \gg q^2$, the matrix element reduces to the product of (g^2/M_W^2) with the product of the currents. If one sets $(g^2/8M_W^2) = G_F/\sqrt{2}$, where G_F is the Fermi weak coupling constant, the intermediate vector boson theory reduces to the Fermi current-current theory.

Many weak processes were worked out in terms of the weak vector boson theory and comparison with experiments set limits on the mass of the W. These early limits were in the several GeV range as the highest energy accelerators available had energies of this order. Since the theory closely parallels quantum electrodynamics (QED), calculations of higher order weak corrections were attempted. Unlike QED, the intermediate vector boson theory, involving as it does massive charged vector particles, was found to be non-renormalizable. This proved to be a significant hurdle which was only overcome many years later with the development of spontaneously broken non-Abelian gauge theories (see section under "Gauge Theories" and "Higgs Mechanism").

Spontaneously broken non-Abelian gauge theories were shown to be renormalizable and Glashow, Weinberg, and Salam developed the theory of electroweak unification with intermediate vector gauge bosons mediating electromagnetic and weak interactions (see also section under "Electroweak Synthesis" and "Standard Electroweak Model"). This theory predicted that charged vector bosons should exist at a mass of about 80GeV, and a neutral vector boson Z^0 should exist at a mass of about 90 GeV, which would play a role in neutral-current processes much like the charged W's play in charged current processes.

These were confirmed with the finding of the W and Z bosons by the UA1 [180] and UA2 [181] collaborations working with the CERN $\bar{p}p$ collider. The experimental observations in these experiments consisted in finding isolated large transverse momentum electrons with associated missing energy in the case of the charged W's, and in the case of the neutral Z, the finding of $Z^0 \rightarrow e^+e^-$ decays. Much precise work on the detailed properties of the Z^0 and the $W^\pm$ particles and their couplings to quarks and leptons has since been carried out at the Large Electron Positron Collider (LEP) working at CERN. The mass of the Z^0 is now known to be $M_{Z^0} = 91.187 \pm 0.007$ GeV and the mass of the charged W's, $M_W = 80.41 \pm 0.10$GeV.

Isotopic Spin

Soon after the neutron was discovered in 1932, one obvious property of the neutron was pointed out by Heisenberg [47]. He noted that, apart from the fact that it is electrically neutral, it has almost the same mass as a proton, only very slightly heavier. Further, as far as the structure of atomic nuclei is concerned, the strong interactions which the protons and neutrons suffer inside the nuclei seem to be similar, regardless of the fact that the proton carries an electric charge (charge independence of the nuclear force). These facts led Heisenberg to suggest that the neutron and proton be treated as two states of a single particle, the *nucleon*. Heisenberg suggested introducing *isotopic spin or isospin* vector $\vec{I}$, whose z-projections $I_z = +1/2$ and $I_z = -1/2$ would serve to label the proton and neutron states of the nucleon. Heisenberg defined charge independence of strong nuclear forces in the language of isospin as: "strong interactions are invariant under rotations in isospin space".

The introduction of isospin is entirely in analogy with ordinary spin. Recall the fact that in the absence of a magnetic field, the two magnetic substates $m_s = \pm 1/2$, of the electron of spin $1/2$, are degenerate. The isotopic spin is different in that it is not a vector in ordinary space, but in a space associated with internal properties of the neutron and the proton, called an *internal space*, in this case, *isospin space*. The isospin vector in this space has three components, I_1, I_2, I_3, just as the ordinary spin vector $\vec{S}$ has three components, S_1, S_2, S_3 in three-dimensional space. Because of the analogy with ordinary spin, all the theoretical machinery that was developed to handle ordinary spin can be used to treat isospin variables. The wave function for a particle with spin and isospin can be written as a product of spatial wave function, the spin wave function, and the isospin wave function. Just as the flip of spin from one state to another is accomplished by the use of operators $S_\pm = (1/2)(S_1 \pm iS_2)$, one can achieve flip of isospin through the operators $I_\pm = (1/2)(I_1 \pm iI_2)$, which will result in a nucleon flipping its state from a proton (isospin "up") into a neutron (isospin "down") and vice versa. These flips are in effect rotations. In the case of ordinary spin they preserve $|\vec{S}|^2$, while the value of $|\vec{I}|^2$ is preserved under rotations of isospin. Thus another way of stating charge independence of nuclear forces is to say: "isospin is conserved in strong interactions". Because a state of spin S can have S_3 projections, $S, S-1, \ldots, 0, \ldots, -S$, there are $(2S+1)$ degenerate states. This is the multiplicity for a particle of spin S. Similarly a particle with isospin I has multiplicity $(2I + 1)$. The nucleon can be said to be an isospin doublet because $I = 1/2$.

The notion of isospin can be extended to other hadrons which have approximately equal masses but different charges. Thus the pion with π^+, π^0, π^- having nearly the same masses can be assigned to an isospin triplet $I = 1$ with $I_3 = +1, 0, -1$, respectively. Likewise, the Σ baryons, which have charge $(+, 0, -)$ states, are a triplet in isospin, $I = 1$, and the Δ isobars, which have four charge states, $(++, +, 0, -)$, are assigned to a quartet in isospin with $I = 3/2$. In assigning isospin to a given set of particles, the value of I_3 is determined by the charge Q carried by the particles, the highest charge state being the one with $I_3 = I$. All the other states of the multiplet are obtained by decreasing I_3 by one unit until one gets to the state of the lowest charge with $I_3 = -I$. To accommodate mesons and baryons, strange and nonstrange, the relation between the charge and isospin is given by the Gell-Mann, Nishijima formula: $Q = I_3 + (1/2)(B + S)$, where B is the baryon number and S is the strangeness quantum number (see also section under "Gell-Mann, Nishijima Formula").

J/ψ meson

The announcement of a new particle, the J/ψ meson, was jointly made in November 1974 by S. C. C. Tings group working at Brookhaven, and B. Richter's group working at SLAC [101]. The experiment at Brookhaven consisted in bombarding a beryllium target with high energy protons and the observation of high mass electron-positron pairs in the resulting product. A strong and a very narrow signal was seen at an invariant mass of the electron-positron pair of mass of 3.1 GeV; Ting named this particle J. The narrow width at this high energy was extremely unusual, for the prevailing thinking at that time was that widths would be very large as the energy increases. Thus, although the particle was seen by Ting and his group in the summer of 1974, they did not rush to announce their results then. They spent their time very carefully checking their experiment. They wanted to make absolutely sure that it was a genuine signal. At about the same time, the same particle was seen in the SPEAR 8.4 GeV electron-positron ring at SLAC by B. Richter and his group. This group observed a very narrow peak in the cross section for $e^+e^- \rightarrow$ hadrons, e^+e^-, $\mu^+\mu^-$, at a center of mass energy of 3.105 ± 0.003 GeV. They quoted a value for the width at half maximum as 1.3 MeV, an incredibly narrow value; the SLAC group named the particle ψ. The two groups decided to make a joint announcement of their J/ψ findings, and this occurred in November of 1974.

Following the discovery, it was soon established that the J/ψ belonged to a family of vector mesons just like the ρ, ω, and ϕ. In the constituent quark model of hadrons, this new particle found a place as a bound state of a new flavor quark and its antiquark. This new flavor was called *charm*. The idea of introducing a fourth quark, the *charm* quark, already had been explored by Bjorken and Glashow in 1964 [102] on the basis of lepton-quark symmetry. They drew attention to the fact that, on the lepton side, the electron and its neutrino and the muon and its neutrino were known at that time, but on the quark side, only three quarks were known, namely, the u, d, and s. A fourth quark, having charge $+2/3$, paired with the s quark (charge $-1/3$), would bring symmetry between leptons and quarks. The charm (c) and the strange (s) quark would form the second generation of quarks after the first generation of u and d.

The proposition that J/ψ was the bound state of the charm quark with its antiquark was vigorously explored and has proven to be correct. Because it is a charm-anticharm bound state, it has "hidden" charm content and is also referred to as a *charmonium bound state*. On the theoretical front, the non-Abelian gauge theory of quantum chromodynamics (QCD) began to be formulated. It was postulated that quarks carry a quantum number called *color* and quarks interact with other quarks through the exchange of colored gluons, which are quanta of the non-Abelian gauge field. A potential model of interquark interaction inspired by QCD is one in which the short distance behavior of the potential is due to one gluon exchange, which reduces to a Coulomb potential between two color charges, while at long distances the potential is linear and is therefore confining. With this model, one can predict the spectrum of charmonium bound states. With parameters in the potential fixed from QCD, the J/ψ is the $n = 1$ $^3S_1$ bound state of $(c\bar{c})$. A bound state is obtained at 3.1 GeV if the mass of the charm quark is chosen in the range 1.1 to 1.4 GeV. This model also predicts a rich spectrum of states for the charmonium system, which have been observed, including the bound state corresponding to the $^1S_0$ state. As one goes up in energy of excited states, one will eventually cross the charm threshold after which one should be able to produce particles with nonzero net charm. Such particles are called *charm particles*, and both mesons and baryons having a charm quark in them have been found (see further in section under "Charm Particles").

K2K Neutrino Experiment

This is a long baseline neutrino oscillation experiment being performed in Japan. The purpose of this experiment is to confirm the occurrence of neutrino oscillations. The muon neutrinos are produced at the KEK 12 GeV proton synchrotron and directed to the SuperKamiokande 50,000 ton Cherenkov water detector some 250 km away. In the period between June 1999 and the end of June 2000, 27 neutrinos from KEK have been detected at SuperKamiokande. They expected this number to be 40.3 ± 5, based on observations with a near detector at KEK. The K2K collaboration considers this to be laboratory evidence for the existence of neutrino oscillations. (See further under "Neutrino Oscillations".)

KamLAND Neutrino Detector

This is an experiment designed by a large collaboration of scientists from Japan and USA. The aim is to test the observed solar neutrino deficit in a terrestrial experiment. If one assumes that CP conservation holds in the lepton sector, the oscillation probability is the same for particles and antiparticles. Hence one can expect that if the deficit in solar neutrinos is due to oscillation of electron neutrinos, the same oscillation effect should be seen with electron antineutrinos also. Nuclear reactors are a good source of electron antineutrinos and the aim of this experiment is to use these. The flux and the energy distributions of the electron antineutrinos are well calculated from the known thermal powers and fuel compositions of the reactors.

The energy of reactor antineutrinos is low, so it can only test what is called the large mixing angle solution for the solar neutrino problem (see further under "Neutrino Oscillations"). The large mixing angle solution implies a maximum for the oscillation probability at a distance of about 250 km from the source. Kamioka is at about this distance from the reactor source. The aim is to set up a 1000-ton liquid scintillator as target and detector at Kamioka. The liquid scintillator consists of 20% trimethylbenzene and 80% paraffin. It measures the e^+ produced by the antineutrinos by charged current reactions in the scintillator. The recoil neutron with a kinetic energy of about 10 keV is moderated and captured on protons after about 170 microseconds. These signals define the neutrino signal. It is crucial to suppress the background at the detector. Here a crucial role is played by the energy spectrum of e^+. The background at the detector can also be inferred directly from the power modulation of the reactor.

At present the detector construction is well under way and data are expected in the next year. The collaboration believes that they will be able to demonstrate convincingly the validity or invalidity of the large mixing angle solution of the solar neutrino problem. This can be achieved in about one to two years of data taking, and the results will be independent of models that power the neutrinos in the Sun.

Kaons: The τ-θ Puzzle

Kaons, or K mesons, are strongly interacting particles with a mass of about 490 MeV. They come in positive, neutral, and negatively charged varieties. The earliest observations of these particles came from work with cosmic rays using Wilson cloud chambers or photographic emulsions as detectors. V-shaped tracks originating from a point in the chamber or the photographic plate would be seen, which led to the name of *V particles* for them [230]. The V was interpreted as arising from the decay of an incident neutral particle. An analysis of the two tracks of the V showed, in many cases, that one of the tracks was a proton and the other a negative pion, with a Q value about 35 MeV. In other cases, both tracks of the V corresponded to pions, with some other Q value. Thus, there were two different neutral V particles, which gave rise to these two kinds of decay products. One of these neutral particles must clearly be heavier than a proton, while the other is lighter than a proton.

With the construction of high energy accelerators, which were able to produce these particles abundantly, clarifications as to their nature emerged. The V particles which were heavier than the proton belong to a family named *hyperons* (Λ^0, Σ, etc.), while the others which were lighter than the proton were called K *mesons* or *kaons*. Further investigations showed that there were two distinct kinds of neutral kaons. One of these was named K^0 and the other $\bar{K}^0$, distinctly different from K^0. Before long, the charged counterparts to these neutral kaons were found, the $K^\pm$ mesons, decaying to $\pi^\pm + \pi^0$. Thus, there were actually four particles: (K^+, K^0) and $(K^-, \bar{K}^0)$, which were antiparticles of one another. In terms of isotopic spin, it was more appropriate to assign isospin $1/2$ to each pair, rather than isospin 1, which would have been appropriate had there been only one neutral K. The kaons and hyperons were assigned a new quantum number called *strangeness*. These particles were produced in association with one another, with opposite values of strangenness, such that in the production reaction the total strangeness was conserved. In their subsequent decays, each particle decays with a change of strangeness, which leads to slow decays. Gell-Mann assigned proper values of strangeness to these new particles in order to extend the relation between isotopic spin and charge for them. (See under "Gell-Mann, Nishijima Formula" for more details).

The masses and other properties of the kaons have been determined. The $K^\pm$ have mass 493.677 ± 0.016 MeV, mean life $\tau = (1.2386 \pm 0.0024) \times 10^{-8}$s. K^+, K^0 are assigned strangeness $S = +1$, $I_3 = +1/2$ and $I_3 = -1/2$, respectively; $K^-, \bar{K}^0$ are assigned strangeness $S = -1$, $I_3 = -1/2$ and $I_3 = +1/2$, respectively. In terms of the quark model, the compositions are $K^+ = u\bar{s}$, $K^0 = d\bar{s}$, $\bar{K}^0 = \bar{d}s$, and $K^- = \bar{u}s$, where s is the strange quark with strangeness -1. The dominant decay modes are

$$K^\pm \to \mu^\pm + \nu_\mu \ (63.5\%),$$

$$K^\pm \to \pi^\pm + \pi^0 \ (21.16\%),$$

$$K^\pm \to \pi^\pm + \pi^+ + \pi^- \ (5.59\%),$$

$$K^\pm \to \pi^0 + \mu^\pm + \nu_\mu \text{ called } K^+_{\mu 3} \ (3.18\%),$$

$$K^\pm \to \pi^0 + e^\pm + \nu_e \text{ called } K^+_{e3} \ (4.82\%),$$

$$K^\pm \to \pi^\pm + \pi^0 + \pi^0 \ (1.73\%).$$

A number of other decay modes are seen at a level below 1% but we do not discuss them here.

The spin of the kaons can be determined from a study of their decays. K^0 has been observed to decay into 2 neutral pions. One uses angular momentum conservation to obtain information on the spin of the K^0. The final state, consisting of two identical bosons, is symmetric under the interchange of the two particles, hence must have even angular momentum $J = 0, 2, 4, \ldots$. Because the two pions are pseudoscalar and the angular momentum is even, the parity P must be even. This suggests J^P for the particle $0^+, 2^+, 4^+, \ldots$. For the charged counterpart, K^+, which decays into $\pi^+ + \pi^0$, all values of J are allowed. The fact that the pion carries zero spin, implies that the parity for a state of angular momentum J has to be $(-1)^J$. Thus 0^+ is a possibility for the charged kaon, also. The three pion decay mode of the kaon provides much more information through a Dalitz plot analysis. That analysis shows that the kaon has spin zero (see also discussion under "Dalitz Plot"). *If parity is conserved*, we may conclude that it has *odd parity* from its three pion decay. Of, course, if parity is not conserved in the decay process, no conclusion can be drawn about the parity of the kaon from its decay modes.

One very important feature is seen from the observed decay modes. Because the kaon is observed to decay into two pions and into three pions, if it is the same particle, which has these two modes of decay, and it has spin zero and a definite parity, then parity cannot be conserved in the decay process. In the early days when the mass determination of these particles was not as precise and absolute belief in parity conservation prevailed, it was thought that these two particles were different particles, having approximately the same mass accidentally. The particle decaying to 2 pions was called the θ particle, while the one decaying into 3 pions was called the τ particle. This became a puzzle, called the τ-θ *puzzle*: whether the θ and the τ particles were different and accidentally had the same mass, or whether they were the same particle, but parity was not conserved in the decay. When experimental data improved to the point where the equality of the masses and lifetimes of these particles could be established more precisely [231], it became clear that they were the same particle, and somehow the parity was not conserved in the decay. This prompted Lee and Yang to investigate the question of parity conservation, in weak interactions in general [141], resulting in the eventual overthrow of parity conservation in weak interaction processes.

Since the parity of the kaon cannot be determined through its decays, one may well ask how does one determine it? The reaction one uses is $K^- + {}^4\mathrm{He} \rightarrow {}^4_\Lambda\mathrm{H} + \pi^0$, where the negative kaons are brought to rest and captured in helium. The capture of the kaon occurs fron an atomic S-state. Some of these captures lead to a final state in which the Λ^0 formed from the capture of the K^- on a proton stays bound in the nucleus, forming the hypernucleus ${}^4_\Lambda\mathrm{H}$. The hypernucleus decays into π^- and ${}^4\mathrm{He}$, both of which are zero spin particles. From detailed studies of this decay, the spin carried by the hypernucleus is determined. It is found from such measurements that the spin of ${}^4_\Lambda\mathrm{H}$ is zero. Going back to the capture reaction, because the capture occurs from an S-state, the orbital angular momentum of the products of the reaction must also be zero. If Λ^0 is assigned the same parity as the nucleon, that is, even parity, then we can conclude that K^- must have the same parity as the final pion and, hence, is odd. In this manner, K^- is established to be a pseudoscalar particle, just as the pion. Although there are no direct experiments which establish the parity for the other charged and the neutral particles, we take the spin-parity for all of them to be $(0, -)$, on the basis that they all form a multiplet with nearly the same mass and, therefore, probably represent the same pseudoscalar particle in different charge states. All the kaons are taken to be pseudoscalar particles just as the pion.

Kaons—Neutral $K_1^0(K_S^0)$, $K_2^0(K_L^0)$

Neutral kaons exhibit certain special properties which we elaborate here. The neutral kaons that are produced by strong interactions are K^0 and $\bar{K}^0$, which have definite strangeness $S = +1$ and $S = -1$, respectively. With strong interactions alone, no transitions of the type $K^0 \leftrightarrow \bar{K}^0$ are possible, because strangeness is conserved. If we include weak interactions, this need not any longer be the case, because of the possibility of decay of either of these particles into π^+ and π^-, in which case the conversion of K^0 into $\bar{K}^0$ can occur through the sequence $K^0 \rightarrow \pi^+ + \pi^- \rightarrow \bar{K}^0$. Thus on including the weak interactions, we find that the $|K^0\rangle$ and $|\bar{K}^0\rangle$ cannot be eigenstates of the full Hamiltonian. One can form linear combinations of these which are eigenstates of CP, (which is respected by the weak interaction Hamiltonian if the phenomenon of CP violation is ignored temporarily),

$$|K_1^0\rangle = \frac{1}{\sqrt{2}}(|K^0\rangle + |\bar{K}^0\rangle), \quad \text{CP eigenvalue} + 1,$$

$$|K_2^0\rangle = \frac{1}{\sqrt{2}}(|K^0\rangle - |\bar{K}^0\rangle), \quad \text{CP eigenvalue} - 1.$$

If CP is conserved in the decay, these two states will be distinguishable by their decay modes. The two pion state has $CP = +1$, and the three pion state has $CP = -1$, and so K_1^0 will decay to 2 pions only, while the K_2^0 will decay into 3 pions only (see also discussion under "CP Violation"). The two pion decay mode occurs much faster because there is more energy released in the decay than in the three pion decay mode, and there is more phase space available. Since CP is known to be violated by a small amount, when that fact is taken into account, the short lived state is called K_S^0, while the long lived state is called K_L^0. The short lived state is predominantly K_1^0, while the long lived state is predominantly K_2^0. Thus if one starts with a beam of K^0, because it is the linear combination $|K^0\rangle = (1/\sqrt{2})(|K_1^0\rangle + |K_2^0\rangle)$, the K_1^0 component (nearly all K_S^0) of the beam will decay away fast, and one will be left with a beam of pure K_2^0 (nearly all K_L^0). The result will be that, near where the K^0 are produced, we will see two pion decays, and a long distance away from the production point, we will see only three pion decays corresponding to K_L^0. This was exactly confirmed in an experiment by Lande et al. [232] who measured the two lifetimes for the first time. Their values are now known to be $\tau_S = (0.8934 \pm 0.0008) \times 10^{-10}$s and $\tau_L = (5.17 \pm 0.04) \times 10^{-8}$s.

There is a very small mass difference between the two states K_S^0 and K_L^0. This has also been measured using the following technique. The amplitudes of these states, in the rest frame of the particles, vary as a function of time according to

$$A(K_S^0, t) = A(K_S^0, 0) \exp\left[-i(m_S - i(1/2)\Gamma_S)t\right],$$

$$A(K_L^0, t) = A(K_L^0, 0) \exp\left[-i(m_L - i(1/2)\Gamma_L)t\right],$$

where $m_{S,L}$ are the masses of the two states and $\Gamma_{S,L}$ are the respective widths. If we start at time $t = 0$ with a beam which is initially all K^0, the amplitudes $A(K_S^0, 0)$ and $A(K_L^0, 0)$ are equal. Because of the mass difference, their relative phases will be shifted as time progresses and the beam will develop a $\bar{K}^0$ component. The appearance of $\bar{K}^0$ can be detected by absorbing these on protons leading to the production of Λ^0. (Note that Λ^0 can be produced only by $\bar{K}^0$ and not by K^0; the appearance of Λ^0 is thus unambiguously related to the appearance of $\bar{K}^0$.) In the rest frame of K^0, after the lapse of time t, the amplitude $A(\bar{K}^0, t)$ for $\bar{K}^0$ as a function of time will be proportional to

$$[A(K_S^0, t) - A(K_L^0, t)].$$

Using the above relations of amplitudes at time t expressed in terms of the amplitudes at time 0, we can work out the probability of observing a $\bar{K}^0$ at time t (neglecting Γ_2 in comparison to Γ_1),

$$P(\bar{K}^0, t) \simeq \frac{|A(K_S^0, 0)|^2}{2}[1 - \exp\left(-\Gamma_S t/2\right)\cos\left(\Delta m t\right) + \exp\left(-\Gamma_S t\right)],$$

where $\Delta m = |m_S - m_L|$. We can plot this quantity, which represents the intensity of $\bar{K}^0$ produced at the time t, as a function of time, for different values of the mass difference. Fitting it to the experimental data, one can determine, the magnitude of the mass difference. In this manner, one can get at the magnitude of this tiny mass difference, the earliest attempt being by Boldt and Caldwell [233] and somewhat later by Camerini et al. [234]. The sign of the mass difference is not determined by such an experiment.

Determination of the sign of the mass difference depends on another type of experiment which involves a *regeneration* of K_1^0 [235]. If we start with a K^0 (or a $\bar{K}^0$) beam and let it propagate for some time t, such that $\tau_S < t < \tau_L$, the K_S^0 component would have decayed, and we would have a pure K_L^0 beam. Now let this beam be passed through some material. Since the $|K_L^0\rangle$ can be thought of as made made up of as a mixture of K^0 and $\bar{K}^0$, the two components K^0 and $\bar{K}^0$ of K_L^0 will be affected differently by the scattering in the medium, which affects the relative phase of these components (depending on the phase shifts suffered by the K^0 and $\bar{K}^0$), and K_S^0 will be regenerated. The probability of regeneration of K_S^0 depends both on the thickness of the material and on the sign of the mass difference. In this way the sign of the mass difference is also determinable.

The quoted value for the mass difference in the recent "Review of Particle Physics" [62] by the Particle Data Group is $m_L - m_S = (3.489 \pm 0.009) \times 10^{-6}$ eV.

KARMEN Experiment

The acronym KARMEN here stands for *KArlsruhe Rutherford Medium Energy Neutrino* collaboration. One major aim of the KARMEN experiment is to measure the neutrino-nucleus interaction cross sections via the charged current and neutral current channels on the nuclei ^{12}C, ^{13}C, and ^{56}Fe. The second major aim is to search for flavor oscillations. Neutrinos are detected in a 56-ton scintillation calorimeter at a distance of 18 m from the source. In the period 1990–1995, data from KARMEN showed no signal for $\bar{\nu}_\mu$ to $\bar{\nu}_e$ oscillations. Since 1997 data have been taken at a new detector confguration. The new data should cover the whole oscillation parameter space available to another experiment, the LSND experiment. The LSND experiment claims to have seen a signal for such oscillations.

$K^*(892)$ Resonance

This particle manifests as a $(K - \pi)$ resonance state in the reaction,

$$K^- + p \rightarrow \bar{K}^0 + \pi^- + p, \text{ or, } K^- + \pi^0 + p, \text{ or, } \bar{K}^0 + \pi^0 + n$$

reported first by Alston et al. [236]. The study was carried out using the 15 in. hydrogen bubble chamber exposed to the K^- beam of momentum 1.15 GeV/c at the LBL-Bevatron. On analysing the kinetic energies of protons resulting from the reaction, they found a strong peak around about 20 MeV. They interpreted this as arising from a $K - \pi$ resonance (or a K^{*-} particle) being produced: $K^- + p \rightarrow K^{*-} + p$, and the K^{*-} subsequently decaying into $\bar{K}^0 + \pi^-$. The mean value for the resonance mass was given by them to be 885 ± 3 MeV and its width as about 16 MeV.

The isospin of this particle was determined by measuring the branching ratio

$$R = \frac{K^{*-} \rightarrow K^- \pi^0}{K^{*-} \rightarrow \bar{K}^0 \pi^-}.$$

These gave a value 0.75 ± 0.35 for R. Theoretically, $R = 1/2$ for isospin $1/2$, while it should be $R = 2$ if the isospin is $3/2$. The measured value is consistent with isospin $1/2$.

To determine the spin of the resonance, they studied the angular distribution of the decay products of the resonance, and found it to be consistent with $J = 0$ or $J = 1$, but excluded $J > 2$. A later experiment, reported by Smith et al. [237] resulted in a spin parity assignment $J^P = 1^-$ for this particle. In terms of quark model assignment, an excited $^3S_1$ bound state of $\bar{u}s$ would represent the vector boson K^{*-}. Many other excited states are also known; for more details, refer to "Review of Particle Physics" [62].

The "Review of Particle Physics" gives for this particle: Mass (K^{*-}) = 891.66 ± 0.26 MeV, full width (K^{*-}) = 50.8 ± 0.9 MeV, $I(J^P) = 1/2(1^-)$. The neutral counterpart has a mass, $M(K^{*0})$ = 896.10 ± 0.28 MeV, full width = 50.5 ± 0.6 MeV. These particles fit into a vector nonet of particles consistent with expectations from SU_3 symmetry and the constituent quark model for mesons.

Kramers Kronig Relations
See section under "Dispersion Relations".

Lagrangian
Lagrangian formulation of mechanics is an alternative formulation to that based on Newton's laws of motion. In particle mechanics, a function called the *Lagrangian L* is introduced, defined by $L = T - V$, where $T = \sum_i (1/2) m_i \vec{v}_i^2$ is the sum of the kinetic energies of the particles. The particles have mass m_i and velocity $\vec{v}_i$ for the ith particle, and V is the sum of the potential energies, due to both any external agents and the mutual interactions between the particles. In general L is a function of the velocities as well as the position coordinates of the particles involved. To be perfectly general, and not be restricted to Cartesian coordinates only, one introduces generalized coordinates $q_i, i = 1, 2, \ldots, 3N$ and velocities $\dot{q}_i = (dq_i/dt), i = 1, 2, \ldots, 3N$ for a system of N particles. Here $\dot{q}_i$ represents the time derivative of the ith coordinate, which is the gen-

eralized velocity corresponding to that coordinate. A quantity called the *action* is defined, which is the time integral of the Lagrangian between fixed end points. The equations of motion are the Euler-Lagrange equations obtained by the variation of the action integral

$$\frac{d}{dt}\left(\frac{\partial L}{\partial \dot{q}_i}\right) - \frac{\partial L}{\partial q_i} = 0, i = 1, 2, 3, \ldots, 3N.$$

For a single particle, $q_1 = x, q_2 = y, q_3 = z$ so $i = 1, 2, 3$ only, and $T = (1/2)m(\dot{x}^2 + \dot{y}^2 + \dot{z}^2)$, $V = V(x, y, z)$. The Lagrangian equations are

$$m\frac{dx}{dt} = -\frac{\partial V}{\partial x}, \qquad m\frac{dy}{dt} = \frac{\partial V}{\partial y}, \qquad m\frac{dz}{dt} = -\frac{\partial V}{\partial z}.$$

Thus the Lagrangian formulation is completely equivalent to the Newtonian formulation of mechanics.

From a theoretical and formal point of view, however, the Lagrangian formulation has some clear advantages. One of these is the clear connection between any symmetries the system possesses and conservation laws associated with those symmetries, known as *Noether's theorem*. To illustrate this in the case of a single particle, we note, if the potential energy of the system is independent of the coordinates x and y, but depends only on z, then the Euler Lagrange equations read

$$\frac{d}{dt}\left(\frac{\partial L}{\partial \dot{x}}\right) = 0, \qquad \frac{d}{dt}\left(\frac{\partial L}{\partial \dot{y}}\right) = 0, \qquad \frac{d}{dt}\left(\frac{\partial L}{\partial \dot{z}}\right) = -\frac{\partial V}{\partial z}.$$

This lets us say $\frac{\partial L}{\partial \dot{x}}$ and $\frac{\partial L}{\partial \dot{y}}$ are constants independent of time, in other words, conserved quantities. These conserved quantities are called *generalized momenta* corresponding to the x and the y coordinates. The potential being independent of x and y, the Lagrangian is invariant under a transformation of coordinates in the x and y directions, the consequence of which is that the corresponding components of the generalized momenta are conserved. This is a very general property of the Lagrangian formulation and is very powerful in identifying conservation laws. Conversely, if we know certain conservation laws hold true for some system for which we do not know the detailed dynamics, it allows us to fomulate a Lagrangian incorporating symmetries associated with those conservation laws.

The Lagrangian formulation for particle mechanics involves a finite number of degrees of freedom. In dealing with continuum mechanics, such as is the case of fluid motion, or in the case of fields, one is dealing with a system possessing an infinite number of degrees of freedom. The Lagrangian formulation has been extended to deal with the dynamics of fields including the effects of relativity and is of great interest to particle physics.

In field theory, one's focus is the field function ϕ, which is a function of space and time coordinates $\phi(\vec{x}, t)$. For simplicity we are considering an example involving only one single scalar field. We start with a *Lagrangian density* $\mathcal{L}$, which is a function of the field and its derivatives with respect to space and time coordinates $\frac{\partial \phi}{\partial x^\mu} \equiv \partial_\mu \phi$. The derivatives with respect to space and time coordinates both occur so as to treat space and time on the same footing as required by special relativity. Integrating $\mathcal{L}$ over all space and time gives the action. The Euler-Lagrange equations of motion in the case of fields is

$$\partial_\mu \left(\frac{\partial \mathcal{L}}{\partial(\partial_\mu \phi)} \right) = \frac{\partial \mathcal{L}}{\partial \phi}.$$

In the equations of motion we have the space and time derivatives appearing in such a way as to guarantee manifest Lorentz invariance. The equations for the scalar field ϕ are just the Euler-Lagrange equations for the ϕ. Similar equations hold for every component of a four-vector field, or for the components of higher rank tensor fields.

We choose a suitable Lorentz invariant Lagrangian density function, so that we obtain the desired field equation from the Euler-Lagrange equation. For example, it is easily verified that, if we want the Klein-Gordon equation to be satisfied by the field function, we choose the Lagrangian density

$$\mathcal{L} = \frac{1}{2}(\partial_\mu \phi)(\partial^\mu \phi) - \frac{1}{2}m^2\phi^2.$$

In this case, the Euler-Lagrange equation gives,

$$\partial_\mu \partial^\mu \phi + m^2 \phi = 0,$$

which is the Klein-Gordon equation.

Lagrangians have been constructed, following and generalizing this approach, for many relativistic field theories, including the electromagnetic field, the electron-positron field, quark fields, chromodynamic fields, and the combined electroweak field. To realize the quanta corresponding to these fields, it is necessary to quantize these theories. In doing so, the field functions become operators, with certain commutation relations imposed on them. In this manner, it is easy to construct quantum field theories of free fields and analyze their contents. The problem of treating fields with interaction between them is, however, in general a difficult one, and exact solutions are not possible except in special cases. Gauge field theories are examples in this category. Readers interested in details will be well advised to consult books and treatises which deal with quantum field theories.

Lamb Shift

This refers to a shift of the $2^2S_{1/2}$ energy level of the hydrogen atom, relative to the $2^2P_{1/2}$ level, first measured by Lamb and Retherford and, hence, called *Lamb shift* [238]. One of the great triumphs of Dirac's relativistic theory of the electron was the prediction of the fine structure of the energy levels of the hydrogen atom. The fine structure for the $n = 2$ levels involves the splitting of the $2^2P_{1/2}$, $2^2P_{3/2}$ states, for which Dirac theory has an exact prediction. Another prediction of the Dirac theory is that the $2^2S_{1/2}$ state must be degenerate with the $2^2P_{1/2}$ state.

Experimental efforts designed to confirm Dirac's theory through a study of the Balmer lines were hampered by the large Doppler effect of the lines in comparison to the small splitting they were trying to measure, giving results which sometimes agreed, and at other times disagreed, with theory. Lamb and Retherford measured these small splittings by using an ingenious method. They bombarded hydrogen atoms with electrons and selected those atoms, which are put in the metastable $2^2S_{1/2}$ state by this process. The metastable atoms are detected using the fact that when they reach a metal surface, the metal ejects electrons. A measure of the ejected electron current gives an idea of the number of metastable atoms reaching the metal. If during their passage to the metal, the metastable atoms are subjected to radio frequency radiation having an energy corresponding to the energy difference between one of the Zeeman components of $2^2S_{1/2}$ and any component of the P levels, the metastable atoms will absorb the radio frequency radiation and make transitions to the P states. This will result in lower numbers of metastable atoms reaching the metal plate, and consequently, the ejected electron current will diminish. Through such measurements, they were able to measure the energy differences between the $2^2S_{1/2}$, $2^2P_{1/2}$ levels and found that the $2^2S_{1/2}$ state lies above the $2^2P_{1/2}$ state by $0.033 cm^{-1}$ (about 1,000 MHz). The shift in the $2^2S_{1/2}$ level relative to the $2^2P_{1/2}$ level was clearly established which is in contradiction to the degeneracy required in Dirac's theory. This measurement indicated that there were further corrections that needed to be made to Dirac theory to get agreement with experiment.

Dirac's theory of the hydrogen atom treats the interaction between the proton and the electron through the static Coulomb potential. The actual interaction which is generated through the exchange of photons between the electron and the proton gives corrections to the static Coulomb potential. These corrections are in general called *quantum electrodynamic corrections*. Efforts to obtain these contributions were beset with divergence difficulties. The first theoretical calculation of the Lamb shift was done by Bethe [239] treating the interaction between the electron and the radiation nonrelativistically. He calculated the correction as a difference between the interaction energies of the bound electron with the radiation field and that of the free electron of the same kinetic energy interacting with the radiation field. For the case of the free electron, one gets a contribution which is interpreted as a contribution to mass of the electron. He found the resulting difference in nonrelativistic radiation theory. It was still logarithmically divergent. He obtained a finite result by cutting off the spectrum of the radiation at the electron rest energy mc^2, anticipating that a full relativistic treatment would give a result not too different from what he got. (This has been borne out by a number of subsequent calculations.) The calculated Lamb shift was in very good agreement with the measured value. The applicability of quantum electrodynamic corrections to physical processes was thus established.

Lambda Hyperon

The Λ hyperon is an electrically neutral unstable particle with a mass higher than the proton, and is observed to decay into a proton and a negative pion. Historically, it was discovered in cosmic ray studies using cloud chambers, in which the decay products left V-shaped tracks and, hence, were called *V particles*. With the building of high energy accelerators in mid to late 1950's, large numbers of these particles could be produced in the laboratory enabling a detailed study of their properties. Such studies showed that these particles are readily produced in association with another particle, known as the kaon (see also under "Associated Production" and under "Kaons"). An example of the production reaction is $\pi^- + p \rightarrow \Lambda^0 + K^0$

The known properties of the Λ particle may be summarized as follows. Its mass is $M_\Lambda = 1115.683 \pm 0.006$ MeV, lifetime $\tau_\Lambda = (2.632 \pm 0.020) \times 10^{-10}$ s. It is assigned a baryon number $B = 1$, an isotopic spin $I = 0$, as there are no other particles which are close in mass to it, and a strangeness $S = -1$; it is produced in association with another particle K^0 (in $\pi^- p$ collisions), which is assigned strangeness $S = +1$. Some of its decay modes are

$$\Lambda^0 \rightarrow p + \pi^- \quad (63.9\%),$$

$$\Lambda^0 \rightarrow n + \pi^0 \quad (35.8\%),$$

$$\Lambda^0 \rightarrow n + \gamma \quad (1.75 \times 10^{-3}),$$

$$\Lambda^0 \rightarrow p + e^- + \bar{\nu}_e \quad (8.32 \times 10^{-4}).$$

The β decay of the Λ^0 is considerably slower than what we would get if we assumed that the matrix elements for the decay were the same as those for the neutron. This was resolved by Cabibbo by introducing the notions of the $\Delta S = \Delta Q$ and the $\Delta I = 1/2$ rules and the Cabibbo angle. (See also discussion under "Cabibbo Angle" and under "Hyperons—Decay Selection Rules".)

Its spin J has been determined by the following general method which is applicable to any particle which decays into a nucleon and a pion. Consider the Λ^0 produced in the reaction $\pi^- + p \to \Lambda^0 + K^0$, with the Λ^0 subsequently decaying to $\pi^- + p$. Let us go over to the rest frame of the Λ^0 and consider its decay in that frame. Let us take the incident pion to be in the x direction. We define a plane given by the direction of the momentum vector of the incident pion and that of the Λ^0, and define the normal to this plane as the z axis. Let us take this axis as the quantization axis. Let θ be the angle between the direction of the decay proton in Λ^0 rest frame and the quantization axis. Then we may use a result derived by Lee and Yang [240] for the case of a particle X of spin J decaying into a nucleon and pion, $X \to N + \pi$. They proved that

$$|\langle\cos\theta\rangle_{av}| \leq \frac{1}{2J+2}.$$

Further, if the decay angular distribution depends only linearly on $\cos\theta$, then

$$|\langle\cos\theta\rangle_{av}| \leq \frac{1}{6J}.$$

Experimentally it was found that in the case of Λ^0 decay, the angular distribution $A(\theta)$ is linear in $\cos\theta$ and $|\langle\cos\theta\rangle_{av}| \simeq 0.19$ [241]. Thus we can immediately conclude that the spin of Λ^0 is 1/2.

The intrinsic parity of Λ^0 is defined, by convention, to be the same as that of the nucleon. Since the nucleon is assigned positive parity, the parity of Λ^0 is positive. Hence it is a $J^P = (1/2)^+$ particle. In terms of the constituent quark model, quark composition of Λ^0 is uds.

$\Lambda(1405S_{01})$ **Resonance**

Evidence for $\Sigma - \pi$ resonance was obtained first by Alston et al. [242], using the hydrogen bubble chamber technique at the LBL-Bevatron using 1.15 GeV/c beam of K^-. The reactions studied were

$$K^- + p \to \Sigma^{\pm} + \pi^{\mp} + \pi^+ + \pi^-,$$

$$K^- + p \to \Sigma^0 + \pi^0 + \pi^+ + \pi^-,$$

$$K^- + p \to \Sigma^{\pm} + \pi^{\mp} + \pi^0.$$

When a plot of the invariant mass distributions of $\Sigma^+\pi^-$ and $\Sigma^-\pi^+$ were examined, a clear peak was found (with no corresponding peak in doubly charged members $\Sigma^+\pi^+$ or $\Sigma^-\pi^-$), at 1405 MeV, with a width of about 35 MeV. Since those original measurements, most recent compilations give a mass of $M = 1407 \pm 4$ MeV and a width $\Gamma = 50 \pm 2$ MeV.

The spin was determined by an analysis of the angular distribution and the polarization of Σ^+ in the decay $\Lambda(1405) \rightarrow \Sigma^+ + \pi^-$, with the Σ^+ subsequently decaying into $p + \pi^0$.

The intrinsic parity of this particle is not fixed by these considerations; it could be odd or even. It is indirectly fixed from the following information. Analysis of $\bar{K}$-N scattering data at low energies suggests the possibility of the occurrence of $\bar{K}$-N bound state in the absence of any other channels. If the Σ-π channel is available to it, the $\bar{K}$-N bound state will not be stable but will appear as a resonance in the Σ-π system. If this resonance is a reflection of the $\bar{K}$-N bound state, the parity should be odd, as the $\bar{K}$ has odd parity and the bound state occurs in the zero angular momentum state. Thus $\Lambda(1405 S_{01})$ has spin 1/2 and is assigned odd parity, $J^P = (1/2)^-$. In terms of the quark model, this resonance is an excited state of the uds system.

Many other Λ resonance states are known: $\Lambda(1520)D_{03}$, $\Lambda(1670)S_{01}$, $\Lambda(1690)D_{03}$, $\Lambda(1800)S_{01}$, $\Lambda(1810)P_{01}$, $\Lambda(1820)F_{05}$, $\Lambda(1830)D_{05}$, $\Lambda(1890)P_{03}$, $\Lambda(2100)G_{07}$, $\Lambda(2110)F_{05}$, $\Lambda(2350)H_{09}$. Details about these states can be obtained from the full "Review of Particle Physics" [62].

Λ_b, Λ_c

These particle states, the Λ-beauty baryon (Λ_b) and the Λ-charm baryon (Λ_c), are best understood in terms of their quark model compositions. If the strange quark s in the uds combination is replaced by the charm quark c, one gets the Λ_c^+ baryon with isospin 0. If one replaces, the strange quark s in the uds combination by the b quark, one gets the Λ_b^0 state. One can envisage there would also be excited states of these quark combinations. Many of these have also been seen in experiments. The spin of the ground states have not been obtained from experimental measurements as yet. A spin of 1/2 is assigned to them on the basis of the quark model. For further details regarding these particle states, reference may be made to the full "Review of Particle Physics" [62].

Large Hadron Collider (LHC)

The Large Hadron Collider, abbreviated LHC, is a new accelerator under construction at CERN to look at elementary particle physics processes at an energy which has not so far been achieved anywhere. The successful operation of the Large Electron Positron (LEP) collider at CERN turned up very accurate data, which were sensitive to phenomena beyond the reach of its highest energies. To explore this region and beyond, one needs even higher energy machines. There are many indications from the data so far obtained that some of the fundamental questions to be answered lie in around the 1 TeV energy region. The Large Hadron Collider is being built to cover this region of energies.

To keep costs as low as possible, the LHC is being designed to share the same 27 km tunnel that the LEP machine used, and will also use all the particle sources and pre-accelerators that were already in place serving other projects. The LHC will incorporate some of the most advanced superconducting magnet technology and accelerator principles and will require great ingenuity on the part of the physicists and engineers. It will be so designed that, in addition to looking for phenomena predicted by current trends in theories, it will also be able to cope with surprises that may arise and be able to probe in new directions.

The accelerator is designed to collide beams of 7 TeV protons on beams of 7 TeV protons, for an energy in center of mass of 14 TeV. At the collision points, the beams must be designed to have brightness exceeding anything so far achieved so that the experiments will have high interaction rates. The accelerator is designed also to be capable of accelerating heavy ions such as lead, with center of mass energies exceeding 1250 TeV. The LHC, in its heavy ion accelerating capabilities, is designed to reach energies a factor of 30 greater than that achieved at the Relativistic Heavy Ion Collider (RHIC) in Brookhaven National Laboratory.

Some of the challenges that the designers will have to face may be summarized as follows.

- The luminosity of LHC must reach 10^{34}cm^{-2}s^{-1}. To achieve this, each of the proton rings will be filled with 2835 bunches, each bunch containing 10^{11} protons. Such large beam currents in an environment of superconducting magnets working at temperatures close to absolute zero will be a big challenge.

- At the collision points, the effects of beam-beam interaction will put limits on the extent to which the beam intensities can be increased without affecting their lifetimes. The LHC will operate at the highest possible limit that this imposes.

- The bunches in the beams leave wake-fields which affect the bunches that follow. Any small disturbance in the bunch will perturb succeeding bunches and under certain conditions these perturbations could get amplified. The instabilities caused by such collective action must be carefully controlled. This has to be done by suitably controlling the electromagnetic fields in the environment surrounding the beams.

- The aim is to store the beams for some ten hours of operation. During this time the particles make many millions of revolutions. The amplitudes of their oscillations about their central orbit must be carefully controlled. Beam-beam interactions may introduce nonlinear components in the guiding and focusing fields which may lead to the onset of chaotic motion. Since no theory can predict chaotic motion accurately, computers are used to track many particles, step by step, through the many thousands of magnets, for a million turns. These results define tolerences for the quality of the magnets under design and production.

- A very efficient collimation system has to be designed so that beam losses to the beam pipe wall will not occur. Otherwise, the beam particle energies converted into heat in the surrounding material will induce a quenching of the superconducting magnets and stop operation of the machine for long periods of time.

- The lattice of magnets in LHC has to be designed so as to have flexibility for undergoing changes and upgrades as may be demanded in the future due to unexpected phenomena.

- The synchrotron radiation loss suffered by the protons in LHC, although not high, still will have to be taken care of. The power emitted is about 3.7 kW, much of which will be absorbed as heat by the beam pipe which is at a very low temperature. The refrigeration system to provide cooling has to take this heat load into account. The ultraviolet light of the synchrotron radiation also releases adsorbed molecules from the beam pipe wall, which will increase the gas pressure in the beam pipe with consequent degrading effects on the beam. The system has to be designed to cope with these effects.

- The LHC magnets will be operated at $1.9°K$. This puts a demand on the coil cable quality and its assembly, which is quite unusual. To meet these demands special cables, capable of carrying 15,000 amps at $1.9°K$, and with the ability to withstand the large magnetic forces as the magnetic fields rise, are being constructed.

Four large detectors will be constructed to look at the products of the collisions at four interaction points. These are: (1) ATLAS (A Toroidal LHC Apparatus), (2) CMS (Compact Muon Solenoid), (3) ALICE (A Large Ion Collider Experiment), and (4) LHCb (CP violation studies in

B-meson decays at LHC). These gigantic detectors are under construction at present.

Left-Right Asymmetry

At SLC, a polarized electron beam is available and has been used to measure precisely the parity violation asymmetry A,

$$A = \frac{\sigma_R - \sigma_L}{\sigma_R + \sigma_L},$$

where σ_R and σ_L are the cross sections for the deep inelastic scattering of a right- and left-handed electrons, respectively, $e_{R,L} + \text{nucleon} \rightarrow eX$. The left-right asymmetry $A_{LR} = -A$. These quantities depend upon the weak mixing angle θ_W through $\sin^2 \theta_W$, and a precision measurement of the asymmetry, leads to a precision determination of the weak mixing angle. For details on the values of $\sin^2 \theta_W$, please see "Review of Particle Physics" [62].

Leptons

Leptons are particles which participate only in processes which involve the electromagnetic and weak interactions. They do not have strong interactions. Presently, three families of leptons are known: (1) the electron family, consisting of the electron (e^-) and its neutrino (ν_e); (2) the muon family, consisting of the muon (μ^-) and its neutrino (ν_μ); and (3) the tau family, consisting of the tau lepton (τ^-) and its neutrino (ν_τ). The electron, the muon, and the tau come in both charge states, negative as well as positive. The postively charged particles, the positron, the positive muon, and the positive tau, along with their associated antineutrinos, are the antiparticles of the lepton families.

The leptons are all established to be particles of spin $1/2$. The masses of the charged leptons are also measured in various experiments: $m_e = (0.51099907 \pm 0.00000015)$ MeV, $m_\mu = (105.658389 \pm 0.000034)$ MeV, $m_\tau = 1777.05$ MeV. The electron is a stable lepton. The muon and the tau are unstable and decay with lifetimes $(2.19703 \pm 0.00004) . 10^{-6}$ s, and $(290.0 \pm 1.2) . 10^{-15}$s respectively. The negative muon is found to decay as $\mu^- \rightarrow e^- + \bar{\nu}_e + \nu_\mu$, while the negative tau has many decay modes involving hadrons in the final state as well as leptonic decay modes. $\tau^- \rightarrow e^- + \bar{\nu}_e + \nu_\tau$ and $\tau^- \rightarrow \mu^- + \bar{\nu}_\mu + \nu_\tau$ each account for about 17% of the decays, the rest being decays to hadrons. The masses of the neutral leptons, the neutrinos, are not very well known experimentally. Only limits are available: $m_{\nu_e} < 3$ eV, $m_{\nu_\mu} < 0.17$ MeV, $m_{\nu_\tau} < 18.2$ MeV. Current theoretical ideas take the mass of every neutrino species to be zero; clearly this is an area in which theory may have to be modified with discovery of new phenomena involving neutrinos, such as neutrino flavor oscillations. Initially, the name *lepton* was given to indicate that these particles are light. The name lepton continues in spite of the fact that the tau lepton is more massive than the proton.

We now give a brief historical survey of how we know the facts mentioned in the previous paragraphs. The first of these leptons, the electron was discovered by J. J. Thompson in cathode rays (see section under "Cathode Rays"). It has negative charge and intrinsic spin $1/2$ as known from atomic spectroscopic measurements. Dirac's relativistic theory of the electron necessitated the introduction of the positron, to give a physical interpretation to the negative energy solutions of the Dirac equation. The positron was discovered by Anderson (see section under "Dirac Equation") in 1932. The muon was discovered by Neddermeyer and Anderson [243] and by Street and Stevenson [244] in 1937. When the muon was found, it was at first thought to be the particle postulated by Yukawa [113], whose exchange between nuclear particles generated the short range strong nuclear force. This idea had to be abandoned because it was soon realized that if the muon was responsible for the strong nuclear force, it should have a lifetime many orders smaller than its actual lifetime. The confusion in this state of affairs was resolved with the discovery by Powell and his collaborators in 1947 of another particle, the pion, in nuclear emulsions exposed to cosmic rays [245]. This particle seemed to be produced in abundance in cosmic rays and could well fit the profile of the particle required by Yukawa. An estimate of its mass put it also around 300 times the electron mass. Evidence that this particle was unstable and decayed into a pair of particles, one of which was charged and the other neutral, was also obtained in the nuclear emulsion studies. The charged particle from the decay had a unique energy, which was evidence for a two body decay. What was seen was interpreted as $\pi^- \to \mu^- + \nu$ decay, where the neutral particle was a neutrino. Other spectacular nuclear emulsion pictures showed, along with the pion decay, the decay of the muon also, leading in the end to an electron and missing energy. In the case of the muon decay, the decay electron was seen to have a distribution of energy, so it had to involve at least three particles in its decay products: $\mu^- \to e^- + \nu + \nu$. So the sequence $\pi^- \to \mu^- + \nu$, $\mu^- \to e^- + \nu + \nu$ was seen. With the advent of accelerators, pions were copiously produced and their properties as well those of the muon could be studied much more accurately than could be done with cosmic rays.

The tau lepton was discovered by Perl et al. in 1976 in e^+e^- annihilation at the SLAC 8.4 GeV electron positron collider [246]. They found anomalous $e\mu$ events with missing energy in electron-positron annihilation, accompanied by no other charged particles or photons. They suggested that these events could be explained, if as a result of the electron-positron annihilation, a pair of heavy leptons were produced, each of which decays into electron (or muon) plus neutrals (neutrinos). They provided a preliminary estimate of the mass of the heavy lepton as in the range 1.6 to 2.0 GeV. Since that first discovery much more precise work has been done on the tau lepton and the mass and lifetime quoted above arise from these further studies.

Now let us turn our attention to the neutral leptons, the neutrinos. Neutrinos are particles which were introduced to explain the continuous energy distribution of electrons from the beta decay of a nucleus. A fundamental beta decay is that undergone by the neutron: $n \rightarrow p + e^- + \bar{\nu}_e$. The shape of the energy spectrum of the electron at its end point is in principle sensitive to the mass of the electron neutrino. In practice, this study is made problematic because of the extremely small number of particles at the end point of the spectrum. The results are consistent with the electron neutrino mass being zero. A very large effort has been put in the measurements of the tritium beta spectrum, and it is these that put the limit on the electron neutrino mass as being less than 2.8 eV (see further under "Neutrino Mass").

We have been distinguishing between neutrinos and antineutrinos and we also know now that neutrinos belonging to different families are different. We now go into the observational reasons which make these necessary. First, how do we know that neutrinos and antineutrinos are distinct? Conventionally, we have associated an antineutrino with e^- beta decay, and the neutrino with e^+ beta decay. From a nuclear reactor we get lots of antineutrinos as a result of the e^- beta decay of fission products. Reines and Cowan [247] performed an experiment to detect the antineutrinos from the reactor by absorbing them in a large tank of water and looking for positrons produced by the inverse beta process: $\bar{\nu}_e + p \rightarrow n + e^+$. To increase the signal to noise ratio, they detected both the positrons by their annihilation pulse and the neutron by the delayed neutron capture gamma rays on protons produced after the known neutron slowing down time of a few microseconds.

The success of this experiment suggests another which is obtained by using crossing symmetry on the inverse beta process considered in the previous paragraph: $\nu_e + n \rightarrow p + e^-$. This should occur at the same rate as the other. Now if the antineutrino and neutrino are the same particle, the reaction $\bar{\nu}_e + n \rightarrow p + e^-$ should also occur at the same rate. An experiment incorporating this idea is the reaction $^{37}\text{Cl}(\bar{\nu}, e^-)^{37}\text{Ar}$, where the initial neutron is bound in the chlorine nucleus. Davies set up a large tank containing a large amount of carbon tetrachloride and irradiated it with antineutrinos from the reactor at Brookhaven National Laboratory and attempted to measure the number of argon atoms produced as a result of the reaction [248]. Knowing the flux of the antineutrinos from the reactor, he could estimate the number of argon atoms to be produced. He found that the cosmic ray backgrounds were such that he could not definitely establish whether the reaction did occur at the expected rate. Even with better shielding from cosmic rays, he did not detect any argon. On the other hand, he was successful in detecting neutrinos from the sun by using the same radiochemical method. This strongly suggested that the neutrino and the antineutrino are distinct particles.

This is what one would expect from the idea of associating a number called the *lepton number* to the leptons and demanding *conservation of lepton number* in reactions [249]. Associate a lepton number $l = +1$ to the electron, the negative muon, and the neutrino, and $l = -1$ to the positron, the positive muon, and the antineutrino, and $l = 0$ to all other particles. Conservation of lepton number allows the reaction $n \rightarrow p + e^- + \bar{\nu}$, because total $l = 0$ on both sides, whereas, $n \rightarrow p + e^- + \nu$ cannot occur, because $l = 0$ on the left-hand side, but $l = 2$ on the right-hand side. The pion decays will be: $\pi^- \rightarrow \mu^- + \bar{\nu}$ and $\pi^+ \rightarrow \mu^+ + \nu$ in accordance with $l = 0$ on both sides. The muon decay which we thought involved two neutrinos, should really be $\mu^\pm \rightarrow e^\pm + \nu + \bar{\nu}$, with the lepton numbers balancing on the two sides. The neutrino and the antineutrino are distinguished by the opposite values of the lepton numbers they carry.

Now we come to the question of why we distinguish between the neutrinos associated with electron, muon, etc. A reaction which can occur is the two body decay of the muon: $\mu^- \rightarrow e^- + \gamma$. This decay has been looked for but has not been found. The earliest effort to look for this decay was made by Hincks and Pontecorvo [250]. Since then many more sensitive experiments have been done and have failed to see this mode. These failures seem to imply that the muon has another property, which the electron does not have. However, the muon does decay into an electron, a neutrino, and an antineutrino—so how can we reconcile these two facts? A reconciliation is possible if we require that the neutrino associated with the electron, ν_e, and the one associated with the muon, ν_μ, are different. Making a variation on the previous assignments, let us associate a lepton number with each family [251]. Thus, if we let $l_e = +1$ for e^- and ν_e, $l_\mu = +1$ for μ^- and ν_μ, $l_e = -1$ for e^+ and $\bar{\nu}_e$, and $l_\mu = -1$ for μ^+ and $\bar{\nu}_\mu$, we can rewrite the muon decay that is observed as: $\mu^- \rightarrow e^- + \bar{\nu}_e + \nu_\mu$. This reaction has the same value of $l_e = 0$ and $l_\mu = +1$ on both sides. The conservation of lepton number must now be refined to the statement of separate conservation of electron lepton number and muon lepton number, respectively. With such a law, the decay $\mu^- \to e^- + \gamma$ does not occur, because the muon lepton number and the electron lepton numbers are not conserved. According to this, the neutrino in the negative pion decay is $\bar{\nu}_\mu$: $\pi^- \to \mu^- + \bar{\nu}_\mu$, while that associated with positive pion decay is ν_μ.

The experimental evidence that the neutrino associated with the electron is different from that associated with the muon, came from the observation of high energy neutrino reactions with matter [252]. The high energy neutrinos (antineutrinos) were obtained from the decay of pions $\pi\pm \to \mu^\pm + (\nu/\bar{\nu})$ at the Brookhaven AGS. If these neutrinos and antineutrinos carry muon lepton number, they should produce only muons and no electrons. The reaction studied was the bombardment of aluminum with the muon neutrinos (or antineutrinos). The expectation was that μ^- (or μ^+) and no electrons (or positrons) should be seen. Indeed, this is exactly what was found. Thus it was established that the electron neutrino is different from the muon neutrino. Extending this, the tau family will also have its own lepton number l_τ which should also be conserved separately.

Lifetime of Particles

Most of the elementary particles seen in nature are unstable and decay into other particles. In the lepton family, only the electron (and the positron in isolation) are known to be stable particles. The nature of the neutrino associated with the electron is at present under intense scrutiny—it is possible that it undergoes oscillations into other flavors of neutrinos. In the hadron family, only the proton (and the antiproton in isolation) are probably stable. All other particles decay into lighter particles, unless some law of conservation forbids the decay. The decay is characterized by a lifetime. The lifetime, the usual symbol for which is τ, is related to the quantity called *half-life*, $T_{1/2}$, familiar from radioactive decays. If we start with N unstable particles which can decay, then $T_{1/2}$ is the mean amount of time it takes for $N/2$ particles to decay. It is related to the lifetime by: $T_{1/2} = (\ln 2)\tau = 0.6932\tau$.

There is a characteristic mean lifetime τ associated with every particle. As examples, for the muon we have a mean lifetime $\tau_\mu = 2.2 \times 10^{-6}$s, for the charged kaons $\tau_{K^+} = 1.24 \times 10^{-8}$ s, and for the neutral pion $\tau_{\pi^0} = 8.4 \times 10^{-17}$ s. When they decay, they may have several modes of decay, the number in each mode being a certain fraction, called the *branching ratio*, of the total number of decays. Thus for example, in K^+ decays, a fraction $\simeq 64\%$ decay into $\mu^+ + \nu_\mu$, and a fraction $\simeq 21\%$ into $\pi^+ + \pi^0$. One of the objectives of the theory is to calculate the branching ratios and the lifetime for each type of particle.

The decays of most elementary particles are governed by three of the four fundamental forces: weak, electromagnetic, and strong. The fourth force, gravitation, does not seem to play a role. Which of these three forces might be involved in a given decay can generally be figured out by looking at the magnitude of the lifetimes. Generally, weak interaction lifetimes fall in the range 10^{-13} s to 15 minutes, while electromagentic lifetimes are of order 10^{-16}–10^{-17} s, and strong interaction lifetimes are of order 10^{-23} s. The longer weak interaction lifetimes, such as 15 minutes, are either a reflection of the fact that the energy difference between the decaying particle and the decay products is very small or that some other conservation law such as conservation of angular momentum is operative.

Theoretically lifetimes are obtained by calculating the probability per unit time that the system will make the transition from the initial to the final state. This probability depends on a product of two factors, the square of the matrix element for the transition and the density of final phase space available for the products of the decay, consistent with kinematic conservation laws. The size of the matrix element for the transition depends on which of the weak, electromagnetic, or strong interaction Hamiltonian is used for providing the decay interaction. The larger the energy difference between the initial and final states, the more phase space is available and therefore the higher the probability for the decay. The lifetime is obtained by taking the reciprocal of the total probability for the decay of the particle into all its modes, and the branching ratio is the ratio of the probability for a particular mode relative to the total probability.

Through the use of the uncertainty principle, a lifetime for a particle can be expressed in terms of a finite width associated with the energy level of the particle. Thus the larger the width, the larger the probability per unit time for the system to decay, and the shorter the lifetime for the particle. One can define partial widths for the different modes of decay, and the total width must be the sum of the partial widths. The branching ratio of a particular mode is then expressible as the ratio of the partial width for that mode relative to the total width. These widths are expressible in energy units of MeV or GeV.

Light-Front Field Theory

A Hamiltonian approach to relativistic quantum field theories based on quantizing on the light-front, instead of with equal time commutators, was proposed by Dirac in 1949 [253]. States in this formalism are defined on the light-cone, instead of on the usual constant time surface ($x_0 =$const), and evolve along the $x_0 = x_3$ direction instead of in x_0, where x_3 is the third component of the coordinate point. Dirac pointed out that there may be several advantages in treating field theories on the light-front. We explain these briefly here.

We know that any dynamical theory which describes interacting particle systems should satisfy the principles of special relativity. Inertial reference frames are related to one another by inhomogeneous Lorentz transformations of the form,

$$x^{'\mu} = a^\mu_\nu x^\nu + b^\mu,$$

where x and x' are the four-dimensional coordinates of the same event in the two different frames. The quantities a^μ_ν and b^μ which represent the transformation form a group that is known as the *Poincaré group*. We will focus attention on the proper subgroup of the inhomogeneous Poincaré transformations which includes only the continuous coordinate transformations and does not contain time reflections or space reflections.

For quantum mechanical systems, the state of the system with unprimed coordinates, $|\Psi\rangle$, is related to the state of the system with primed coordinates, $|\Psi'\rangle$, by a unitary transformation $U(a,b)$: $|\Psi'\rangle = U(a,b)|\Psi\rangle$. Focusing attention on infinitesimal transformations, we may write $a_{\mu\nu} = g_{\mu\nu} + \epsilon_{\mu\nu}$, and $b_\mu = \epsilon_\mu$, where $g_{\mu\nu}$ is the metric and the ϵ's are infinitesimals. The infinitesimal form for the unitary transformation is

$$U(a,b) = 1 + i\epsilon^\mu P_\mu - \frac{i}{2}\epsilon^{\mu\nu} M_{\mu\nu},$$

where P_μ are the four-momentum generators and $M_{\mu\nu}$ are the angular momentum (and boost) generators. Considering the fact that two successive transformations, (a,b) followed by (a',b'), are equivalent to a single transformation, (a'',b''), the U's satisfy the multiplication law,

$$U(a'',b'') = U(a',b')U(a,b).$$

This implies that the generators satisfy the following commutation relations:

$$[M_{\mu\nu}, M_{\rho\sigma}] = i(g_{\nu\rho}M_{\mu\sigma} - g_{\mu\rho}M_{\nu\sigma} + g_{\mu\sigma}M_{\nu\rho} - g_{\nu\sigma}M_{\mu\rho})$$

$$[P_\mu, P_\nu] = 0, \qquad [P_\mu, M_{\rho\sigma}] = i(g_{\mu\rho}P_\sigma - g_{\mu\sigma}P_\rho).$$

It was pointed out by Dirac that for interacting fields there is no unique way for including interactions in these generators. The only requirement is that, whatever way the interactions are added to the generators belonging to free fields, the above commutation relations must still be satisfied. In the nonrelativistic case, it is possible to add an interaction to the generator of time translations (Hamiltonian) only and have all the commutation relations satisfied. In the relativistic case, however, the commutation relations above can be satisfied only by adding interaction terms to more than one generator. It is a matter of choice which generators we choose to call dynamical (that is, which contain interactions) and which kinematical (that is, which do not contain interactions).

Based on these considerations, there are several possible kinematical surfaces which can be constructed. We consider two of these: (1) instant form surface, x_0 =const, and (2) the light-cone surface, $x_0 + x_3 = 0$.

Of these more familiar choice is the instant form (equal time dynamics) in which one adds the interaction to the Hamiltonian P_0. The system evolves in the time variable x_0. The physical quantities defined at an instant of time that are left invariant are the generators of space-translations (three-momenta), $P_i, i = 1, 2, 3$, and the generators of rotations (angular momenta), $J_1 = M_{23}, J_2 = M_{31}, J_3 = M_{12}$, for a total of six kinematical quantities. The remaining four generators are dynamical, which evolve the system away from the surface x_0 =const: the three rotationless boosts, $K_i = M_{0i}, i = 1, 2, 3$, and the Hamiltonian $H = P_0$.

Dynamics based on the light-cone as the invariant surface is referred to as *light-front dynamics*. Dirac showed that this form has the largest number of kinematical generators, namely seven. This is one of the advantages mentioned previously. The light-front coordiantes are defined by

$$(x^+, x^1, x^2, x^-) \equiv \left(\frac{x^0 + x^3}{\sqrt{2}}, x^1, x^2, \frac{x_0 - x_3}{\sqrt{2}} \right).$$

The seven kinematic generators are

$$P^+, P^1, P^2, J_3, K_3, \frac{K_2 - J_1}{\sqrt{2}}, \frac{K_1 + J_2}{\sqrt{2}}.$$

Here the transverse components of angular momenta are mixed with transverse boosts. Hence, unfortunately, manifest rotational invariance is lost. The dynamical generators, which evolve the system away from the initial surface $x^+ = 0$, are

$$P^-, \frac{K_1 - J_2}{\sqrt{2}}, \frac{K_2 + J_1}{\sqrt{2}}.$$

P^- is the light-front Hamiltonian, canonically conjugate to the light-front time x^+.

Besides having seven kinematic generators, a major advantage of the light-front formulation of field theory is that the vacuum state of the theory is the *physical* vacuum state for massive particles. If M is the mass of a state, then since $P^2 = 2P^+P^- - P_1^2 - P_2^2 = M^2$, both P^+ and P^- are positive for positive energies, so momentum conservation forbids pair creation from the vacuum. This means the vacuum does not have to be "renormalized" by effects of disconnected pair-creation, pair-annihilation processes as in the instant form.

Another connection to employing light-front coordinates is found in the use of infinite-momentum frames, where the coordinate system is boosted along the third axis (the longitudinal direction) to the speed of light. In these frames the motion of the particles is essentially along the $x^0 = x^3$ direction. However, it must be emphasized that use of light-front coordinates is not equivalent to boosting the system to the infinite momentum frame.

A formulation of non-perturbative QCD, based on the light-front formulation and applicable to bound-state problems, has been given in a paper by Wilson et al. [254]. It outlines a broad program of deriving effective Hamiltonians from QCD at any low momentum scale and treating non-perturbative effects. There are formidable problems to be solved in connection with the infrared problems inherent in this method.

Lightest Supersymmetric Particle

If there is conservation of R-parity, one expects that the heavy supersymmetric particles will decay to lower mass supersymmetric particles, and the lowest mass state cannot decay any further without violating the R-quantum number. Hence, it will be stable. (See further under "Supersymmetry".)

Livingston Plot

Since the early 1950's, when efforts were made to construct charge particle accelerators in the laboratory, there has been a phenomenal growth in the highest energy achieved by the accelerators. It has been noticed that roughly every 7 years, the highest energy reached by the accelerators has increased an order of magnitude. This can be dramatically exhibited in a plot called the *Livingston Plot* in which the abcissa represents the years and the ordinate represents the energy reached by the accelerators. A Livingston Plot for the years starting from 1960 onward is shown in Figure 4.15.

One might wonder how long such a spectacular growth can be maintained. It is interesting to note that, with the tremendous progress made in a succession of technologies, in the same time as the energy of the

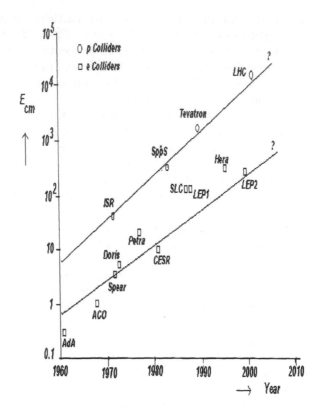

Figure 4.15: The highest energy E_{cm} achieved (in GeV) plotted as a function of the year, called the *Livingston Plot*. (Adapted from *Introduction to the Physics of High Energy Accelerators* by D. A. Edwards, M. J. Syphers. Copyright © 1993 by John Wiley & Sons, Inc. Reprinted by permission of John Wiley & Sons, Inc.)

accelerators have increased, the cost per unit energy to produce these higher energies has actually decreased. With every new accelerator or collider that has been built, great progress has been made in our quest toward the understanding of the ultimate structure of matter. At the same time new questions open up, the answers to which depend on the construction and operation of even higher energy accelerators. Thus, there are very good physics reasons why this growth in the energy of the accelerators must be maintained. Of course, the associated question is whether further developments in technologies will bring down the cost per unit energy so that new ventures may still be economically feasible.

The Large Hadron Collider (LHC) at CERN is the highest energy collider that is under construction at present and is expected to go into operation by the year 2005. There are also discussions taking place on the feasibility of building electron linear colliders reaching center of mass energies of 500 GeV or 1 TeV.

Local Gauge Invariance
See section under "Gauge Theories".

LSND Neutrino Experiment
The acronym LSND stands for *Liquid Scintillation Neutrino Detector*. The experiment was performed at the Los Alamos laboratory and was designed to search for oscillations of low energy neutrinos. The source of neutrinos was the 800 MeV proton accelerator with 1 mA intensity. The neutrinos were detected by 167 tons of liquid scintillator at a distance of 30 m from the source. LSND discovered an excess of events, which if they were due to neutrino oscillations, would correspond to to an oscillation probability of 0.3%.

MACRO Detector
This is a tracking detector for magnetic monopoles and for atmospheric neutrinos.

Magnetic Monopole
In Maxwell's electromagnetic theory, while isolated electric charges are admitted, there are no corresponding isolated magnetic poles. Electric charges are the source of electric fields, while magnetic fields have electric currents as their sources. An electric current loop of area A, carrying a current I in it, has a magnetic moment IA, and produces a magnetic field equivalent to a bar magnet with a north pole and a south pole of suitable strength, oriented perpendicular to the plane of the area of the current loop. This basic asymmetry between electric charges and magnetic poles is a feature of the basic equations of electromagnetism. Another observational fact about electric charges is that it comes in integral multiples of a fundamental charge unit. There is no theoretical explanation for why the electric charge seems to be quantized.

Dirac put forward a new idea concerning the quantization of electric charge. He showed that the quantization of electric charge can be understood, if isolated magnetic monopoles exist [255], although he could not obtain a value for the unit of the monopole. He further showed that quantum mechanics does not forbid the existence of isolated magnetic poles. He also discussed why these particles are not observed in nature.

The idea that electric charge is quantized if an isolated magnetic pole exists can be understood from the following considerations. Let us consider the motion of a spinless charged particle of charge e and mass m, moving far away from a stationary isolated magnetic pole taken as the origin. The magnetic field at the position of the electric charge a distance $\vec{x}$ from the origin is $\vec{B} = g(\vec{x}/x^3)$, where g is the strength of the magnetic pole. The equation of motion of the charge in the field of the magnetic monopole is

$$m\frac{d\vec{v}}{dt} = e(\vec{v} \times \vec{B}) = eg\vec{v} \times \frac{\vec{x}}{x^3},$$

where $\vec{x}$ and $m\vec{v}$ are the operators of position and momentum, respectively, and the right-hand side of this equation must be understood to mean the symmetrized product. The operators of position and momentum have the usual commutation relations,

$$[x_k, mv_l] = i\delta_{kl}.$$

The conserved angular momentum vector $\vec{J}$ can be derived to be,

$$\vec{J} = \vec{x} \times m\vec{v} - eg\frac{\vec{x}}{x}.$$

It is the operator for the generation of rotations.

Let us now consider the wave function of the electric charged particle in some state represented by the coordinate space wave function $\psi(\vec{x}, t)$. Let us carry out an infinitesimal rotation of the coordinate system about the axis given by the direction of $\vec{x}$, $\delta\vec{\omega} = \hat{x}\delta\phi$, where $\delta\phi$ is the infinitesimal rotation angle. The operator generating rotation on the wave function is $\delta\vec{\omega} \cdot \vec{J} = -eg\delta\phi$. For a finite rotation through an angle ϕ, this results in the transformation of the wave function,

$$\psi(\vec{x}, t) \rightarrow e^{-ieg\phi}\psi(\vec{x}, t).$$

Now it is apparent that the demand of single (or double) valuedness of the wave function under rotation through 2π implies that the product eg has to be an integer (or an integer plus $1/2$). This has the consequence that, if a magnetic monopole of strength g exists, the charge e is quantized.

More generally, the topological properties in grand unified theories can be such that when the gauge group G is broken to a lower group H at some mass scale M, monopoles may result. Polyakov [228] and 't Hooft [227] have discovered that magnetic monopoles exist if H contains a U_1 factor. The mass of such monopoles is of the order M/α, where α is the square of the relevant gauge coupling. In SU_5 GUTS, for example, monopoles of mass 10^{16} GeV exist. In this case, the Higgs field which breaks SU_5 to $SU_3 \times SU_2 \times U_1$ leads to knots in the gauge fields, which manifest as monopoles. The presence of such monopoles has an impact on the kinds of grand unification schemes and cosmology, which have been studied.

Majorana Neutrino

We have seen in Dirac theory of the electron, the electron and the positron are clearly different particles, being antiparticles of one another. They are distinguishable by the sign of their charge. There are neutral bosons, such as the π^0 meson, for which the antiparticle is the same as the particle. The question naturally arises as to whether there is a theory for a neutral fermion, such as the Dirac theory, in which the antiparticle is the same as the particle. Indeed, that such a theory is possible was discovered by Majorana [256] in 1937. A choice of representation of the Dirac matrices is possible so that the Dirac equation is purely real. Such an equation is suitable for describing a neutral fermion and in this situation, the charge conjugate (Majorana) fermion is the same as the fermion.

The question arises as to whether neutrinos are Dirac fermions or Majorana fermions. Massless fermions have helicity (the projection of spin in the direction of motion) as a good quantum number and are described by the *Weyl equation*. For massive fermions, one can construct chiral projections of left-handed and right-handed chiralities, and the mass terms in the equation mix the chiralties. Thus for a massive fermion, helicity eigenstates are different from the chirality states which are involved in weak interactions. In the standard model of electroweak interactions, only neutrinos of left-handed chirality ν_L and antineutrinos of right-handed chirality ν_R^c interact with gauge fields. Components of opposite chirality ν_R and ν_L^c are sterile, in the sense that they do not participate in weak interactions. If neutrinos are massless, chirality and helicity are the same, and in weak interaction processes, we cannot produce neutrinos and antineutrinos of the same helicity in the standard model. If they have a mass, however, the helicity can be changed by going into a Lorentz frame by a boost which changes the sign of the momentum. In this case, we have means of producing neutrinos and antineutrinos with the same helicity and comparing their activities [257]. If these interact differently, they are Dirac fermions, while if they interact identically, they are Majorana fermions.

At present it seems that the only feasible experiment to settle whether neutrinos are Majorana neutrinos is neutrinoless double beta decay. This is a process unlike the ordinary double beta decay with emission of two neutrinos ($\beta\beta2\nu$), which can occur in second order weak interaction $(A, Z) \to (A, Z + 2) + 2e^- + 2\bar{\nu}_e$ and is, therefore, rare. The process of neutrinoless double beta decay, $(A, Z) \to (A, Z + 2) + 2e^-$, is a process which can occur only if the lepton number is violated by two units, a departure from the standard model. A mass term which violates lepton number by two units is called a *Majorana mass term*. Such a term can convert the ν_{eL}^c emitted with one of the electrons, into a ν_{eL}, which is absorbed at the second vertex where the second electron is emitted. The amplitude for the process with lepton number change by two units suffers a strong suppression relative to the two neutrino double beta decay ($\beta\beta2\nu$) but has a larger available phase space for the final particles. Further, the signal for the final state in the neutrinoless double beta decay is clean—it will have one peak at the sum of the energies of the two electrons. (In the case of two neutrino double beta decay, one has a less clean signal, and the continuous electron spectrum has to be distinguished from the background.) The chances of seeing a neutrinoless double beta decay require that the single beta decay of the initial isotope must be absent or suppressed, and there be a large energy difference (Q value for the transition) between (A, Z) and $(A, Z+2)$ isotopes. Typical lifetimes for double beta decays are in the range of 10^{19} to 10^{24} years.

A number of experiments have pursued the neutrinoless double beta decay mode. ^{76}Ge has been examined for $\beta\beta0\nu$ decays and has not been observed. A constraint on the half-life of $> 1.9 \times 10^{24}$ years has been set from these data at 90% confidence level. This translates into an effective Majorana neutrino mass < 1.1 eV [258]. An experiment looking at ^{136}Xe [259] has produced a result $> 3.4 \times 10^{23}$ years which implies an effective Majorana neutrino mass < 2.8 eV. Improved experiments are in progress with ^{76}Ge, ^{136}Xe, ^{100}Mo, and ^{116}Cd to reach effective Majorana mass limits < 0.1 eV.

Mandelstam Representation

Amplitudes for scattering processes are in general complex functions of energy and momentum transfer. It helps to know the analyticity properties of the amplitude in both of these variables. There is a conjecture by Mandelstam [196], based on analysing nonrelativistic scattering problems and also analyzing a number of Feynman diagrams of perturbation theory, on the behavior of the amplitude for the process: $A+B \to C+D$, with masses $M_A = M_1$, $M_B = M_2$, $M_C = M_3$, $M_D = M_4$ and four-momenta $P_A = P_1$, $P_B = P_2$, $P_C = P_3$, $P_D = P_4$.

Before we state the conjecture below, let us introduce the Lorentz invariant variables,

$$s = (P_1 + P_2)^2, \qquad t = (P_1 - P_3)^2, \qquad u = (P_1 - P_4)^2,$$

which are called the *Mandelstam variables* (with $P_1^2 = M_1^2$, etc.). Only two of these three variables are independent because of the relation, $s + t + u = \sum_{i=1}^{4} M_i^2$ between them.

In the center of mass system for the reaction $A + B \to C + D$, s is equal to E_{cm}^2, where E_{cm} is the sum of the energies of particles A and B, t represents the square of the momentum transfer between particles A and C, and u (which is not an independent variable) represents the square of the momentum transfer between particles A and D. This is called the *s-channel reaction*. In the s-channel, s is positive, while, t and u are negative.

From this reaction, we can form another reaction, $A + \bar{C} \to \bar{B} + D$, by taking the antiparticle of C to the left-hand side and the antiparticle of B to the right-hand side. The antiparticles have four-momenta which are the negatives of the momenta of the particles, $P_2 \to -P_2, P_3 \to -P_3$, relative to the s-channel reaction. Hence here $s = (P_1 - P_2)^2$, $t = (P_1 + P_3)^2$, and $u = (P_1 - P_4)^2$. This channel is called the *t-channel reaction*. In this channel, t is positive and represents the square of center of mass energy of the $A, \bar{C}$ system, while s and u are squares of momentum transfers and are negative.

We can form yet another reaction from the above, $A + \bar{D} \to \bar{B} + C$, by taking the antiparticle of D to the left-hand side and the antiparticle of B to the right-hand side. Correspondingly here, $s = (P_1 - P_2)^2$, $t = (P_1 - P_3)^2$, and $u = (P_1 + P_4)^2$. This is called the *u-channel reaction*. In this channel, u is positive and represents the square of center of mass energy of the $A, \bar{D}$ system, while s and t are squares of momentum transfers and are negative.

As an example, let us consider the s-channel pion nucleon reaction: $\pi^+ + P \to \pi^+ + P$. The t-channel reaction is $\pi^+ + \pi^- \to \bar{P} + P$, and the u-channel reaction is $\pi^+ + \bar{P} \to \pi^+ + \bar{P}$. The t-channel reaction can be realized in the laboratory by running it in the reverse direction: $\bar{P} + P \to \pi^+ + \pi^-$. The u-channel reaction is difficult to realize in the laboratory; we do not have a target of antiprotons to bombard with a pion beam.

Now we come to the statement of the conjecture. For any "two particle to two particle" process, the amplitude can be written as a sum of terms, each of which involves products of certain factors, say, Dirac spinors, Dirac matrices, etc., and some functions which are Lorentz invariant functions. The Mandelstam conjecture is stated for the invariant functions. Consider one of these invariant functions represented by

$F(s, t, u)$ of the variables s, t, u. Mandelstam conjectured that these functions are analytic in the s, t, u plane, with cuts in these variables determined by generalized unitarity conditions. The Mandelstam representation incorporating these cuts is given by,

$$
\begin{aligned}
F(s, t, u) = & \frac{1}{\pi} \int_{s_0}^{\infty} ds' \frac{f_1(s')}{s' - s} + \frac{1}{\pi} \int_{t_0}^{\infty} dt' \frac{f_2(t')}{t' - t} \\
& + \frac{1}{\pi} \int_{u_0}^{\infty} du' \frac{f_3(u')}{u' - u} + \frac{1}{\pi^2} \int_{s_0}^{\infty} \int_{t_0}^{\infty} ds' dt' \frac{g_{12}(s't')}{(s' - s)(t' - t)} \\
& + \frac{1}{\pi^2} \int_{t_0}^{\infty} \int_{u_0}^{\infty} dt' du' \frac{g_{23}(u't')}{(t' - t)(u' - u)} \\
& + \frac{1}{\pi^2} \int_{s_0}^{\infty} \int_{u_0}^{\infty} ds' du' \frac{g_{13}(s'u')}{(s' - s)(u' - u)}.
\end{aligned}
$$

Here s_0, t_0, u_0 are the starting points for the cuts in the s, t, u variables, determined by the generalized unitarity conditions. Such a representation for the amplitude is very handy in applying dispersion relations to the physical processes under consideration and the corresponding t- and u-channel processes.

Mass Generation, Gauge Fields
See sections under "Higgs Mechanism" and "Higgs-Kibble Mechanism".

Massive muon pairs
See section under "Drell-Yan Mechanism".

Mesons
Mesons are strongly interacting particles. The name *meson* was coined to describe a particle with mass intermediate between that of the electron and the proton. The meson was first introduced by Yukawa [113] in an attempt to explain the short range nature of the strong nuclear force. According to Yukawa, quanta with masses about 300 times the electron mass, exchanged between the nuclear particles, could generate a nuclear force with a range of about 10^{-13} cm. When the muon (originally named the μ-meson) was discovered in cosmic rays with a mass of about 200 times the electron mass, it was first wrongly identified with the particle postulated by Yukawa [113]. The discovery of the "pi" (π) meson in cosmic rays using the nuclear emulsion technique by Powell and collaborators, who also obtained pictures showing the sequential decay of the π meson into the muon and that of the muon into an electron, showed clearly that there was more than one particle with mass between that of the electron and the proton. Soon it was established that it was

the π meson and not the muon that had the right properties to partic-ipate in the generation of short range nuclear interactions as suggested by Yukawa.

Further studies of cosmic ray events showed that there were other mesons with masses larger than those of the π mesons. They had masses of about 1,000 electron masses. Some decayed into two π mesons while others decayed into three, leaving V-shaped tracks in the record-ing medium. For this reason, they were referred to as V *particles* at the time. Some V particles were found to be even heavier than the proton and to decay into a proton and a π meson and, hence, could not be identified as mesons. Clarifications as to the nature of these V parti-cles had to await the building of high energy accelerators, with which one could produce, much more copiously, the events seen in cosmic rays. Such studies eventually lead to an understanding that there were two kinds of V particles. The ones that decayed into a proton and π meson (classified as baryons or heavy particles) and those that decayed into π mesons only (classifed as mesons) were produced in association with one another. This property of association was encoded in terms of a property of these particles known as *strangeness*. The π mesons were called *non-strange mesons*, while the new V particles, which decayed into π mesons (and lighter than a proton), were called *strange mesons*. The strange mesons were then known as K *mesons*. With further de-velopment of high energy accelerators and detection techniques, literally many hundreds more strongly interacting particles, both baryons and mesons, were produced. Gradually the names *pions* and *kaons* came to be adopted for the π and the K mesons, respectively.

Clearly not all these particles could be classed as fundamental. A way had to be found to understand them in terms of excitations in a more fundamental system. The founding of the constituent quark model by Gell-Mann and Zweig showed that this could indeed be done. According to this model, the baryons are composed of three quark combinations, while the mesons are composed of quark-antiquark combinations. The three quark and the quark-antiquark combinations could exist in a num-ber of states of excitation and these could be the hundreds of strongly interacting particles being seen. Initially, only three types of quarks were introduced, the u, d, and s ("up", "down", and "strange"). Now the stan-dard model envisages the existence of six quarks: (u, d) (c, s) (t, b), where c is the charm quark, t the top quark, and b the bottom quark. Briefly, we review the properties of these quarks: u, c, and t carry charge $+(2/3)e$, while d, s, and b carry charge $-(1/3)e$, and each quark carries a baryon number $(1/3)$. Bound state combinations can be formed from these and the antiquarks. Because quarks carry a spin of $1/2$, the lowest bound states would be classified as $^1S_0$ and the $^3S_1$ states with total baryon number zero. These could be identified with the pseudoscalar and vector mesons. Higher excited states could come from higher orbital angular momenta between the quark and antiquark. The quarks also carry a quantum number called *color*, and the color dependent forces are such that only color neutral (color singlet) combinations manifest as mesons. Thus now, in addition to nonstrange and strange mesons, we can have charm mesons, bottom mesons, etc., if the quarks in them are charm, bottom, etc., respectively. The "Review of Particle Physics" [62] lists many of these meson states and their quark constitutions and various detailed properties such as rates of transition between various states. It is a rich source of information on mesons.

Minimal Supersymmetric Standard Model (MSSM)

This model is an extension of the standard model [260]. It assumes $B - L$ conservation, where B and L are baryon number and lepton number, respectively. The supersymmetric partners of the standard model particles are added to the standard model particles. The Higgs sector for the standard model has one complex doublet with hypercharge $Y = 1$ capable of giving mass to u-type quarks. (There the conjugate doublet, with hypercharge $Y = -1$, gives rise to mass for the d-type quarks.) In the supersymmetric theory, the conjugate doublet is not allowed, and one needs to have two complex Higgs doublets, one with hypercharge $Y = +1$ and another with $Y = -1$, in order to give mass to both u-type quarks and d-type quarks. There is a total of eight fields, of which three are the would-be Goldstone bosons, providing masses to the $W^\pm$ and Z^0. The remaining five lead to physical particles in this model. There are two physical charged Higgs scalars $H^\pm$, two neutral scalars H_1^0, H_2^0, and one neutral pseudoscalar H_3^0.

There are many more parameters in this model than in the standard model. The MSSM model has as many as 124 independent parameters. Of these 19 have correspondence to the standard model parameters. The remaining 105 parameters are entirely new parameters. One of the new paramters is expressed as the ratio of the vacuum expectation values of the neutral scalars (v_1/v_2) (with $\sqrt{(v_1^2 + v_2^2)} = 246$ GeV) and is usually denoted by $\tan \beta$. If at the new machines, supersymmetric particles are found, there will be a lot of work to be done to determine these parameters and see how the MSSM model fares.

Among the particles of interest, special attention is focused on the Higgs sector of MSSM. Of the two neutral Higgs, H_1^0 and H_2^0, H_1^0 has the lower mass. When this mass is computed in the theory including radiative corrections at the one loop level, one finds that the mass of

H_1^0 is less than about 125 GeV for a top quark mass of 175 GeV and top-squark mass of about 1 TeV. This prediction opened an exciting prospect for LEP II, since it has enough energy to look for this Higgs at around 100 GeV to 115 GeV. To date LEP II has found no clear signal corresponding to this Higgs particle. This will certainly be one of the main concerns at LHC when it operates.

Among many other features of the MSSM model, another particular result we would like to mention here is about the running of the different coupling constants as a function of the energy. It is found that the three coupling constants, the strong, the weak, and the electromagnetic, meet at 10^{16} GeV energy, representing a unification scale which is somewhat less than the Planck scale.

A lot of theoretical and experimental work has already been done on the consequences of the MSSM model. For fuller details, we refer the reader to the "Reviews of Particle Physics" [62] and to the other references mentioned there.

MINOS Neutrino Experiment

The acronym MINOS stands for *Main Injector Neutrino Oscillation Search.* The neutrino beam is produced at Fermilab by 120 GeV protons from the main injector impinging on a carbon target. Secondary particles, mostly pions and kaons, are focused down an 800 m evacuated decay pipe. The decays of pions and kaons, produce a beam of muon neutrinos with a very small admixture of electron neutrinos. The muon neutrino beam is directed toward the Soudan mine a distance of 730 km from the Fermilab, where the Soudan2 1000 detector is located. This detector is a gas ionization time projection calorimeter consisting of 224 independent modules, each module being of dimensions 1m × 1m × 2.5m and filled with a mixture of 85% argon and 15% carbon dioxide gas. The new MINOS 10,000 detector will be used to detect the neutrinos and search for neutrino oscillation effects.

MuNu Experiment

This is an experiment designed to measure electron antineutrino scattering on electrons at the BUGEY reactor in France. The aim is to study the feasibility of measuring the magnetic moment of the neutrino.

Muon

The first evidence for the muon came from the study of cosmic rays with a cloud chamber by Neddermeyer and Anderson [261] and was confirmed in a paper by Street and Stevenson [244]. The method of identifying these particles was based on a comparison of the measured energy loss by radiation suffered by particles in cosmic ray showers with the theoretical expressions derived for them by Bethe and Heitler. They showed that the energy loss was consistent with those of particles of mass between the electron and the proton. The mass was estimated from these measurements to be about 300 times the electron mass. All the early experiments were done studying cosmic rays with a cloud chamber.

It was soon discovered that these particles were unstable in experiments performed by Rossi et al. [262] and Williams and Roberts [263]. The first measurements of the mean lifetime for these particles were presented by Rasetti [264] and Rossi and Nereson [265]. The result they obtained was approximately 2 microseconds. There was already evidence in the work of Williams and Roberts that the muon decayed into an electron. The first evidence that the electron in the muon decay had a continuous energy spectrum came from the work of Hincks and Pontecorvo [266]. They found the energy of the electrons in excess of 25 MeV present in the decay electron spectrum. Confirmation of this result came from the work of Leighton et al. [267]. Their measurements showed that the decay spectrum extended from 9 to 55 MeV with a continuous distribution in between. They also pointed to the fact that the shape of the spectrum and the end point was strong evidence for the muon to have a spin of $1/2$. From the observed end point energy, they deduced the mass of the muon to be 217 ± 4 electron masses. The observation of the continuous energy distribution of the electron shows that in the decay products, along with the electron, two other neutral particles must be involved.

With the development of accelerators and electronic methods of detection, these early results on the muons have been amply confirmed and the energy spectrum and the mean lifetime have been measured very accurately.

A critical experiment performed by Conversi, Pancini, and Piccioni tried to obtain information on whether the muon could be the mediator of strong interaction [268] as proposed by Yukawa [113]. The difference in behavior of positive and negative muons stopped in dense materials was the focus of interest of these experiments. The effect of the Coulomb field of the nucleus on the negative muon would tend to increase the probability of capture by the nucleus, while for the positive muon the opposite would be the case. If this were true, very few negative muons would decay and all the decays could be attributed to the positive muons. They found, however, that slow negative muons undergo nuclear absorption for sufficiently large atomic number Z as seen from the fact that there were no decay electrons. When Z decreases below $\simeq 10$, decay electrons start appearing. This implies that the lifetime

for nuclear absorption and the lifetime for electron decay of the muon
become nearly equal at $Z \simeq 10$. In carbon with $Z = 6$, the emergence of
decay electrons indicated that the slow negative muon undergoes decay
rather than being absorbed by the nucleus. The fact that at $Z \simeq 10$, the
lifetime for nuclear absorption becomes nearly equal to the lifetime for
decay, 2.2×10^{-6} s, shows that the muon does not interact very strongly
with nuclear particles. If it had been a strongly interacting particle, its
lifetime for absorption should have been much smaller, about 10^{-19} sec,
a factor of 10^{13} smaller. This striking discrepancy led to the proposal
of the two meson hypothesis by Marshak and Bethe [269], according
to which a meson other than the muon was responsible for the nuclear
forces. This conclusion came even before the π meson was discovered
by Powell and collaborators in cosmic rays. The observation of the π
mesons put the two meson hypothesis on a sure footing.

A further step in the mechanism for the nuclear absorption of muons
was taken by Pontecorvo [270], who noted the fact that the lifetime for
the capture of the negative muon by the nucleus is of the order of the
lifetime for electron K-capture process, once allowance is made for the
difference in the disintegration energy and the difference in the size of
the orbit of the electron and the muon. He was thus led to propose that
the muon decays by an interaction of the same strength as that involved
in β decay, coupling the electron and neutrino currents with the muon
and neutrino currents. This is the first time the idea of the universality
of Fermi weak interactions was proposed.

Before it was firmly established that the electron in muon decay had
a continuous energy spectrum, it was thought that the muon decayed
into an electron and a neutrino, in conformity with an idea of Yukawa,
and that the muon was the mediator of nuclear forces. However, the
experiment of Conversi, Pancini, and Piccioni disagrees with this picture.
So an alternative mode of decay of the muon was sought. Could it be
that $\mu^- \rightarrow e^- + \gamma$ decay occurs? The signal for such a mode of decay
would be the appearance of monoenergetic electron and photon, each
of energy about 50 MeV. An experiment looking for the monoenergetic
electron and photon was performed by Hincks and Pontecorvo [250].
They did not find any. The nonappearance of this mode of decay is now
understood in terms of a separate conservation law of lepton number for
the electron and the muon.

Muonic Atoms

A muonic atom is formed when one of the electrons of the atom is replaced by a negative muon. The muon being about 200 times more
massive than the electron, the radii of the states of the muonic atom are

about 200 times smaller. When the muon makes a transition between states, there is emission of radiation, which typically lies in the X-ray region of the spectrum. A high precision measurement of the X-ray transition energies was possible with the development of Ge(Li) detectors. This was a highly fruitful activity in many cyclotron laboratories in the late 1960's and early 1970's.

The equivalent of the Lamb-shift correction in muonic atoms is dominated by corrections due to vacuum polarization in QED. The precision measurements of the transition energies provided another independent check of QED radiative corrections and electron-muon universality. In heavy muonic atoms, fine structure and hyperfine structure measurements are sensitive to the properties of the nucleus, such as the electric charge distribution and the magnetic moments. These measurements gave independent measurements of properties of nuclei, which could be compared with those obtained from electron scattering measurements. A further experiment, μ-e conversion in a muonic atom, tested the conservations of muon and electron lepton numbers to high precision, providing data supplementing what was known from the lack of occurrence of $\mu \rightarrow e + \gamma$.

Nambu, Jona-Lasinio Nonlinear Model of Hadrons

The model proposed by Nambu and Jona-Lasinio was a very interesting one based on an analogy with the theory of superconductivity [271]. They suggested that the nucleon mass arises as a self energy of some primary fermion field through the same mechanism that gives rise to the energy gap in the theory of superconductivity.

This is a model in which an important vital step was taken which introduced the notion of spontaneous breakdown of gauge symmetry. They considered a simplified model of four-fermion nonlinear interaction having γ_5 gauge symmetry. As a consequence they found zero mass pseudoscalar bound states of the nucleon-antinucleon system which could be identified with an idealized pion of zero mass (what is now identified as the *Goldstone boson*). They further found that they could generate finite mass pseudoscalar bosons, provided a symmetry breaking term appeared in the Lagrangian of the theory. The hypothesis that the symmetries of strong interactions are spontaneously broken was very fruitful in obtaining many properties of strong interactions of hadrons at low energies. The idea of spontaneous breaking of gauge symmetry has had a profound influence on the development of a unified theory of electroweak interactions.

Neutral Intermediate Vector Boson
The idea that a neutral intermediate vector boson could play a role in weak interactions was first noted by Glashow [272]. He pointed out that with a triplet of leptons interacting with a triplet of vector bosons (two charged intermediate vector bosons and the photon), the theory possesses no partial-symmetries. On the other hand, he suggested that a simple partial-symmetric model, capable of reproducing all the weak and electromagnetic interaction data of the time, could be obtained by adding a neutral intermediate vector boson to the two charged ones, and having at least four intermediate vector boson fields including the photon. He showed that such a theory exhibits partial-symmetries which are the leptonic analogues of strangeness and isotopic spin of the strong interactions. The subsequent discovery of the existence of weak neutral current effects and the discovery of the Z^0 shows the correctness of these early ideas on the structure of weak and electromagnetic interactions of leptons.

Neutral Weak Current
The first experimental evidence for the existence of weak neutral currents came from neutrino experiments performed at CERN with the Gargamelle bubble chamber [273].

In the purely leptonic sector, this bubble chamber group observed a reaction which was interpreted as $\bar{\nu}_\mu e^- \to \bar{\nu}_\mu e^-$. They evaluated a mixing angle, the so-called *weak mixing angle* θ_W with $0.1 < \sin^2 \theta_W < 0.6$. This reaction proceeds through the exchange of the neutral intermediate vector boson (Z^0) between the muon-antineutrino and the electron.

Another reaction studied was $\nu_\mu + \text{nucleus} \to \nu_\mu + X$, the final state in the reaction producing no electrons or muons but hadrons. This is exactly what would be expected from a neutral current weak interaction. The theory would involve the product of the neutral current formed from the neutrinos with the neutral current formed from the nucleons in the nucleus. A consequence of this will be that in the final state one will have a neutrino rather than an electron or a muon. (The charged current process would involve $\nu_\mu + \text{nucleus} \to \mu^- + X$ and, hence, will have a muon in the final state.) The rate of the neutral current process was compared with that from the charged current process. The observed neutral current process could be pictured as due to the exchange of the neutral intermediate vector boson (Z^0) between the neutrino and the nucleons of the nucleus, just as the charged current process is viewed as due to the exchange of charged intermediate vector bosons $(W^\pm)$.

The neutral weak currents also play a role in atoms. In addition to the photon exchange interaction between the electron in the atom and the nucleus, there is also the exchange of the neutral intermediate vector boson Z^0 between them. The effect due to the neutral intermediate vector boson is many orders of magnitude smaller than the electromagnetic effects. However, there is one distinction between these two types of interactions. The electromagnetic interaction is parity conserving, whereas the weak interaction does not conserve parity. Hence, in the transitions between atomic energy levels, there should be small, but observable, parity violating effects. Among these parity violating effects is one where a rotation of the plane of polarization occurs for a polarized wave traversing the medium containing the atoms. This is the so-called *optical rotation*. The optical rotation is related to the amount of parity violating amplitude in the state and can be calculated in terms of the parameters of coupling of the neutral weak current. Experiments looking for this optical activity have been done in atomic bismuth. The specific transition was the $^4S_{3/2} \rightarrow^2 D_{5/2}$ $M1$ transition at 648 nm [274]. The measurements agreed with those expected from theory for a value of the weak mixing angle given by $\sin^2 \theta_W = 0.25$.

There are also other precise more recent measurements of atomic parity violation in cesium, thallium, lead, and bismuth [275]. Further confirmation of the existence of the neutral weak current comes from measurements of parity non-conservation in inelastic electron scattering. Parity violating asymmetry arises in inelastic scattering of longitudinally polarized electrons on a deuterium target, $e^- + $ deuteron $\rightarrow e^- + X$. The asymmetry parameter A is measured by finding the ratio of the difference in the cross sections for left-handed (σ_L) and right-handed (σ_R) polarized electrons, to their sum. The polarization asymmetries are very small, typically of order 10^{-5}. The measurement of such a small asymmetry was accomplished by using special techniques involving a Pockels cell with which the longitudinal polarization could be switched, and the cross sections measured in the same setup [276]. The deduced value for $\sin^2 \theta_W$ from such measurements was 0.224 ± 0.020. (See also under "Parity-Violating Asymmetry in Deep Inelastic Scattering".)

Neutralino

These are mixtures of photinos, z-inos, and neutral higgsinos, which are the supersymmetric partners of the photon, the Z^0, and the neutral Higgs bosons, respectively. (See under "Supersymmetry".)

Neutrino

In beta decay of nuclei, electrons are emitted when a nucleus in a given energy and angular momentum state transforms into a daughter nucleus in a definite state of energy and angular momentum. One would expect the electrons to exhibit a monochromatic energy spectrum corresponding to this transition. Experimentally one finds the energy spectrum of the electrons is continuous. Faced with the unsavoury prospect of abandoning energy conservation in beta decay, a way out was suggested by Pauli in an open letter in 1930 [72]. The idea he put forward was that, along with the electron, another neutral unseen particle is emitted, which carries away the balance of energy and angular momentum in the transition. Fermi gave the name *neutrino* to this particle. Systematic studies of a number of beta transitions suggested that the neutrino, besides being electrically neutral, must carry an intrinsic spin (1/2) just like the electron and must have nearly zero rest mass. The neutrino, just like the electron, should be described by Dirac's relativistic equation, which in turn implies that it must have an antiparticle, the *antineutrino*.

Since those early days of the proposal of neutrinos as a particle emitted in beta decay, much work has been done, establishing that there really is such a particle and measuring its various properties. As a result of these further studies, it is now known that it is the antineutrino that is emitted along with the electron in beta decay, while the neutrino is emitted in positron beta decays or in K-capture processes.

Neutrino Beams
The studies of weak interactions at high energies require that one have high energy neutrinos (or antineutrinos) available. It was Pontecorvo [277] who first suggested that production of intense neutrino beams by using accelerators may be feasible. The motivation for producing such beams was to perform experiments which could show whether the neutrinos associated with electrons in beta decay and the neutrinos associated with muons in pion decays are the same or different. If $\bar{\nu}_\mu$ is the same as $\bar{\nu}_e$, then it should be possible for it to induce the reaction $\bar{\nu}_\mu + p \rightarrow e^+ + n$. The signature will be the production of positrons in the final state. Schwartz [278] made a concrete proposal for producing neutrino beams and his proposal was later executed at the Brookhaven (AGS) proton synchrotron. He proposed that a natural source for high energy neutrinos are pions. On their decay, they can produce neutrinos with energies in the laboratory ranging from zero to 45% of the pion energy, and the direction of these neutrinos will be dominantly in the pion direction. Thus, one could produce ν_μ beams from π^+ decays and $\bar{\nu}_\mu$ beams from π^- decays.

Neutrino Deep Inelastic Scattering
With the production of narrow band neutrino and antineutrino beams obtained from decays of collimated beams of pions and kaons, the deep

inelastic scattering processes involving neutrinos and antineutrinos on nucleons have been studied experimentally. In these, the charged current (CC) process involves $\nu_\mu(\bar\nu_\mu)+p \to \mu^-(\mu^+)+X$, and neutral current (NC) process involves $\nu(\bar\nu) + p \to \nu(\bar\nu) + X$. The initial neutrino direction is known and the four-momentum of the final charged leptons can be measured for the (CC) process. Using the variables introduced to describe deep inelastic scattering of electrons on nucleons, $x = Q^2/(2M\nu)$, $y = [(q{\cdot}P)/(p{\cdot}P)]$ (see under "Bjorken Scaling"), the (CC) cross sections for neutrinos (antineutrinos) on nucleon works out to be

$$\frac{d^2\sigma^{CC}}{dxdy}(\genfrac{}{}{0pt}{}{\nu}{\bar\nu}\ p) = \frac{G_F^2 s}{4\pi}[xy^2 F_1^{CC}(x) + 2(1-y)F_2^{CC}(x)$$
$$\pm\ xy(2-y)F_3^{CC}(x)],$$

where the $+$ sign holds for neutrinos and the $-$ sign for antineutrinos, and where, in addition to the form factors F_1 and F_2, which appeared in deep inelastic electron-nucleon scattering, a third form factor F_3 appears in neutrino reactions because of parity violation in the weak interactions of neutrinos with nucleons. Here also Bjorken scaling is observed: in the limit of large Q^2 and ν, but x fixed, the form factors are functions of x only and do not depend upon Q^2. The Callan-Gross relation is also verifed to be valid here. Just as in the electron-nucleon deep inelastic scattering, violations from Bjorken scaling do occur, which can be accommodated in QCD.

The case of the neutral current process is somewhat less precisely handled experimentally. Here, because the outgoing neutrino is not observable and the incident neutrino direction is known but not its energy, the energy of the recoil hadron jet represented by X has to be used to measure the variable ν. The angle of jet recoil is also measurable. These measurements, without any additional information, are not sufficient to extract ν, Q^2, and incident E_ν for each event. In such a situation one has to derive cross sections averaged over the neutrino spectrum, integrated over Q^2 or over Q^2 and ν. This difficulty may be somewhat overcome if one is working with a narrow band neutrino beam arising from the decays of collimated beams of pions and kaons.

Neutrino Helicity

It has been mentioned that the study of angular momentum conservation in beta decay suggests that the neutrino spin is $1/2$. We now present the experimental evidence that bears on this statement. If the mass of the neutrino is zero, it can be shown that the Dirac equation for the neutrino can be reduced to an equation with just two components. These two states are labeled by the helicity of the neutrino (the spin projection

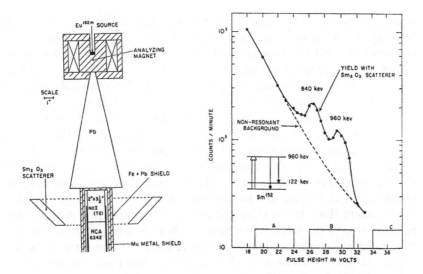

Figure 4.16: Experiment to determine the helicity of the neutrino. (Figure from M. Goldhaber, L. Grodzins, A. W. Sunyar, *Physical Review* 109, 1015, 1958. Copyright 1958 by the American Physical Society.)

along the direction of the neutrino momentum). States of given helicity are not parity conserving, and hence neutrinos of zero mass, described by Dirac equation, naturally lead to parity violation. Detailed study of the question of parity conservation in beta decays has found that parity is indeed not conserved (see section under "Parity Nonconservation in Nuclear β Decays"). The question arises whether the helicity of the charged leptons and the neutrino involved in beta decay can be directly determined experimentally. This has indeed been possible using some ingenious techniques. We describe these briefly.

A combination of clever techniques was put to use in the measurement of the helicity of the neutrino by Goldhaber et al. [117]. The experimental arrangement used by them is shown in Figure 4.16. The experiment measured the helicity of the neutrino arising from the K-capture in the isotope ^{152}Eu leading to the emission of a neutrino and a recoiling (excited) daughter nucleus ^{152}Sm*. The excited daughter nucleus decays by emitting a 961 keV γ ray to the ground state of ^{152}Sm. It is known that ^{152}Eu as well as the ground state of ^{152}Sm has spin 0. Thus by angular momentum conservation, the angular momenta carried off by the neutrino and the gamma ray must equal the angular momentum brought in by K-capture, the magnitude of which is 1/2. Let us take the direction of the gamma ray as the axis of quantization. Let m_γ, m_ν, m_K represent respectively the projection of the angular momentum of the gamma ray, the neutrino, and the capture electron, in this direction. The possible values for the projection are: $m_\gamma = \pm 1$, $m_\nu = \pm 1/2$, $m_K = \pm 1/2$, and these must satisfy, $m_\gamma + m_\nu + m_K = 0$. Two possible ways of satisfying this requirement are: $m_\gamma = +1, m_\nu = -1/2, m_K = -1/2$ or $m_\gamma = -1, m_\nu = +1/2, m_K = +1/2$. To decide which of these possibilities holds, one can measure the circular polarization of the emitted 961 keV gamma ray using the analyzing magnet shown in Figure 4.16. It was found that $m_\gamma = -1$, so that m_ν has to be $+1/2$. Now to determine the helicity of the neutrino from this information, we also have to know the direction of the neutrino momentum. It was ascertained by a further measurement that the neutrino momentum *was indeed in a direction opposite to that of the gamma ray*. Since we have just established that m_ν, the projection of the neutrino angular momentum along the direction of the photon momentum is $+1/2$, if the neutrino momentum is in a direction opposite to that of the gamma ray, it follows that the helicity of the neutrino is negative, -1. The neutrino is thus established to be left-handed. (The helicity assigned to the antineutrino is right-handed as the antineutrino is the antiparticle to the neutrino.)

The establishment of the fact that the neutrino momentum is opposite to the gamma-ray momentum was made by observing the nuclear resonance scattering of the emitted gamma rays on a target containing ^{152}Sm. Nuclear resonance scattering can occur only if the gamma ray energy is equal to the difference in energy between the excited and ground states of ^{152}Sm nuclei. This condition is usually not met because the emitting nucleus recoils, and the recoil energy given to the nucleus lowers the gamma ray energy by a small amount. However, if the nucleus, as it emits the gamma ray, recoils in the direction that the gamma ray is emitted, the gamma ray as seen by the scattering ^{152}Sm nucleus is Doppler shifted (blue shifted), compensating for the recoil energy loss. Thus the nuclear resonance scattering of the gamma rays occurs only if the source of the gamma rays (the excited daughter ^{152}Sm* nucleus) recoils toward the scatterer when it emits the gamma ray. The neutrino momentum that gives this recoil is in a direction opposite to that of the emitted gamma ray. The resonantly scattered gamma rays from the scatterer are detected by a suitably shielded NaI crystal. In Figure 4.16, the signal from the resonantly scattered gamma rays is clearly seen, which establishes that the emitting nucleus is recoiling in the direction of the emitted gamma ray, and the neutrino momentum is opposite to that of the gamma ray.

Measurements of the longitudinal polarization of the charged leptons (electron or positron) in beta decays have also been done. One method for measuring the longitudinal polarization of the decay leptons is by scattering the leptons of known momentum from a thin magnetized iron foil. It is known that two electrons of each iron atom are oriented parallel to the magnetizing field. The direction of the magnetizing field can be chosen parallel or antiparallel to the momentum of the incident leptons. The electron-electron scattering (called *Møller scattering*) depends on the relative orientation of the spins. It is greater for antiparallel orientation than for parallel. Thus, by appropriately choosing the magnetization direction in the iron, one can determine the amount of longitudinal polarization of the beta particles. In this way it has been found that electrons from beta decay carry helicity $-\beta$, while positrons carry helicity $+\beta$, where β is the velocity of the beta particle (in units of the the velocity of light). For high energy beta particles ($\beta \simeq 1$), the electron helicity is -1 (left-handed), while for the positron, the helicity is $+1$ (right-handed). In the high energy limit where the lepton mass can be ignored, the helicities for the charged leptons (anti-leptons) go over smoothly into those for the neutrinos (antineutrinos).

These observations allow one to associate a quantum number, called the *lepton number*, with these particles and demand conservation of lepton number. It is assigned a value $l = +1$ for the electron and its neutrino, while it is -1 for the positron and the antineutrino, and 0 for all other particles such as the proton or the neutron. Thus in neutron decay, conservation of lepton number requires that, along with the electron, only an antineutrino (and not a neutrino) will be emitted (see details in section under "Leptons").

Neutrino—Leptonic Quantum Numbers

We restate here that leptonic quantum numbers are associated with the neutrinos and their corresponding charged lepton partners. The ν_e and e^- are allocated an electron lepton number $l_e = 1$; thus the antiparticles $\bar{\nu}_e$ and e^+ have the opposite value $l_e = -1$. Associated with ν_μ and μ^- is the muon lepton number $l_\mu = +1$, and $\bar{\nu}_\mu$ and μ^+ have $l_\mu = -1$. Similar allocations of l_τ can be made for the tau lepton and its corresponding neutrinos. In any reaction, each type of leptonic quantum number seems to be conserved separately. Thus in positive muon decay, we will only have $\mu^+ \rightarrow e^+ + \nu_e + \bar{\nu}_\mu$ occuring. This decay has $l_\mu = +1$, and $l_e = 0$ on both sides.

Neutrino—Majorana

See section under "Majorana Neutrino".

Neutrino Mass

First evidence on the mass of the (anti)neutrino emitted in electron beta decay came already from the study of the energy spectrum of the beta electrons near the end point of the energy spectrum (see section under "Beta-Decay Theory"). Fermi, who developed a field theory of beta decay and derived an explicit expression for the continuous energy spectrum of electrons emitted in beta decay, showed that the energy spectrum has a different shape at the end point depending on the mass of the (anti)neutrino. At the end point, it would have nearly a horizontal tangent if its mass was zero, while it would have a vertical tangent if its mass was nonzero. This is true if there is perfect energy resolution. Finite experimental resolution makes clear cut statements regarding the mass of the neutrino from end point measurements problematical.

An experiment by Lyubimov et al. [279] has suggested that 17 eV $< m_{\bar{\nu}} < 40$ eV, but the finite lower bound has not been confirmed by any other experiments. A large amount of effort has been expended on precision measurements of tritium β decay. It has a small end point energy, $W = 18.6$ keV. At the end point, effects of leaving the daughter ^{3}He atomic system in excited states have also to be taken into account. Until very recently, the best fits to the data with the square of the (anti)neutrino mass as a free parameter required negative values, which is an indication that there are probably some unexplained systematic effects distorting the data at the end point [280]. A new method for investigating the tritium beta spectrum using a solenoid retarding spectrometer, consisting of an electrostatic spectrometer with a guiding magnetic field, was proposed by a group working in Mainz, Germany, in 1992 [281]. They had been perfecting this technique over a number of years and measured the end point of the tritium beta spectrum at high resolution. The source is molecular tritium frozen on to an aluminum substrate. The most recent data obtained in 1999 showed that they could be fitted with positive values of squared mass for the neutrino. If they fit using data in the last 15 eV from the end point, they obtained a limit of $m_\nu < 2.8$ eV (95%C. L.), while if they used data in the last 70 ev from the end point, the limit obtained was $m_\nu < 2.2$ eV (95%C. L.). It seems that the Mainz group with its new technique has managed to get the systematic effects under control and they are getting sensible results which are fitted with positive mass square for the neutrinos.

It turns out that the existence of a nonzero mass for the neutrino (antineutrino) has profound cosmological significance. With a neutrino mass of the order of 40 eV, the relic neutrinos from the big bang pervading the universe could give enough of a gravitational effect so as to cause the universe to be closed (that is, lead to a "big crunch") or give rise to an oscillating universe with repeated big bangs and big crunches.

Neutrino Oscillations

Pontecorvo speculated on the possibility that oscillations may occur between neutrino and antineutrino states if the lepton number were not conserved [282]. He argued for such a possibility based on what was known about the K^0-$\bar{K}^0$ system (see section under "Kaons—Neutral $K_1^0(K_S^0), K_2^0(K_L^0)$"). Second order weak interactions induce transitions in which $K^0 \leftrightarrow \bar{K}^0$ are possible in the kaon system. If neutrinos and antineutrinos are also mixed states as far as the lepton number is concerned, then oscillations might occur between these states just as in the neutral kaon system.

Since that early suggestion, other oscillation possibilities have been envisaged. If we call the states with definite lepton numbers l_e, l_μ, l_τ as states of definite flavor, then these represent eigenstates which are the basis for weak interactions. If the neutrinos possess nonvanishing masses, then the eigenstates of definite mass are mixtures of flavor eigenstates, and vice versa. These two sets of states are related by unitary transformations.

To illustrate the quantum mechanics of the flavor oscillations (cyclic transformation of flavors), we consider just two flavors and restrict to propagation in vacuum. Let the kets representing the flavor states be $|\nu_e\rangle$ and $|\nu_x\rangle$, and the mass eigenstates be $|\nu_1\rangle$ and $|\nu_2\rangle$ with energies $E_1 = \sqrt{(p^2 + m_1^2)}$ and $E_2 = \sqrt{(p^2 + m_2^2)}$, respectively, where m_1 and m_2 are the masses of these eigenstates, and the particle has a momentum p. At any time t, the flavor eigenstates are

$$|\nu_e(t)\rangle = \cos\theta|\nu_1\rangle e^{-iE_1 t} + \sin\theta|\nu_2\rangle e^{-iE_2 t},$$

$$|\nu_x(t)\rangle = -\sin\theta|\nu_1\rangle e^{-iE_1 t} + \cos\theta|\nu_2\rangle e^{-iE_2 t},$$

where θ is the mixing angle between the flavors (assumed less than 45 degrees). The probability that a neutrino starting at time $t = 0$ with electron flavor will retain its electron flavor at time t is obtained from the amplitude,

$$\langle\nu_e(0)|\nu_e(t)\rangle = \cos^2\theta e^{-iE_1 t} + \sin^2\theta e^{-iE_2 t}.$$

The probability P_{ee} for it to retain its electron flavor is

$$P_{ee}(t) = |\langle\nu_e(0)|\nu_e(t)\rangle|^2 = 1 - \sin^2 2\theta \sin^2 \frac{(E_1 - E_2)t}{2}.$$

If the masses m_1 and m_2 are very small and the neutrinos are relativistic, then we may write $E_{1,2} \simeq p + m_{1,2}^2/(2E)$ (we have used $E \simeq p$ in the second term). Hence we have, $E_2 - E_1 = (m_2^2 - m_1^2)/(2E)) \equiv \pm \Delta m^2/(2E)$, $\Delta m^2 = |m_1^2 - m_2^2|$ with plus sign applying when $m_2 > m_1$ and minus sign when $m_1 > m_2$ and Δm^2 is a positive quantity. For either sign we have

$$\sin^2 \frac{(E_1 - E_2)t}{2} = \sin^2 \frac{\Delta m^2 t}{4E}.$$

Thus we have

$$P_{ee}(t) = 1 - \sin^2 2\theta \sin^2 \frac{\pi t}{T_v},$$

where we have introduced the vacuum oscillation time $T_v = 4\pi E/\Delta m^2$. We can also write this in terms of distances; if in a time t, the distance traveled is $t \simeq R$ (with c the velocity of light $= 1$), and we introduce the vacuum oscillation length $R_v = T_v = 4\pi E/\Delta m^2$ (because we take $c = 1$), we have

$$P_{ee}(R) = 1 - \sin^2 2\theta \sin^2 \frac{\pi R}{R_v}.$$

Thus, if we start with an electron flavor at some location and let the particle travel a distance R, the probability to retain its flavor oscillates with the distance R. The oscillation amplitude is determined by the mixing parameter $\sin^2 2\theta$. Because R_v is a function of the energy E, if one has a distribution of energies for the neutrinos, one should average the above probability for flavor retention over the energies. Averaging gives $P_{ee}|_{ave} = 1 - \frac{1}{2}\sin^2 2\theta$ for vacuum oscillations at distances R which are large compared with R_v. Thus there may be at most a decrease of a factor of two, independent of the precise values of the neutrino masses or, for example, the earth-sun distance. From the above, it can also be verified that the probability P_{ex} for starting in electron flavor at $R = 0$ and ending up in flavor "x" at R is $P_{ex}(R) = \sin^2 2\theta \sin^2 \frac{\pi R}{R_v}$. Thus in the two flavor case we will have, $P_{ee}(R) + P_{ex}(R) = 1$, as it should be.

The vacuum oscillation probability depends on two parameters, $\sin^2 2\theta$ and Δm^2. A convenient expression to give R_v in meters is

$$R_v = 2.5 \left(\frac{(p/\text{MeV})}{(\Delta m^2/(\text{eV})^2)} \right)$$

For a value of $\Delta m^2 \simeq 10^{-4}$ eV^2 and $p \simeq 1 - 10$ MeV, $R_v \simeq 250 - 2,500$ kilometers. For smaller values of Δm^2, the values of R_v are correspondingly larger. Δm^2 of order 10^{-11} eV^2 gives a value of $R_v \simeq 10^{11}$ meters, which is approximately equal to the radius of the earth orbit.

Probabilities for oscillations when the neutrinos propagate in matter rather than in vacuum have also been worked out [283,284]. In matter, a difference arises between neutrinos of different flavors. The electron neutrino ν_e (and not ν_μ or ν_τ) interacts with electrons of the medium through W-boson exchange, thus altering the oscillation patterns. (Neutral current interactions, which involve Z^0 exchange, affect all neutrino flavors equally and hence need not be considered as they will only give an overall phase change and not relative phase differences.)

What Mikheyev and Smirnov pointed out (building on the earlier work of Wolfenstein) was that the effect of matter can significantly affect the neutrino oscillations and can lead to substantial enhancements of the effects over the vacuum oscillations. Such matter enhancements are called *Mikheyev-Smirnov-Wolfenstein (MSW) effect*. The presence of matter can be expressed neatly in terms of two parameters which are generalizations of the vacuum oscillation parameters. The vacuum mixing angle θ is replaced by a matter mixing angle θ_m, and the square of the mass difference, Δm^2, is replaced by a new quantity, $\Delta_m = (\Delta m^2/(2E))\sqrt{[(\pm A - \cos 2\theta)^2 + \sin^2 2\theta]}$, where $A = 2\sqrt{2}G_F N_e E/\Delta m^2$, G_F being the Fermi weak coupling constant, and N_e the electron density in the medium ($+$ sign $m_2 > m_1$, and $-$ sign $m_1 > m_2$). We can write $A = (R_v/R_e)$, where $R_e = \sqrt{2}\pi/(G_F N_e$ may be called the *neutrino-electron interaction length*, and R_v is the vacuum oscillation length introduced previously. For definiteness we take $m_2 > m_1$ in the work below.

The matter mixing angle parameter, $\sin^2 2\theta_m$ can be expressed in terms of the vacuum mixing angle parameter as (for $m_2 > m_1$)

$$\sin 2\theta_m = \frac{\sin 2\theta}{\sqrt{[1 - 2A\cos 2\theta + A^2]}}.$$

This expression shows that even if the vacuum mixing angle parameter $\sin^2 2\theta$ is small, a resonance situation can arise when $A = \cos 2\theta$, and the matter mixing angle parameter $\sin^2 2\theta_m$ can become unity (that is, $\theta_m = \pi/4$). The resonance has a width corresponding to densities satisfying $|(A - \cos 2\theta)| = |\sin 2\theta|$. One can introduce a *matter oscillation length R_m*. It can be expressed in terms of the vacuum oscillation length as $R_m = R_v/D_m$, where $D_m = \sqrt{[1 - 2A\cos 2\theta + A^2]}$. At resonance, $D_m|_{res} = \sin 2\theta$, and $\sin 2\theta_m|_{res} = 1$. Note that for electron antineutrinos, the sign of A will change, and for the same $\Delta m^2 > 0$, no resonance occurs. Very different oscillation effects occur for neutrinos and antineutrinos in matter which contains only electrons.

The flavor conversion probability, in the case of matter of constant density, takes the form, ($t \simeq R$)

$$P_{ex}(R) = \sin^2 2\theta_m \sin^2 \frac{\pi R}{R_m}.$$

In the limit when $A \ll 1, R_v \ll R_e, \theta_m \to \theta$, and $R_m \to R_v$, and the expression reduces to the expression for vacuum oscillations. In the opposite limit when $A \gg 1, R_v \gg R_e$, $\sin 2\theta_m \to \sin 2\theta / A$, $R_m = (R_v/A) = R_e$, the conversion probability is suppressed by A^2 factor, so it is small. For the resonant case when $A = \cos 2\theta$, $\sin 2\theta_m \to 1$, $R_m \to (R_v / \sin 2\theta)$ and $P_{ex}(R) = \sin^2(\frac{\pi R \sin 2\theta}{R_v})$. Even in this resonant case, because of the energy dependence of the oscillatory sine squared function, on averaging over energy, we will get a factor $1/2$. Thus for constant density, the mixing can give at most a reduction of a factor of 2 and no larger.

One can obtain somewhat larger reductions than the factor of two, if one treats the case of varying density and if this variation with distance is slow. When the density varies slowly, the mass eigenstates propagate through the region of varying density without making any transitions, and the effect of variation in the density is felt only through the behavior of the matter mixing angle parameter $\sin 2\theta_m$. The expressions for the flavor states at any point are

$$|\nu_e\rangle = \cos\theta_m |\nu_1\rangle e^{-iE_1 t} + \sin\theta_m |\nu_2\rangle e^{-iE_2 t},$$

$$|\nu_x\rangle = -\sin\theta_m |\nu_1\rangle e^{-iE_1 t} + \cos\theta_m |\nu_2\rangle e^{-iE_2 t}.$$

Let us consider that $m_2 > m_1$ and start from a region of high density and move toward the region of low density, corresponding to an electron neutrino being born at the center of the sun and propagating out through a region of slowly decreasing density. This electron neutrino will encounter the resonance mentioned above, provided its energy is larger than that required for the central density of the sun. From the resonance condition, $A = \cos 2\theta$, it can be seen that this minimum energy is $E_c = 6.6 \cos 2\theta (\Delta m^2 / (10^{-4} \text{ eV}^2))$ MeV. So, for $\Delta m^2 = 10^{-4} \text{ eV}^2$, the minimum value is about 6 MeV. Higher energy neutrinos will encounter the resonance at some point on their path where the density is lower than the central density.

In the region of high density, ν_e is mostly ν_2, because here $\sin\theta_m$ is close to unity. As it proceeds into lower density regions, if the density varies slowly enough, the ν_2 will propagate as ν_2. Eventually it exits from the sun into vacuum, where $\theta_m \to \theta$ which is small. Now ν_2 is mostly in ν_x because $\cos\theta_m \to \cos\theta$ is now large in this region. Thus a complete conversion of flavor can occur in the adiabatic case when the density varies slowly. To treat the problem fully, one needs to carry out a numerical solution of the propagation equations. In this way one can obtain regions of the $\Delta m^2 - \sin^2 2\theta$ parameter space where equal values of P_{ex} occur. Diagrams which show plots of P_{ex} in the $\Delta m^2 - \sin^2 2\theta$ parameter space are called *Mikheyev-Smirnov diagrams*. Typically, the contour of equal P_{ex} is in the shape of a triangle, with one side parallel to the abscissa ($\sin^2 2\theta$), the other parallel to the ordinate (Δm^2), the third side being the sloping line joining the end points of the other two lines to complete the triangle. In such a diagram the horizontal and vertical portions of the contour are where the adiabatic approximation is valid. The sloping line of the contour represents regions of the parameter space where the adiabatic approximation is not valid and one has to take into account the transitions between the mass eigenstates during their propagation in regions of varying density. The more general problem has been tackled by Parke [285], to which we refer the reader and we do not go into those details here.

Neutrino Oscillations—Solar Neutrino Problem

There have been two types of experiments designed to detect neutrino oscillations. In the first type, the probability that neutrinos of a certain flavor, originating from some location with a certain flux, survive as neutrinos of the same flavor at the detector, which is some distance away from the source, is measured. If measurement at the detector shows a lower flux for the original flavor, the reduced survival probability is measured. Such experiments may be called *disappearance experiments*. In the second type of experiment, one tries to detect directly, the neutrino of a diferent flavor, by looking for the associated charged lepton in the detector. Such experiments may be called *appearance experiments*.

In disappearance experiments, one would have to locate detectors at different distances from the source and compare the fluxes measured at the different locations to get an idea of the reduced flux arising from oscillations. These locations will allow one to put limits on the minimum and maximum values of the parameter Δm^2 to which these detectors are sensitive. The minimum value of Δm^2 corresponds to oscillation lengths which are much larger than the distance from the source to the farthest detector location, while there is also a maximum Δm^2 which will correspond to smaller oscillation lengths leading to averaging out of the oscillations in the nearer detector. Solar neutrino experiments belong mainly in this category. Other examples are nuclear reactor experiments, which are a good source of electron antineutrinos. Since their energies are low, they will not be able to produce muons or tau's, precluding appearance experiments. Thus, nuclear reactor experiments will also have to be of the disappearance type.

Appearance experiments, on the other hand, need only a few clear events to establish that oscillation has occurred. Examples are accelerator experiments, which are a source of muon neutrinos, so that one looks for electrons above some background level in the detector. Such a signal will establish that they are produced due to flavor oscillation of the muon neutrino.

Experiments looking for neutrino oscillations are the following: solar neutrino experiments, reactor neutrino experiments, accelerator neutrino experiments, and atmospheric neutrino experiments. We give a brief view of what has been learned about the parameter space of neutrino oscillations from these experiments.

Solar neutrino experiments were the earliest to be mounted. They attempted to measure the neutrino flux reaching the earth from the sun [286]. The neutrino flux was measured by looking for captures of the neutrinos in ^{37}Cl (Homestake experiment) contained in a large tank of dry cleaning liquid, resulting in the production of ^{37}Ar, which was then extracted by radio-chemical methods and assayed. This reaction has a threshold of about 7 MeV and, hence, is sensitive only to the ^{8}B neutrinos from the sun, which extend in energies all the way from a few MeV to about 15 MeV. Theoretical calculations based on a model of the sun, now called the *Standard Solar Model (SSM)* [287], give the flux of ^{8}B neutrinos originating from the core of the sun. Davis measured a much lower neutrino flux than was predicted by the SSM, and repeated measurements have confirmed the early discrepancy. This discrepancy was the origin of the term *solar neutrino problem*.

As it happens, the sun puts out neutrinos with a wide distribution of energies, and the Davis experiment was sensitive only to the highest energy neutrinos which are not the most abundant in flux. About 98% of solar energy is produced by the *p-p* reaction cycle, in which 4 protons are converted into ^{4}He, generating two positrons and two electron-neutrinos in the process. The maximum energy of these electron-neutrinos is 0.42 MeV. They are called the *pp neutrinos*. Some heavier nuclei are also produced in the *pp* cycle, notably Li, Be, and B with very little contribution to the energy production from the sun. ^{8}B decays and gives rise to neutrinos with a continuous spectrum stretching to 14 MeV, and ^{7}Be produces a discrete neutrino line at 0.80 MeV due to electron capture. There is also a small flux of (pep) neutrinos of about 1.44 MeV and (hep) neutrinos with a continuum of energies up to 18.7 MeV [288].

SAGE [289] and GALLEX [290] experiments were designed to measure the flux of the neutrinos arising from the *pp* cycle. Kamiokande [291] and SuperKamiokande [292] are water Cerenkov detectors, which were designed to observe ν-e scattering, the scattered electrons giving Cerenkov light which is detected. It has a threshold of 7 MeV and, hence, is capable of looking only at the ^8B neutrinos. All these experiments have reported results. The results for the SAGE, GALLEX, and Homestake experiments are given in terms of a unit called *Solar Neutrino Unit (SNU)*,

Experiment	Measurement	Theory
Homestake(SNU)	$2.56 \pm 0.16 \pm 0.16$	$7.7^{+1.2}_{-1.0}$
GALLEX(SNU)	$77.5 \pm 6.2^{+4.3}_{-4.7}$	129^{+8}_{-6}
SAGE(SNU)	$66.6^{+7.8}_{-8.1}$	129^{+8}_{-6}
Kamiokande(Flux)	$2.80 \pm 0.19 \pm 0.33$	$5.15^{+1.0}_{-0.7}$
SuperKamiokande(Flux)	$2.44 \pm 0.05^{+0.09}_{-0.07}$	$5.15^{+1.0}_{-0.7}$

Table 4.3: Results of solar neutrino measurements compared with theoretical expectations according to SSM. Statistical as well as systematic errors are given separately for the experiments. The errors in theory reflect the uncertainties of some of the underlying nuclear reaction cross sections and assumptions of the SSM.

where 1 SNU$=10^{-36}$ events/scatterer/s, while the results of Kamiokande and SuperKamiokande experiments are expressed as the observed flux of ^{8}B neutrinos above 7.5 MeV in units of No·cm^{-2}s^{-1} at the earth. Table 4.3 presents the results of these experiments and theoretical expectations. Homestake, GALLEX, and SAGE use radio-chemical methods of measurement, while Kamiokande and SuperKamiokande rely on direct observation of neutrino electron scattering in water. These latter experiments also demonstrate that the neutrinos are coming directly from the sun by showing that the recoil scattered electrons are in the direction of the line joining the sun and earth. SuperKamiokande results represent the most precise measurements of the flux of the high energy neutrinos (> 7.5 MeV) from the sun. From an examination of the table, we see that all the measured results are lower than the theoretical expectations.

Much effort has been devoted to understanding the solar neutrino deficit in terms of neutrino oscillations in vacuum, or including MSW effects. We present here only the work by Bahcall, Krastev, and Smirnov on this subject [293]. They present analyses of the data from all the solar neutrino experiments paying special attention to the SuperKamiokande data. The analysis strongly indicates that the large mixing angle (LMA) MSW solution for active flavors fits the data well with $\Delta m^2 = 1.8 \times 10^{-5}$ eV^2 and $\sin^2 2\theta = 0.76$. For vacuum oscillations, they find the data are best fitted with $\Delta m^2 = 6.5 \times 10^{-11}$ eV^2 and $\sin^2 2\theta = 0.75$. We do not present here the large number of effects they have taken into account in arriving at their conclusions. Persons interested in the details should consult the full paper.

Another new experiment studying neutrinos has come on stream at the time of this writing. The Sudbury Neutrino Observatory (SNO) in Sudbury, Ontario, Canada, uses 1 kilotonne of heavy water to detect neutrinos, besides scattering off electrons as in SuperKamiokande, reactions such as, $\nu_e d \rightarrow ppe$, and $\nu_x d \rightarrow pn\nu_x$ also. By observing the neutrons from the second reaction, and their later capture, one can measure neutral current rates from all flavors x of neutrinos. The inverse beta decay from the first reaction will help construct the neutrino spectrum above 5 MeV accurately. The comparison of the charged current and neutral current rates will give signals which are independent of any solar models. The spectral shapes will help differentiate between the different solutions. (See further under "Sudbury Neutrino Observatory (SNO)".)

Searches for neutrino oscillations using reactor and accelerator neutrinos have not provided any conclusive evidence for neutrino oscillations. Among the many efforts afoot in reactor studies, the work at the Goesgen reactor in Switzerland and at the Bugey reactor in France may be mentioned. Accelerator experiments either in progress or to be done at Brookhaven National Laboratory, Fermilab, CERN, Rutherford Laboratory with Karmen detector, and the Liquid Scintillation Neutrino Detector (LSND) at the Los Alamos Meson Physics Facility (LAMPF) may all produce convincing evidence for neutrino oscillations in the foreseeable future.

Neutrinos—More Than One Kind
Danby et al. [294] carried out an experiment designed to test whether the ν associated with the electron is the same as the ν associated with the muon. They produced ν_μ and $\bar{\nu}_\mu$ from the decay of $\pi^\pm$, and they let the ν_μ impinge on a target of aluminum and studied the products of the reaction. They found only muons produced in the final state; no electrons were produced. Thus it was clearly demonstrated that the neutrino associated with the muon is different from the neutrino associated with the electron in beta decay. For further details see section under "Leptons".

Neutrinos—Number of Species
With the advent of the high energy electron-positron collider rings and the use of these in producing an abundant supply of Z^0, much precise work has been performed on measuring the mass and the width of the Z^0 resonance. The first evidence that the number of light neutrino species equals three comes from such measurements at SLC [295]. The measurements show the mass of the Z^0 to be 91.14 ± 0.12 GeV and its width to be $2.42^{+0.45}_{-0.35}$ GeV.

In the standard model, the couplings of the leptons and the quarks to the Z^0 are specified, and one can calculate the widths for the decays of

Z^0 into various modes, e^+e^-, μ^+, μ^-, $u\bar{u}$, $d\bar{d}$, etc. If one constrains the widths to all these visible modes to that given by the standard model, by subtraction from the total width, one can find the width to modes involving decays to neutral final particles, the neutrinos associated with the electron, muon, and τ. For this they found the value 0.46 ± 0.10 GeV. The standard model also gives the value of 0.17 GeV for the partial width to a given neutrino species. The total width for the neutrals then translates into no more than 2.8 ± 0.6 neutrino species. In other words, there are only three light neutrino species consistent with the width data on the Z^0. This has been further confirmed by other groups, L3, OPAL, DELPHI, and ALEPH collaborations working with the large electron-positron collider (LEP) [296].

Neutron

The discovery of the neutron has a very interesting history. Bothe and Becker [297] in Germany observed that, when beryllium was bombarded by alpha particles emitted from polonium, it emitted a radiation of high penetrating power. Curie and Joliot [298] tried to measure the ionization produced by this radiation by letting this radiation pass through a thin window into an ionization chamber. They noticed that if they placed any material containing hydrogen, the ionization inside the chamber increased, and the ionization appeared to be due to the ejection of protons of velocities nearly the tenth of the speed of light. Since at that time the only other neutral particle that was known was the photon, they thought the beryllium radiation was gamma rays, and that the basic process by which the energy was transferred to the proton was a Compton scattering process of the gamma ray. Using this assumption, they estimated the energy of the gamma rays to be about 50 MeV.

At this point Chadwick [299] undertook further observations of the nature of the radiation emanating from beryllium. With an ionization chamber and an amplifier he measured the ionization produced by the sudden entry of a proton or an alpha particle into the chamber with the aid of an oscilloscope. He found that the beryllium radiation ejects particles, not only from hydrogen, but also from helium, lithium, nitrogen, beryllium, carbon, etc. The ionizing power of the ejected particles from hydrogen was like that of protons with speeds about one tenth of the speed of light. The ejected particles from the other elements also had large ionizing power, suggesting that they were recoil atoms of those materials. Chadwick found that the recoil atoms had ranges and ionizations which could not possibly have arisen by Compton process involving a gamma ray of 50 MeV energy on the basis of energy momentum conservation in the collision process. He found that all the inconsistencies disappear if he assumed that the radiation emanating from beryllium consisted of neutral particles of protonic mass and dubbed such particles *neutrons*. Assuming incident velocity for the neutron to be about one tenth of the speed of light, he found that the observed energies of the recoil atoms were in good agreement with what would be expected from collisions of the neutron with atoms of the material. Even from these early measurements, he was able to conclude that the neutron was heavier than the proton and suggested that the mass was probably between 1.005 and 1.008 times the mass of the proton. He even proposed a model for the neutron as a bound state of a proton and an electron with a binding energy of about 1 to 2 MeV.

Neutron Decay

As the neutron is more massive than the proton, it was expected that it would decay into a proton by the beta decay process, $n \to p + e^- + \bar{\nu}_e$. First observation of the beta decay of the neutron came from the work of Robson [300]. Working at the Chalk River nuclear reactor, he identified the positively charged particle arising from the decay of the neutron to be a proton by a measurement of the charge to mass ratio. He also showed that the proton signal disappears when a thin boron shutter was placed in the path of the neutron beam thereby absorbing the neutrons. This established that the protons arise from the decay of the neutrons. From the measurement he quoted a value for the lifetime of the neutron as between 9 and 18 minutes. The modern value for the mean lifetime of the neutron is 886.7 ± 1.9 s.

Neutron—Magnetic Moment

The earliest estimate of $-1/2$ nuclear magneton for the magnetic moment of the neutron came from studies of hyperfine structure of the atomic spectra of a number of elements by Altshuler and Tamm [301]. Direct measurement of this from a Stern-Gerlach experiment [150] on the deuteron gave a value for it between 0.5 and 1 nuclear magneton. Knowing from another measurement on the proton, that its magnetic moment was about 2.5 nuclear magnetons, and assuming the additivity of the magnetic moments of the proton and the neutron in the deuteron, Altshuler and Tamm came up with a value for the magnetic moment of the neutron between -1.5 and -2 nuclear magnetons. A much more accurate determination of the neutron magnetic moment was made by Alvarez and Bloch [302] who used an extension of the magnetic resonance method of Rabi and collaborators. Their value for the magnetic moment was -1.93 ± 0.02 nuclear magnetons. The modern value for this quantity is -1.9130428 ± 0.0000005 nuclear magneton.

Neutron—Mass
An early measurement of the mass of the neutron was made by the observation of the nuclear photoeffect of the deuteron, the analogue of the atomic photoeffect in which measurements of excitation and ionization potentials gave information about atomic energy levels and binding energies. Nuclear photoeffect was studied by Chadwick and Goldhaber [148] in which they bombarded deuterium nuclei with gamma rays from ThC″ with an energy of 2.62 MeV; the reaction studied was $\gamma + d \rightarrow p + n$. Deuteron was chosen because it was the simplest nucleus next to the proton and had a very low binding energy. With a cloud chamber, which was used to study this process, they could not get good accuracy in measurements. Still they managed to extract a mass for the neutron from these early measurements. The modern value for the neutron mass is $m_n = 1.008664904 \pm 0.000000014$ u, where "u" is the atomic mass unit defined with mass of $^{12}C/12 = 1.6605402 \times 10^{-27}$ kg. On the same scale the proton mass is $m_p = 1.007276470 \pm 0.000000012$ u, giving $m_n - m_p = 0.001388434 \pm 0.000000009$ u.

Neutron—Spin
The establishment of neutron spin to be 1/2 was first done by Schwinger in 1937 [303]. His reasonings went as follows. The fact that the proton and the deuteron have spin 1/2 and 1, respectively, suggests that the neutron spin can be 1/2 or 3/2. To decide between these possibilities, he suggested looking at experiments on the scattering of neutrons by para- and orthohydrogen. The experiments showed that the scattering cross section of orthohydrogen is much larger than that of parahydrogen. He showed that assuming a spin 3/2 for the neutron, one would expect theoretically that the para- and ortho- cross sections would be comparable in magnitude. Assuming a spin 1/2 for the neutron, on the other hand, gives much larger ortho- cross section than para- cross section in agreement with the experimental data. Thus, the spin of the neutron was determined to be 1/2.

Neutron—Statistics
The spin of the neutron has been measured to be 1/2. In accordance with the spin-statistics theorem of Pauli, the neutron must obey Fermi-Dirac statistics.

Noether's Theorem
This theorem represents an important tool to deal with symmetries associated with physical systems. Physical systems possess various symmetries, for example, symmetry under space-time transformations, internal

symmetry such as isospin and chiral symmetry, which is like an internal
symmetry acting differently on left-handed and right-handed chirality
parts of the Dirac field. In its simplest version Noether's theorem applies to these three mentioned symmetries. The theorem states that for
each generator of symmetry, there exists a conserved current associated
with that generator.

To establish the connection between a symmetry and a conservation
law, we consider two examples from the above. First, we consider a
Lagrangian density $\mathcal{L}(\phi, \partial^\mu \phi)$ for a scalar field theory and consider its
change under an infinitesimal space-time transformation, $x^\mu \to x'^\mu =
x^\mu + \epsilon^\mu$, with ϵ^μ infinitesimal. The change in the Lagrangian density
$\delta\mathcal{L} = \mathcal{L}' - \mathcal{L}$ is

$$\delta\mathcal{L} = \frac{\partial\mathcal{L}}{\partial\phi}\delta\phi + \frac{\partial\mathcal{L}}{\partial\alpha^\mu}\delta\alpha^\mu, \qquad \alpha^\mu = \partial^\mu\phi.$$

Here $\delta\phi = \phi' - \phi = \epsilon^\nu \alpha_\nu$ and $\delta\alpha^\mu = \partial^\mu(\epsilon^\nu\alpha_\nu)$. Now from the Euler-
Lagrange equation of motion we have $\frac{\partial\mathcal{L}}{\partial\phi} = \frac{\partial}{\partial x^\mu}\frac{\partial\mathcal{L}}{\partial\alpha^\mu}\partial^\mu(\epsilon^\nu\alpha_\nu)$. Using this
in the above we can write

$$\delta\mathcal{L} = \frac{\partial}{\partial x^\mu}\frac{\partial\mathcal{L}}{\partial\alpha^\mu}(\epsilon^\nu\alpha_\nu) + \frac{\partial\mathcal{L}}{\partial\alpha^\mu}\partial^\mu(\epsilon^\nu\alpha_\nu)$$

$$= \epsilon^\mu\frac{\partial\mathcal{L}}{\partial x^\mu}.$$

This can be written as

$$\frac{\partial}{\partial x^\mu}\left[-\epsilon^\mu\mathcal{L} + \frac{\partial\mathcal{L}}{\partial\alpha_\mu}\epsilon^\nu\alpha_\nu\right] = 0.$$

This can be cast in the form $\frac{\partial J^{\mu\nu}}{\partial x^\mu} = 0$, where

$$J^{\mu\nu} = -g^{\mu\nu}\mathcal{L} + \frac{\partial\mathcal{L}}{\partial\alpha_\mu}\alpha^\nu.$$

The $J^{\mu\nu}$ represents a set of conserved quantities, of which $J^{0\nu} = P^\nu$ are
the energy-momentum four-vector densities which satisfy a conservation
law. This is the familiar conservation law following from symmetry under
space-time translations.

The second example we consider in which $\mathcal{L}$ is invariant under a
global transformation involving some internal symmetry, $\phi(x) \to \phi' =
e^{-i\epsilon}\phi(x)$. The transformation is a phase transformation of the field
ϕ. In infinitesimal form we have $\delta\phi = \phi' - \phi = -i\epsilon\phi$. We see that
there is a set of conserved quantities $J^\mu = -i\frac{\partial}{\partial\alpha_\mu}\phi$, satisfying $\frac{\partial J^\mu}{\partial x^\mu} = 0$.
The quantities $J^\mu(x)$ are a set of currents which satisfy the conservation

law in this case. The zero component J^0 when integrated over three-dimensional volume is $Q = \int d^3x\, J^0$ and satisfies, $\frac{\partial Q}{\partial t} = 0$, that is, the charge Q is conserved.

NOMAD Neutrino Experiment

The acronym NOMAD stands for *Neutrino Oscillation MAgnetic Detector*. The NOMAD collaboration has set up the NOMAD detector such that every 14 s 10^{13} muon neutrinos from the CERN SPS go through the 3 ton detector. The detector can register up to half a million neutrino interactions per year. The aim is to study ν_μ-ν_τ oscillations. The track detector and the electromagnetic calorimeter are situated in a 0.4 T magnetic field.

Non-Abelian Gauge Theories

See section under "Gauge Theories".

Nuclear Forces

Historically, the forces between nuclear particles were considered one of the four fundamental forces of nature. Heisenberg's theory of atomic nuclei included forces between nuclear particles, which had exchange properties; the forces depended upon relative spins, relative positions, and relative positions and spins. Hence, such forces were called *exchange forces*. The specifically nuclear interactions between two protons, a proton and a neutron, and two neutrons are the same regardless of the electric charge carried by the nuclear particles. Such a property was called *charge independence of nuclear forces*. This fact taken together with the fact that the proton and neutron have a very small mass difference, the concept of *isotopic spin* was introduced, according to which the proton and the neutron are considered degenerate isotopic spin substates of a particle called the *nucleon*, just like the two ordinary spin degenerate substates of an electron in the absence of a magnetic field. The concept of charge independence now enters the theory with isotopic spin as the fact that the nuclear forces are invariant under rotations in isotopic spin space.

With developments in particle physics in the last three decades, we now know that quantum chromodynamics involving non-Abelian color gauge fields which couple to colored quarks, much like the Abelian electromagnetic field, which couples to electrons in quantum electrodynamics, generates the fundamental strong interaction force. The quarks interact with one another through the mediation of the color gauge fields, the quanta of which are the gluons. The nucleons are colorless bound states of three quarks. Nuclei are bound states of nucleons in much the same way as molecules are bound states of neutral atoms. The binding of atoms into a molecule occurs due to van der Waals forces between neutral atoms, which owe their origin to the fundamental electrical forces between electrons and nuclei. In much the same way, the nuclear forces between nucleons in nuclei may originate as the equivalent of the van der Waals forces from quantum chromodynamics. Details of this are not clear as yet because bound states belong in the non-perturbative region of the underlying theory.

Nuclear Reactor

A nuclear reactor is a device in which a self-sustaining fission chain reaction occurs which leads to the production of energy, other fissile materials, and isotopes. The first construction and operation of a chain reacting pile was carried out under the direction and supervision of Fermi at the University of Chicago during the months of October and November of 1942. Many people contributed to the project under Fermi's supervision. It was put into operation on December 2, 1942, a date which announced the arrival of the nuclear era [304].

The first reactor used high purity enriched uranium (enriched in the amount of ^{235}U with respect to its natural abundance) and graphite in it. When the uranium nuclei suffer fission, they give rise to neutrons along with the fission products. These neutrons are slowed down in the moderator, graphite in this case, after which they are highly effective in causing further fissions. In each succeeding fission this process repeats with further neutrons generated, slowed down, and causing further fissions. There is a certain critical mass of uranium at which the process becomes self-sustaining. The fission processes release energy which can be extracted from the system by a suitable system of heat exchangers. Neutron captures on the isotope ^{238}U lead eventually to the production of ^{239}Pu. Plutonium suffers fissions as readily as ^{235}U under slow neutron bombardment and is a product of the reactor operation. It is extracted by chemical methods from the spent fuel in reactors. Many isotopes for medical and other uses are also produced in the reactor. In many countries, production of energy by nuclear means contributes substantially to their total energy needs.

Octet Multiplet

Baryons of spin and parity, $J^P = \frac{1}{2}^+$, are observed to form a multiplet of eight particles with approximately the same mass. Such a multiplet, called a *baryon octet*, naturally lends itself to a description in terms of the octet representation of the group SU_3. Many of the characteristics of the data can be systematically understood through the detailed workings of SU_3.

Mesons of the same spin and parity are also observed to form multiplets containing eight particles of approximately the same mass. These form the *meson octet*.

If one uses the three-dimensional fundamental representation (3) of SU_3 and its conjugate $\bar{3}$, the basic particles of the representations being the quarks and antiquarks, respectively, we can form all the baryons from three quark bound states and the mesons from quark-antiquark bound states. Such a description is the basis for the constituent quark model for baryons and mesons. For details, see section under "Eightfold Way" and "Constituent Quark Model of Hadrons".

ω Meson

This particle manifests as a three-pion resonance at a mass of 782 MeV, in isospin 0 and $J^P = 1^-$ state. It was discovered first in the study of $\bar{p}p$ annihilation products, using the hydrogen bubble chamber at the LBL Bevatron [305]. The reaction studied was $\bar{p}+p \rightarrow \pi^+ + \pi^+ + \pi^0 + \pi^- + \pi^-$. When the invariant mass distributions of combinations of any three pions were plotted, they found no peaks in the mass distributions for total charge 1 and 2, while when the charge was zero, a sharp peak was found with a mass of about 782 MeV. This is called the *neutral ω meson*. The width of the peak was determined to be between 9 and 10 MeV. These results imply that the isospin is zero for this state and that the decay involves strong interactions. The G parity for this state is -1. Taken together with isospin zero, this means that the the charge conjugation parity for the neutral ω meson is $C = -1$.

The spin and parity of the state was determined by looking at the Dalitz plot (see section under "Dalitz Plot"). It can be shown that if the spin associated with this state is zero ($J = 0$), there should be no events in which the two charged pions have the same energy. There should be fewer events near the symmetry axes of the Dalitz plot. There was no such depletion found. Hence, $J = 0$ is excluded. Now consider a configuration in which the three pion momenta are all in the same line, and let this line make an angle θ with respect to some axis of quantization. Such a configuration will be described in the wave function with the angular function $Y_J^M(\theta)$, the parity of which is $(-1)^3(-1)^J$, the first $(-1)^3$ factor coming from the intrinsic parity for the three pions. Since parity is conserved in the strong decay, if the parity of the ω meson were $(-1)^J$, there should be no such events. Such events should occur on the boundary of the Dalitz plot. An examination of the plot in the boundary region shows that there are far fewer events there, which leads one to conclude that spin-parity assignments $1^-, 2^+, 3^-, \ldots$ are possible. One can also show that, if the spin-parity assignment was 2^+, there should be far fewer events in the center of the Dalitz plot where all the pions have the same energy. This is not observed. Rejecting values of spin greater than 2 as they would give much narrower widths, one takes 1^- as the spin-parity assignment for this state. The modern value for the mass of this state is 781.94 ± 0.12 MeV, with a width 8.41 ± 0.09 MeV, with $J^{PC} = 1^{--}$ and $I^G = 0^-$. In terms of the quark model, this vector meson is formed from the spin triplet combination of $u, \bar{u}$ and $d, \bar{d}$ in the isospin zero state $|\omega\rangle = (1/\sqrt{2})(u\bar{u} + d\bar{d})$.

Ω^- Hyperon

With the assumption of SU_3 symmetry, a formula for the masses of the baryonic resonances, known as the *Gell-Mann Okubo mass formula*, predicts that the separation of levels with hypercharges $Y = 0, -1, -2$ should be constant in the decuplet representation. (See further in the section under "Eightfold Way".) Based on this, and knowing the masses of the relevant $Y = 0$ and $Y = -1$ states, one can predict the mass of the $Y = -2$ member of the decuplet. The prediction turns out to be a mass of 1679 MeV. Such a particle has indeed been found [306] (see Figure 3.1 in Chapter 3). The modern value of its mass is 1672.45 ± 0.29 MeV, remarkably close to where it was predicted to be found. Because its mass is less than that of $\Xi + \bar{K}$, it cannot decay by strong interactions into this product. It can decay only by weak interactions into $\Lambda + K^-$, $\Xi^0 + \pi^-$, $\Xi^- + \pi^0, \ldots$. Its mean lifetime turns out to be $0.822 \pm 0.012) \times 10^{-10}$ s. The branching ratios for the three mentioned modes of decay are 67.8%, 23.6%, and 8.6%, respectively. The spin-parity assigned for this state, $J^P = (3/2)^+$, is based on SU_3 symmetry; no direct experimental determination of this is available. The quark model assignment for this state is a bound state of sss with spin-parity $(3/2)^+$.

$\bar{\Omega}^+$ Particle

The antiparticle of the Ω^-, the $\bar{\Omega}^+$ particle has also been seen produced in the reaction $K^+ + d \rightarrow \bar{\Omega}\Lambda\Lambda p\pi^+\pi^-$ and decays via the mode $\bar{\Lambda} + K^+$. The mass that fits the data on this particle has the value 1673.1 ± 1.0 MeV [307].

Paraquark Model

The idea of introducing another degree of freedom to get out of the difficulties with Pauli principle for the ground state baryons was proposed by Greenberg [308]. He proposed that quarks obeyed para-statistics of order 3, thus providing three additional degrees of freedom. These extra degrees of freedom have since been called *color*. See section under "Colored Quarks and Gluons".

Parity Conservation

The concept of spatial parity as a conserved quantity in a quantum mechanical system owes its origin to Wigner [309]. The parity operation is a discrete transformation involving the behavior of a physical system under inversion of spatial coordinates $\vec{r} \to -\vec{r}$. As this transformation means that we are going from a right-handed system of coordinates to a left-handed one, we might say that we are investigating whether the system has "left-right symmetry". It was tacitly assumed that, at a fundamental level, nature does not distinguish between "right" and "left" in all fundamental laws. It was quite a surprise when it was discovered that weak interactions violate this symmetry principle.

If the quantum mechanical system is described by the wave function $\psi(\vec{r})$, and P represents the parity operator, we have $P\psi(\vec{r}) \to \psi(-\vec{r})$. Applying P twice we see that $P^2\psi(\vec{r}) = \psi(\vec{r})$, or that $P^2 = 1$. This implies that the eigenvalues of the parity operator are ± 1. Parity will be a conserved quantity if the operator P commutes with the Hamiltonian of the system: $[H, P] = 0$. If the Hamiltonian has the property that $H(-\vec{r}) = H(\vec{r})$, such as when the Hamiltonian is spherically symmetric, then clearly, $[P, H] = 0$, and the system described by the wave function $\psi(\vec{r})$ will have definite parity. For systems which are described by a spherically symmetric Hamiltonian, the wave functions $\psi(\vec{r})$ have the form $R(r)Y_{lm}(\theta, \phi)$, where r, θ, ϕ are the coordinates of the point in spherical polar coordinates, and the values of l label the orbital angular momentum, and m its projection on the z-axis. The operation of spatial inversion in spherical coordinates is achieved by $r \to r$, $\theta \to \pi - \theta$ and $\phi \to \pi + \phi$. Since under these operations

$$Y_{lm}(\theta, \phi) \to (-1)^l Y_{lm}(\theta, \phi),$$

the wave function of the system $R(r)Y_{lm}(\theta, \phi) \to (-1)^l R(r)Y_{lm}(\theta, \phi)$, the eigenvalues $+1(-1)$ of the parity operator are associated with even (odd) values of l. Thus, $s, d, g, \ldots$ states have even parity, while the $p, f, h, \ldots$ states have odd parity. These parities are called *orbital parities*, since they arise from a consideration of the orbital angular momentum associated with the state. Since the wave function associated with a composite system can be expressed as a product of the wave functions for the individual systems, parity for the composite system is a product of the parities of the individual systems.

This may be applied to an atomic system in interaction with electromagnetic radiation. The system's energy levels are characterized by angular momentum (in addition to other quantum numbers), and when radiation is emitted, the highest probability occurs for electric dipole radiation (E1) with the selection rule $\Delta l = \pm 1$. Thus, the parity associated with the energy levels involved in the transition must change from even to odd or odd to even. If parity is conserved in electromagnetic interactions of atoms, the emitted E1 radiation must have parity -1 associated with it, so that the parity of the total system is conserved. This means that electric dipole radiation will be emitted when the atom goes from f to d state, d to p state, p to s state, etc. All experimental results are in conformity with parity conservation in electromagnetic interactions.

Every elementary particle also carries an *intrinsic parity*. By convention, neutrons and protons are assigned an intrinsic parity which is $+1$. Studies of nuclear reactions at energies where no new particles are produced show that parity is conserved in strong interactions; that is, the parity of the initial and final states are the same. When as a result of nuclear reactions between known particles, a new particle is created, if parity is conserved in the production reaction, the intrinsic parity of the newly produced particle is determined so that the initial and final states have the same parity. An example may serve to illustrate this point. Consider the reaction $p + p \rightarrow \pi^+ + d$ (which is a reaction involving strong interactions) producing the new particle π^+. The two protons and the deuteron all have intrinsic parity $+1$. The π^+ must have a suitable intrinsic parity associated with it such that parity is conserved on the two sides of the reaction.

In relativistic collisions antiparticles are produced and the question arises as to what intrinsic parity is to be associated with the antiparticle of a given particle. The answer to this question depends on whether we are considering particles which are bosons or fermions. For a boson, the antiparticle has the same parity as the particle. For a fermion, a consequence of the Dirac equation is that the intrinsic parity of the antifermion is opposite to that of the fermion. This is amenable to experimental tests in positronium decays.

Positronium is a bound state of an electron and a positron. It is like a hydrogen atom in which the reduced mass of the electron is half the electron mass. The lowest bound S states are ones in which the spins of the electron and positron are parallel ($^3S_1$ state) or antiparallel ($^1S_0$ state). The $^1S_0$ state decays into two photons with a mean lifetime of about $10^{-10}s$, while the $^3S_1$ state decays into three photons with a mean lifetime of about $10^{-7}s$. Let us consider the two photon decay: $e^+ + e^- (^1S_0) \rightarrow \gamma + \gamma$. Because the orbital parity associated with the S state is $+1$, the parity on the left-hand side is determined by the intrinsic parities of the electron and the positron. It will be $+1$ if they have the same intrinsic parity, and -1 if they have opposite intrinsic parity. If we can determine experimentally the parity carried by the two photon system on the right-hand side, using parity conservation, we would have determined the parity on the left-hand side. In the rest frame of the positronium, the two photons have equal and opposite momenta, $\pm\vec{p}_\gamma$. Let their polarization vectors (the direction of their electric vectors) be $\vec{e}_1$ and $\vec{e}_2$. Two possible forms for the wave function of the two photon system (with total angular momentum 0) can be written down which satisfy Bose symmetry and which involve labels of both photons: $\psi_1 \propto (\vec{e}_1 \cdot \vec{e}_2)$ and $\psi_2 \propto (\vec{e}_1 \times \vec{e}_2) \cdot \vec{p}_\gamma$. Of these ψ_1 has even parity because it is a scalar product of two vectors, each of which changes sign under space inversion. On the other hand, ψ_2 has odd parity, because, it involves an additional vector $\vec{p}_\gamma$ in the scalar product, which changes sign under space inversion. These functions are such that, for even parity, the polarization vectors of the two photons must be parallel to one another, while for odd parity, they must be perpendicular to one another. Whether the polarizations are parallel or perpendicular can be determined by experiment. Just such a determination was made in an experiment performed by Wu and Shaknov [310]. In their experiment, they used the dependence of Compton scattering of photons in a magnetized material on the polarization of the photon to determine the polarization direction of each of the photons. In this way, they determined the polarizations of the two photons to be perpendicular to one another. The two photon system has odd parity which implies that the intrinsic parity of the positron is opposite that of the electron. This is true for other fermions, too. Thus, the antiproton and antineutron have opposite intrinsic parities to those of the proton and the neutron, respectively.

Parity Nonconservation in Nuclear β Decays

The observation of the decays of kaons into two pion and three pion final states, the final states having opposite parities, opened the question of whether parity is conserved in weak interactions. One place where lots of experimental data were available was in beta decay. The question was first posed whether one could detect parity nonconservation in beta decays.

An experiment to detect parity nonconservation in beta decays was carried out by Wu et al. [311]. They studied β decay of oriented nuclei to see if they could find any asymmetry in the direction of emission of the electrons with respect to the spin direction of the nuclei. Any such correlation between the spin of the nucleus and the momentum of the

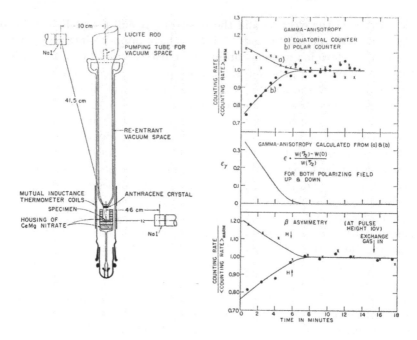

Figure 4.17: The experimental arrangement which established parity violation in nuclear beta decay. (Figure from C. S. Wu, E. Ambler, R. W. Hayward, D. D. Hoppes, R. P. Hudson, *Physical Review* 105, 1413, 1956. Copyright 1956 by the American Physical Society.)

emitted electrons would signal the presence of a pseudoscalar $\vec{J}\cdot\vec{p}_e$, which changes sign under reflections. If electrons were emitted preferentially in the direction in which nuclear spins were pointing (or preferentially in the opposite direction), such an observation would constitute parity violation in β-decay. To do such an experiment, they chose to work with ^{60}Co, which has a nuclear spin $J = 5$ and β-decays by electron emission to an excited state of ^{60}Ni, which has a nuclear spin, $J = 4$. The β transition is a pure Gamow-Teller transition. To orient the ^{60}Co nuclear spins, they mounted the sample inside a solenoid and put the assembly inside a cryostat maintained at a temperature of 10 millikelvin (see Figure 4.17). The magnetic field generated by the solenoid oriented the cobalt nuclear spins in the direction of the magnetic field. On the left portion of Figure 4.17 the portion of the cryostat containing the ^{60}Co nuclei is shown. The anthracene crystal just above the specimen counts the beta particles by sending the scintillations from the crystal through a lucite rod to a photomultiplier tube. A measure of the degree of orientation of the cobalt nuclei was obtained by looking at the angular distribution of the gamma rays emitted in the decay of the excited ^{60}Ni. Two NaI counters, one level with the source and the other located close to the axis of the lucite rod, aided in the determination of the angular anisotropy of the gamma rays and thus the degree of orientation of the cobalt source. On the right side of Figure 4.17, the top part shows the gamma ray counting rates in the two NaI counters. The calculated gamma-anisotropy is shown in the middle part. In the bottom, the beta asymmetry is shown for the two magnetic field directions. It should be noted that the beta and the gamma asymmetry disappear at the same time, which is the warm-up time (about 6 minutes). Notice also that the beta counts become independent of the field direction when the sample becomes warm, and the cobalt nuclei lose their orientation.

Let θ represent the angle between the spin direction of the cobalt nuclei and the direction of motion of the β-electrons. The experiment measured the number $N(\theta)$ of electrons emitted as a function of the angle θ. Wu and collegues found that this number distribution could be expressed in the form

$$N(\theta) = 1 - (\hat{n} \cdot \vec{p}_e/E),$$

where $\hat{n}$ is a unit vector in the direction of the nuclear spin (magnetic field), $\vec{p}_e$ the vector momentum of the β-electron, and E its total energy. It is clear from this expression that more electrons are emitted in a direction opposite to the nuclear spin than in a direction parallel to it. As mentioned above, this is clear violation of front to back symmetry with respect to the nuclear spin and, hence, violation of parity in nuclear beta decay.

In the Gamow-Teller transition, nuclear parity does not change, so detection of parity nonconservation in the decay implies that the emitted leptons are in a mixture of orbital s and p states. Since $N(\theta)$ is a maximum for $\theta = \pi$ and $L_z = 0$ in this direction, the electron and neutrino spins must be aligned in this direction, or in other words, they are longitudinally polarized. This longitudinal polarization has been measured in electron and positron decays, by scattering them against electrons in a magnetized material, and has been found to be $-\beta$ and $+\beta$, respectively, where β is the velocity of the decay lepton (in units of the velocity of light). The longitudinal polarization of the neutrino (helicity) has also been determined by an ingenious experiment (see section under "Neutrino Helicity").

Parity-Violating Asymmetry in Deep Inelastic Scattering
In the standard model of electroweak synthesis, one of the predictions is for the existence of the neutral intermediate vector boson Z^0, with

a mass of about 90 GeV, with prescribed couplings to the leptons and quarks. The reality of the effects of the Z^0, called *neutral current effects*, were measured a considerable time before the actual production of the Z^0 in the laboratory. One of these effects occurs in the deep inelastic scattering of polarized electrons on deuterium target $e_{R,L}d \rightarrow eX$. If parity is violated in the couplings of Z^0, a parity-violating asymmetry, defined by

$$A = \frac{\sigma_R - \sigma_L}{\sigma_R + \sigma_{L|}}$$

must have a nonzero value, where $\sigma_{R,L}$ are the cross sections for the deep inelastic scattering of right- and left-handed electrons on the deuteron. In the quark-parton model this asymmetry can be shown to have the form

$$\frac{A}{Q^2} = c_1 + c_2 \frac{1 - (1 - y)^2}{1 + (1 - y)^2},$$

where Q^2 is the negative of the square of the four-momentum transfer, and y is the fraction of the energy transferred from the electron to the hadrons. For the deuteron target, one can show that

$$c_1 \sim \frac{3G_F}{5\sqrt{2}\pi\alpha}\left(-\frac{3}{4} + \frac{5}{3}\sin^2\theta_W\right),$$

and

$$c_2 \sim \frac{9G_F}{5\sqrt{2}\pi\alpha}\left(\sin^2\theta_W - \frac{1}{4}\right),$$

where G_F and α are the Fermi weak coupling constant and the fine structure constant, respectively. In a very beautiful experiment performed at SLAC, this parity-violating asymmetry was measured [312] in 1979, about four years before the Z^0 was directly produced in the laboratory with the CERN Sp$\bar{p}$S collider. This not only established the reality of the effect of the Z^0 boson but also gave a good measurement of the weak mixing angle θ_W.

Particle-Antiparticle Conjugation
See section under "Charge Conjugation".

Parton Model
See also section under "Bjorken Scaling—Explanation".

The study of the deep inelastic scattering of electrons on protons shows that the form factors depend only on the dimensionless variable $x = Q^2/(2M\nu)$, when one takes the limit $Q^2 \rightarrow \infty$ and $\nu \rightarrow \infty$ but with x finite (Bjorken scaling), where $q^2 = -Q^2$ is the square of the four-momentum transferred by the electron and ν is the energy transferred by the electron in the proton rest frame. The parton model explanation of Bjorken scaling was proposed by Feynman [92]. He suggested that the proton may be pictured as a collection of pointlike free particles called *partons*, and that the scattering of the incident high energy electron by the proton, can be evaluated by calculating first the scattering from these free pointlike constituents.

The reason why this is possible has to do with the time duration of the collision of the electron in comparison to the time during which the proton dissociates into a virtual state of free partons. If $q^\mu(q^0, \vec{q})$ is the virtual photon four-momentum, q^0 is very large for deep inelastic scattering. The collision time (t_{coll}) with the proton is of the order of the time taken for the proton to absorb the photon, $\simeq (1/q^0)$. If δE is the magnitude of the difference in energy between the proton state and its state in virtual dissociation into partons, then the dissociation time (t_{diss}) is $\simeq (1/\delta E)$. One can show that t_{diss} is very large compared with t_{coll} at high energies, in which case the partons can be considered as free particles during the collision. We can easily estimate $q^0, \delta E$ as follows. Working in the center of mass frame of the electron-proton system, let the incident proton of mass M have initial four-momentum P of $(P^0 = \sqrt{\vec{P}^2 + M^2}, \vec{P})$, then the electron's initial four-momentum is $(|\vec{P}|, -\vec{P})$ where we have neglected the electron mass. After scattering, the electron four-momentum is $(|\vec{P}| - q^0, \vec{P} - \vec{q})$. Again if we neglect the mass of the final electron, we must have

$$(|\vec{P}| - q^0)^2 - (\vec{P} - \vec{q})^2 = 0,$$

which simplifies to $q^2 - 2q^0|\vec{P}| - 2\vec{P} \cdot \vec{q} = 0$. We also have $P \cdot q = q^0\sqrt{\vec{P}^2 + M^2} - \vec{P} \cdot \vec{q}$. Eliminating $\vec{P} \cdot \vec{q}$ between these two expressions and simplifying, we get

$$q^0 \simeq \frac{2P \cdot q + q^2}{4|\vec{P}|}.$$

If the proton dissociates into a parton of mass M_1 and other partons of mass M_2 with momenta $x|\vec{P}|$ and $(1-x)|\vec{P}|$, then

$$\delta E = \sqrt{x|\vec{P}|^2 + M_1^2} + \sqrt{(1-x)^2|\vec{P}|^2 + M_2^2} - \sqrt{\vec{P}^2 + M^2}.$$

This simplifies to

$$\delta E \simeq \left(\frac{M_1^2}{2x} + \frac{M_2^2}{2(1-x)} - \frac{M^2}{2} \right) / |\vec{P}|.$$

In the deep inelastic process, remembering $q^2 = -Q^2$ and $P \cdot q = M\nu$ are very large (compared to M^2), we have for the ratio of the collision time to the dissociation time,

$$(t_{coll}/t_{diss}) = [(2M_1^2/x) + (2M_2^2/(1-x))]/(2M\nu - Q^2).$$

Clearly this ratio is very small when both Q^2 and $2M\nu$ are very large compared to M^2. Thus these considerations show why in the deep inelastic process one can treat the partons in the proton as not interacting with one another within the duration of the collision with the electron. The partons scatter the electron incoherently, and the electron-proton cross section is the sum of the parton contributions. However, as the fractional momentum x carried by the parton is continuously distributed between 0 and 1, the sum over partons is replaced by an integral over x with a function $f_i(x)$, which is the probability distribution function for a parton of type i to carry a fractional momentum x of the proton (and summed over all parton types i). These are called *parton distribution functions* and they characterize the proton in the parton model. After the scattering process, over a long time scale, the partons are assumed to recombine into the final state hadronic fragments arising from the proton.

Parton Model—Bjorken Scaling Violation

In the interests of economy, let us suppose that the partons are identified with quarks. Then one can use quantum chromodynamics (QCD) to govern the interactions of quarks. In the Bjorken scaling limit, the proton is just a collection of free quarks. In QCD, the quarks emit and absorb gluons, the coupling between them tending to zero at infinite momentum transfers ($Q^2 \rightarrow \infty$) due to the asymptotic freedom of the theory. At any finite Q^2, the parton distribution functions acquire Q^2 dependent corrections due to the possibility of emission of gluons, so the parton distribution functions $f_i(x)$ are replaced by the quark and gluon distributions, $q(x, Q^2)$ and $g(x, Q^2)$, respectively, in the proton. The virtual photon which probes the proton now views it not merely as a collection of quarks but as the more complicated (quark plus gluon) system which leads to Bjorken scaling violations (explicit dependence on Q^2), departing from the simple parton model.

A set of differential equations to give the Q^2 evolution of the quark and gluon distribution functions was derived by Gribov and Lipatov, and Altarelli and Parisi using perturbative QCD [313]. These are usually called the *Altarelli-Parisi* evolution equations, although Gribov and Lipatov had derived these equations earlier independently for Abelian theories. The evolution equations are, introducing a scale parameter Λ and $t = \ln(Q^2/\Lambda^2)$, into the so-called *leading logarithmic approximation*,

$$
\begin{aligned}
\frac{dq_i(x,t)}{dt} &= \frac{\alpha_s(t)}{2\pi}[P_{qq} \bowtie q_i + P_{qg} \bowtie g], \\
\frac{dg(x,t)}{dt} &= \frac{\alpha_s(t)}{2\pi}[P_{gq} \bowtie \sum_i q_i + P_{gg} \bowtie g],
\end{aligned}
$$

where the symbolic notation

$$
P_{qq} \bowtie q_i = \int_x^1 \frac{dw}{w} P_{qq}(\frac{x}{w}) q_i(w,t),
$$

and similar notations in the other terms have been used. Here, P_{qq}, P_{gq}, P_{gg} are quark-quark, quark-gluon, and gluon-gluon splitting functions, respectively, for which expressions are derived in perturbative QCD. $\alpha_s(t)$ is the "running coupling constant" expressed as a function of t. Using these, one can solve the integro-differential equations and obtain the quark and gluon distribution functions at any t (hence at any Q^2), given some initial values at t_0 (Q_0^2), where t_0 is chosen such that perturbation theory is applicable for $t > t_0$. The initial parton distributions at t_0 are obtained from fits to experimental data at $Q_0^2 \simeq 2 - 2.5$ GeV^2. The Q^2 dependence of the functions depends on the value of Λ, called the *QCD scale*. It is determined by fits to experimental data on the Q^2 dependence of the form factors. Such work establishes that perturbative QCD is capable of describing high energy phenomena involving hadrons very well, over a very wide range of Q^2 values, and goes a long way toward establishing QCD as a fundamental theory for strong interactions.

Path Integral Formalism

In 1948, Feynman [314] proposed an alternative formulation of non-relativistic quantum mechanics. This formulation and its relativistic generalizations are found necessary in dealing with many problems in perturbative quantum field theories, especially non-Abelian gauge field theories.

The basic idea behind this formulation is the following. In quantum mechanics, we know that the probability of an event, which is capable of occurring through several different ways, is obtained by taking the sum of the complex amplitudes, one for each such way, and finding the absolute square of the sum. Thus the probability that a particle will be found to have the path $x(t)$ in some region of space will be the square of a sum of contributions, one from each path in the region. Feynman proposed that the contribution from any single path is given by $\exp iS$, where S is the classical action in units of $\hbar$ for that path. He also proposed that the total contribution from all paths reaching a given point (x, t) from earlier times is the wave function $\psi(x, t)$. He showed that this function satisfies the Schrödinger equation. He also studied the relation of this method to matrix and operator algebra and considered applications in particular in quantum electrodynamics.

Peccei-Quinn Symmetry

Conservation of CP seems to hold for strong interactions. Strong interactions of hadrons have their origins in the interaction of quarks and gluons, which make up the hadrons. The theory describing the interactions of quarks and gluons is based on the color gauge field theory, called *Quantum Chromodynamics (QCD)*. The question we may well ask is, does quantum chromodynamics give CP conservation?

It has been found that the non-Abelian QCD fields possess Euclidean solutions which carry a topological quantum number. This requires a much more complicated vacuum state than is normally included in the perturbative treatment of field theories. There are infinitely many vacua labeled by a parameter called θ. The Hilbert space of states factor into subspaces of states built on these distinct θ vacua. The vacuum to vacuum transition amplitude, when calculated in a particular θ vacuum, involves a sum over field configurations labeled by the topological quantum number q defined by

$$\frac{g^2}{32\pi^2} \int d^4 x F_a^{\mu\nu} \tilde{F}_{\mu\nu}^a = q,$$

where g is the coupling constant, $F_{\mu\nu}^a$ is the non-Abelian gauge field, and $\tilde{F}_{\mu\nu}^a = \epsilon_{\mu\nu\rho\sigma} F_a^{\rho\sigma}$, called the *dual non-Abelian gauge field*. Here the ϵ tensor is totally antisymmetric in its indices, with $\epsilon^{0123} = 1$. This constraint on the field configurations imposed through q modifies the Lagrangian density of the field theory by adding a term,

$$i\theta \frac{g^2}{32\pi^2} F_{\mu\nu}^a \tilde{F}_a^{\mu\nu},$$

to it. Such a term clearly violates both P and CP. It will give violation of CP in strong interactions, which will be disastrous, for there is no experimental evidence for such a violation. Peccei and Quinn [315] suggested a way out of this problem. They suggested that if the full Lagrangian possesses a chiral U_1 invariance, changes in θ are equivalent to changes in the definitions of the fields in the Lagrangian and will lead to no physical consequences. The theory with the chiral U_1 symmetry (called *Peccei-Quinn symmetry*) is equivalent to a theory with $\theta = 0$, and P and CP will be conserved in strong interactions.

A value of zero for θ is a natural result of Peccei-Quinn symmetry, which is exact at the classical level, and which will give rise to a massless Goldstone boson. Such a particle is called the *Axion* (see further in the section under "Axions"). Quantum mechanical effects, such as the triangle anomaly with gluons, give a nonzero value to the mass of the axion arising from the spontaneous breaking of Peccei-Quinn symmetry; thus the axion is really a pseudo-Goldstone boson.

ϕ Meson

The first evidence for this particle came from a study of the reactions involving K^- from the Brookhaven AGS proton synchrotron incident on protons in a hydrogen bubble chamber [316]. There were many final state products as a result of the K^--p interactions. Out of these, the final states involving $\Lambda^0 + K^+ + K^-$ and $\Lambda^0 + K^0 + \bar{K}^0$, when analyzed, showed a peak in the $K - \bar{K}$ mass distribution at a mass of about 1020 MeV and a width of about 3 MeV. The small width is a reflection of the fact that the phase space available for the decay of this state is rather small (the threshold being about 988 MeV for the K^+K^- mode and 995 MeV for the $K^0\bar{K}^0$ mode) and is due to a weak decay process. This particle is called the ϕ *meson*.

One can also study reactions in which the final state particles are $\Sigma^0 + K^+ + K^-$ and $\Sigma^+ + K^- + K^0$. A peak in the K^+K^- mass distribution is seen but no such corresponding peak is seen in the K^-K^0 mass distribution. This is consistent with the assignment of isospin $I = 0$ for this resonant state. The decay $\phi \rightarrow K^+K^-$ allows us to conclude that its parity is $(-1)^J$, where J is the spin of the ϕ. Under the charge conjugation operation C, we have $C\phi C^{-1} = (-1)^J \phi$, and because $I = 0$, the G parity operator $G = Ce^{i\pi I_2}$ is the same as C. The G-parity for the ϕ particle is $(-1)^J$. It has also been observed that the branching ratio for ϕ decaying into a pair of pions, relative to decay to a $K\bar{K}$, is very small, leading to the conclusion that the G-parity is (-1) for this particle, and hence its spin J must be odd. J values higher than 1 are excluded from a measurement of the branching ratio of $\phi \rightarrow K^0\bar{K}^0$ to $\phi \rightarrow K^+K^-$, which depends on the assumed spin of the ϕ; odd $J > 1$ give a poor fit to this ratio. Hence, $J = 1$ is assigned to this particle. Thus the ϕ meson is a vector meson and, in the constituent quark model, is almost completely an $s\bar{s}$ bound state due to dominance of the $K\bar{K}$ decay mode.

The decay of the ϕ meson into a pair of charged leptons has also been observed [317]. In the constituent quark model, the ratio of the decays of the ρ^0, ω^0, and ϕ^0 can be worked out on the basis of their quark compositions. These also show that the ϕ is almost entirely $s\bar{s}$ bound state.

Photino
This is a hypothetical particle of spin $1/2$, which is the supersymmetric partner of the spin 1 photon. (See under "Supersymmetry".)

Photoelectric Effect
The idea of the quantum of radiation made great strides toward believability with Einstein's explanation of the photoelectric effect based on Planck's ideas of quanta of radiation [318]. Hertz at first, and later Millikan in more detailed studies, had observed that when metal surfaces were illuminated by light, they emitted electrically charged particles. They established that (1) these emitted charged particles were negatively charged electrons and not positive ions; (2) there was a threshold frequency of light ν_0, characteristic of each metal, below which frequency there was no emission of electrons; (3) the magnitude of the current of electrons emitted was proportional to the intensity of the light and independent of the frequency of the light for $\nu > \nu_0$; and (4) the energy of the emitted electrons, as measured by a retarding potential applied to slow them down, showed that the energy was proportional to the frequency of the light and independent of its intensity.

These observations cannot be explained on the basis of the classical theory of light waves. Classically, the energy associated with the light waves is related to its intensity, and so the energy of the photoelectrons should vary with the intensity of light and not have the frequency dependence that is actually observed. There is also no explanation classically why the photoelectric current should vary with the intensity of the light. With Einstein's use of the quantum picture for light, where the energy of light photons is $E = h\nu$, higher frequency of light means higher energy for the quantum, so that the emitted photoelectron will have higher kinetic energy. The threshold frequency ν_0 for each metal implies a characteristic threshold energy $E_0 = h\nu_0$ below which it may be energetically impossible to free an electron from the metal. The equation for the kinetic energy acquired by the electron may be written as kinetic energy $= h\nu - h\nu_0$. The variation of the photoelectric current with the intensity of the light is understood because a low intensity light source puts out fewer quanta than a high intensity source for the same frequency. More photoelectrons will be emitted when a higher intensity source is used. This explanation put the photon hypothesis for the light quantum on a sound footing.

Photon—Mass, Spin, Statistics
A number of phenomena advanced the corpuscular nature of light. Most notable among these was the behavior of the energy density of radiation

from a black body as a function of the frequency of radiation. Expression for the energy density, derived by Rayleigh and Jeans, on the basis of classical theory of radiation showed a quadratically increasing function of the frequency. This behavior, besides being physically absurd, disagreed with experimental measurements at high frequencies. Planck solved the problem by making a radical new assumption regarding the nature of light. He proposed that the energy E of light came in discrete packets, or quanta, $E = h\nu$, where ν is the frequency of the radiation, and h is a constant with dimension energy $\times$ time. The expression for the energy density of black body radiation, using the idea of quanta of radiation, showed a decrease for high frequencies and was found to agree very well with experimental results. Thus was born the quantum of the electromagnetic field, the photon.

For a particle to be recognized as a particle, it must have, along with its energy, an associated momentum, such that in collision processes, one can see the effect of momentum conservation. Establishment of the fact that the photon, in addition to having an energy $E = h\nu$, has a momentum $p = h\nu/c$ associated with it (where c is the velocity of light), did not take long. Stark [319] was the first to make manifest the reality of associating momentum and energy to the photon in the X-ray region. Compton effect, discovered in 1923, established the idea of the photon as an elementary particle endowed with energy and momentum beyond any doubt (see further under "Compton Effect").

The Compton effect has some additional consequences. By establishing that the momentum of a photon is given by $p = h\nu/c$, it also determines that the rest mass of the photon is zero. This follows from Einstein's relation between energy and momentum: $E = \sqrt{c^2 p^2 + m_0^2 c^4}$, for a particle of rest mass m_0. It is seen from here that $E = cp$ or $p = E/c = h\nu/c$ follows, when the rest mass m_0 is set to zero. The Compton effect also shows that photons must be treated individually in any processes of collisions in which they participate. It shows that a photon cannot be split; a part of the energy of the photon cannot be scattered.

Another property associated with a particle is its intrinsic spin. The fact that the photon has polarization suggests that, when we associate a wave fuction for the photon in quantum mechanics, it must have one component for each state of polarization of the photon. The photon as a particle with rest mass zero and spin 1 can be shown to have just two states of polarization. Thus the photon is an elementary particle with spin 1. The direct establishment that the photon spin is 1 was done by Raman and Bhagavantham [320]. Raman was aware of the relationship of photon spin to selection rules in spectroscopy—the possibility of the quantum being either right- or left-handed around an axis parallel to the direction of its motion, corresponding to the two alternative projections of its angular momentum on this axis. They considered the spectrum of light scattering by molecules of a gas, which has three components: the unshifted Rayleigh line, and two rotational bands, one on the high frequency side and another on the low frequency side of the Rayleigh line. They applied energy and angular momentum conservation to the scattering of photons by the rotating molecules and deduced that the high frequency and low frequency scattered components in the forward direction would consist of photons which have experienced a reversal of spin. That is, if the incident light be circularly polarized, the Rayleigh line would be circularly polarized in the same sense, while the rotational Raman scattered lines would be reverse circularly polarized. They showed that the observations in their experiment were in accord with these deductions and hence deduced that the photon had an intrinsic spin of one unit.

The statistics observed by a collection of light quanta was discovered by Bose [321]. Photons obey Bose statistics; that is, a collection of photons is described by a wave function, which is a totally symmetric function of the coordinates and other attributes of all the photons in the assembly.

Pion

The pion, originally named the π *meson*, was among the earliest elementary particles to be discovered in cosmic rays by Powell and collaborators working with photographic emulsion plates [322]. That such a particle might exist was predicted by Yukawa [113] in 1935 based on theoretical considerations for generating the strong short range nuclear force between the nuclear constituents. He based his considerations on an analogy with electromagnetic forces between charged particles which are generated by the exchange of photons between them. The basic electric force, which is the Coulomb force, has an infinite range of interaction because the exchanged photons are massless. He showed that exchange of massive quanta between nuclear particles would give rise to a short range force, the range being inversely proportional to the mass of the exchanged particle. By taking the range of the nuclear force to be about 1.3×10^{-13} cm, he estimated that the exchanged particles would have a mass about 300 times the electron mass (energy equivalent $\simeq 150$ MeV). At that time such particles could not be produced in the laboratory because no accelerators were available with enough energy to produce them. Cosmic rays were the only source known at the time which had high energy particles among them.

The search for particles with mass intermediate between those of the electron and the proton were undertaken by a number of researchers working in cosmic rays, in the early period using cloud chambers and later using photographic emulsion plates. A particle, now called the *muon*, was the first to be found in such studies (see further under "Muon"). At first it was thought to be the particle that Yukawa had predicted. For a number of reasons explained in the section on muons, it turned out that it was not the particle which could be the mediator of strong nuclear forces. Powell and collaborators [322] exposed photographic emulsion plates to cosmic rays at high altitudes. When the plates were developed, they found them to contain tracks of new particles which had not been seen before, which were named π *mesons*. They saw events, which they interpreted as the decay of a π meson coming to rest in the emulsion and decaying into a muon. The decay muon was always found to have the same range in the emulsions, namely about 600μ. This is consistent with the kinematics of a two-body decay in which the decay products will have monochromatic energies. The other member of the two-body decay product did not leave a track in the emulsion, probably because it was neutral. The kinematics were consistent with the neutral particle having zero mass and they considered it to be a neutrino. They interpreted these decays as examples of $\pi^+ \to \mu^+ + \nu_\mu$. There were also events recorded where, in addition to the π decay to μ, they saw the μ decay to e. From the π-decay measurements, they came up with a figure for the mean lifetime of the π as 10^{-8} s.

Progress of such work with cosmic rays was slow, but with the building of high energy accelerators at various institutions, rapid progress could be made in this field. This fueled the growth of an era of good measurements with which one could find precise values for the mass and lifetime, and the spin, parity, etc., in addition to discovering new particles (K mesons and hyperons). Hints for the presence of these new particles were already coming from cosmic rays work. Efforts were also afoot to measure the scattering cross sections for pions on protons and neutrons.

π^+ Mass, Lifetime, Decay Modes
The mass and mean lifetime of the π^+ has been measured in a variety of experiments since it was first found. We give only the modern values here: $m_{\pi^+} = 139.56995 \pm 0.00035$ MeV and $\tau_{\pi^+} = (2.6033 \pm 0.0005) \times 10^{-8}$ s. The dominant decay mode is $\pi^+ \to \mu^+ + \nu_\mu$ with a branching ratio of $(99.98770 \pm 0.00004)\%$. All other decay modes have branching ratios of the order of less than 10^{-4}. An interesting mode is one leading

to β decay of the pion $\pi^+ \to \pi^0 + e^+ + \nu_e$ with a branching ratio $(1.025 \pm 0.034) \times 10^{-8}$ [323].

π^+ Spin Determination

The spin of the π^+ has been determined by using a clever idea. The reaction $p + p \to \pi^+ + d$ and its inverse are measured at the same center of mass energy. Using the principle of detailed balance between these two reactions one can determine the spin of the π^+. The total cross section for the forward process is calculated to be

$$\sigma_{\text{forward}} = \left(\frac{1}{2}\right)^2 \sum_{if} |M_{if}|^2 \frac{E_\pi E_d p_\pi}{4\pi p_f}.$$

Here the first factor $(1/2)^2$ comes from averaging over the spins of the two initial protons, M_{if} is the matrix element for the transition, and the last factor comes from the final phase space with p_π, E_π, E_d and p_f being the momentum of pion, energy of the pion, energy of the deuteron, and the momenum of the proton, respectively, in the p-p center of mass frame. The cross section for the reverse process, at the same center of mass energy, is

$$\sigma_{\text{reverse}} = \frac{1}{6(2J+1)} \sum_{if} |M_{if}|^2 \frac{p_f E_\pi E_d}{4\pi p_\pi}.$$

Here, of the first factor $1/6(2J+1)$, $1/3(2J+1)$ comes from averaging over the spin of the π^+ and of the deuteron, the other factor of $1/2$ comes from taking account of the identity of the two protons in the final state, and the rest of the factors have the same meaning as in the previous expression. Taking a ratio of these two expressions we get

$$\frac{\sigma_{\text{reverse}}}{\sigma_{\text{forward}}} = \frac{2}{3(2J+1)} \left(\frac{p_f}{p_\pi}\right)^2,$$

where the unknown matrix element factor cancels out. The reverse cross section for the absorption of π^+ has been measured at a pion laboratory energy of 24 MeV [324]. The forward cross section has also been studied at a proton incident laboratory energy of 341 MeV [325]. The center of mass energy here is approximately that involved in that of the reverse reaction. Thus, all the factors in the ratio are known except the $2J+1$ which is determined by the ratio. It is found that $J = 0$ best fits the data. There is no direct experimental determination of the parity of the π^+.

π^- Parity Determination

There is no direct experimental determination of the spin of the π^-. However, the spin of π^- must be zero because it is just the antiparticle of π^+. (Likewise, the mass and the lifetime of π^- must be the same as that of π^+.)

There is a direct experimental determination of the parity of π^-. We describe how that was done. It can be shown that, if the reaction $\pi^- + d \to n + n$ occurs for π^- at rest, then the parity of the π^- must be odd. The reasoning behind this conclusion is as follows. It turns out that after the π^- slows down in the material, it gets captured into high atomic orbits from which it cascades down to lower states. It also turns out that there is an appreciable probability for it to be in an atomic S state. Suppose capture occurs from these atomic S states. In such a case the initial angular momentum and parity of the system is $1^\pm$, the + sign holding for even parity, and the − sign for odd parity for the π^-. The final state two neutron system, by Pauli principle, can only be in $^1S_0, ^3P_{0,1,2}, ^1D_2, \ldots$. Of these the only state that has angular momentum 1 is the $^3P_1$ state, and this state has odd parity. So, if the reaction occurs at all, the parity of π^- has to be odd.

The capture reaction has indeed been observed to occur by Panofsky et al. [326]; hence the parity of π^- is determined to be odd. Thus, both π^+ and π^- are spin zero particles with odd parity, that is, they are pseudoscalars.

In showing that the above captures do indeed occur from an S state, the following considerations are involved. The whole slowing down process for the π^- and the formation of the π-mesic atom takes a time much shorter than the mean lifetime of the π^-. It is captured in an atomic orbit with a large value of the principal quantum number n from which by Auger processes it cascades down to $n \simeq 7$. Then a mechanism discovered by Day, Snow, and Sucher [327] decides the further fate of the pionic atom. The small radius neutral pionic atom wanders about and soon penetrates the electron shell around a deuterium atom. The intense electric fields in this region cause Stark mixing of the n^2 degenerate levels and populate the S levels. These populations are much more than what they would be if the pionic atom came down to the $2P$ and $1S$ states via radiative transitions. The capture probability rapidly decreases as the angular momentum of the orbit increases, so S state capture is favoured over capture from higher angular momentum states, a fact which was used above.

π^0 Mass, Lifetime, Spin, Parity

The neutral π has been detected by its decay mode, $\pi^0 \to \gamma + \gamma$. Precision measurements on this decay allow one to determine the mass and mean lifetime of this particle. They are $m_{\pi^0} = 134.9764 \pm 0.0006$ MeV and $\tau = (8.4 \pm 0.6) \times 10^{-17}$ s. The branching ratio for the 2 gamma decay mode is $(98.798 \pm 0.032)\%$. The $e^+ e^- \gamma$ decay [328] mode has a branching ratio $(1.198 \pm 0.032)\%$. Other branching ratios for two electron pairs and one electron pair, which can be viewed as the internal conversion of the two photons into two pairs or a single pair, respectively, are about 3×10^{-5} and 7×10^{-8}, respectively.

The spin of the π^0 is not measured directly. Its closeness in mass to the charged members suggests that they could form an isotopic triplet if its spin were also zero. Additional evidence for its spin being zero comes from the occurrence of the 2 gamma decay mode. It can be theoretically proven that a system with spin 1 cannot decay into 2 gammas while a system with spin 0 can.

The parity of the π^0 is determined from the following consideration. Assuming spin 0 for the π^0, the decay amplitudes for the two gamma decay mode, which incorporate the requirements of Bose symmetry (linearity in the polarization vectors of the two photons) have the forms

$$\vec{e}_1 \cdot \vec{e}_2 \text{ (for even parity)},$$

$$\vec{p}_\gamma \cdot \vec{e}_1 \times \vec{e}_2, \text{ (for odd parity)},$$

where $\vec{e}_1, \vec{e}_2, \vec{p}_\gamma$ are the polarization vectors of the photons, and the momentum of the photon in π^0 rest frame, respectively. These amplitudes predict that, for even parity of π^0, the polarization vectors of the two photons must be parallel, while for odd parity, they must be perpendicular. In the modes in which these photons are internally converted, these polarization correlations are carried over. If the angle between the planes of the two pairs is ϕ, then for even (odd) parity, the distribution in this angle is of the form $1 + 0.48 \cos 2\phi (1 - 0.48 \cos 2\phi)$ [329]. The planes of the two pairs are found to be perpendicular to one another, $\phi = \pi/2$. Thus, the parity of π^0 is determined to be odd.

Pion-Nucleon Scattering

In the 1950's a very large amount of effort was expended in understanding the properties of pions and their interaction with nuclear particles. Of special interest was the scattering of pions on nucleons. As soon as it was established that pions were pseudoscalar particles, the dynamics of pion nucleon interaction, based on the Yukawa coupling between pseudoscalar mesons and nucleon (pseudoscalar) currents with a coupling constant g, was intensely investigated. Theoretical developments were based on the use of perturbation theory for the Yukawa interaction patterned on calculations in quantum electrodynamics developed by Feynman, Schwinger, Tomonaga, and Dyson. Experimental data on the scattering of pions on nucleons revealed that the magnitude of the cross sections were quite large and showed a rapid rise with energy. These features indicated that $g^2/4\pi$ would have to be of order 15 to provide a fit to the data. With such a large coupling constant it was clear that perturbation theory was inadequate to deal with the problem. No general nonperturbative methods of treating the problem were known.

For a brief period, theoretical effort was put into the so-called Tamm-Dancoff approximation to solve for the pion-nucleon amplitude, limiting the coupling of this state to other nearest neighbouring states. This generated an integral equation for the pion-nucleon amplitude [330] with a kernel limited to order $g^2/4\pi$. Using angular momentum decomposition, equations were developed for various angular momentum states, and expressions for the phase shifts were derived. Numerical solutions were obtained and the phase shifts indicated a rapid rise in the $I = 3/2, J = 3/2$ state of the pion-nucleon system, indicating a possible resonance in this state. The whole program was not developed further because (1) the method was non-covariant, and (2) a consistent renormalization program could not be properly developed to proceed with kernels of higher orders.

Methods of analyzing the experimental data by performing partial wave analysis of the data were developed around this time. Through such analyses, one could extract the behavior of phase shifts as a function of energy and other quantum numbers associated with the state. Many ambiguities encountered in the process had to be resolved. Through such analyses resonances could be found and classified as to their degree of inelasticity in the various channels.

Another approach called the method of *Dispersion Relations* was developed. Here the central focus was on the analyticity properties of scattering amplitudes stimulated by the work of Gell-Mann, Goldberger, and Thirring [155]. Dispersion relations relate the real part of the scattering amplitude to the imaginary parts through dispersion integrals. By the optical theorem, the imaginary part of the forward scattering amplitude is related to the total cross section, which is measured in experiments. Thus using a measurement of the total cross section as a function of the energy to derive the imaginary part of the forward scattering amplitude, one can determine the real part by evaluating the dispersion integral. Regions of analyticity of the scattering amplitudes with positions of cuts and poles specified in different channels give relations between physical processes which are related to each other by certain transformations called *crossed-channel transformations*.

For pion-nucleon scattering the method starts from relativistic expressions for the scattering amplitude, which are written in terms of Dirac covariants (to take care of spin components) multiplied by Lorentz invariant scalar functions of energy and invariant four-momentum transfers. The invariant functions also depend on the isotopic spin of the state. Using this technique, Hamilton carried out extensive dispersion relations analyses of pion nucleon scattering [331]. We do not go into details on this subject here, because the development of QCD as the theory of strong interactions between quarks, which are the constituents of hadrons, has provided us with a new field theoretical approach for the problem. (For some further information on dispersion relations, see section under "Dispersion Relations".)

Pion-Pion Interactions

At energies in pion-nucleon scattering which are sufficient to produce an extra pion, it is possible that the two pions in the final state suffer interactions. Study of the effect of such final state interactions, which would give results different from assuming pure phase space distributions for the two pions, would help in understanding the nature of the pion-pion interactions. Such methods were advocated by Goebel [332] and further extended by Chew and Low [333]. Pursuance of this work led to the prediction of spin 1, isospin 1, resonance in the two pion system, which has since been termed the ρ *meson* [334]. More information may be seen in section under "Rho Meson (ρ) Resonance".

Planck Mass

An interesting comment was made by Dirac in connection with the Newtonian gravitational constant G_N. Its measured value is given to be 6.67259×10^{-11} m^{3}kg^{-1} s^{-2}. He noticed that, if one divides $\hbar c$ by G_N, one gets a quantity of the dimensions of square of a huge mass. When one works out what the energy equivalent of this mass is, one gets a value 1.221048×10^{19} GeV. He commented on the size of this mass when compared with the mass of the electron, proton, etc. The mass given by $\sqrt{\hbar c/G_N}$ is denoted by M_P and is called the *Planck mass*. In terms of grams, it is roughly 10^{-5} grams. This mass now obtains a special significance in terms of the energy scale at which all the fundamental forces become unified and, accordingly, represents the energy at which quantum effects in gravity become important.

Pomeranchuk Theorem

This theorem relates the interaction cross sections for particles on a given target with that of antiparticles on the same given target at asymptotically high energies. It was derived by Pomeranchuk, on the basis of dispersion relations, and is called *Pomeranchuk theorem* [335]. It may be stated as follows:

$$\lim_{s \to \infty} \left(\sigma_{tot}(s)^{XY} - \sigma_{tot}(s)^{\bar{X}Y} \right) = 0,$$

where s is the Mandelstam variable equal to the square of the center of mass energy in the $X(\bar{X}), Y$ system, and σ_{tot} is the total cross section.

As we have seen in the section under "Dispersion Relations", the real part of the forward scattering amplitude for a given process is related to the imaginary part through a dispersion integral over the imaginary part of the forward scattering amplitude. The imaginary part of the forward scattering amplitude is related to the total cross section by the optical theorem and is, hence, measurable. The total cross section contains contributions from elastic as well as inelastic processes at high energies.

Let us consider a dispersion relation in which the differences of the imaginary parts of the forward scattering amplitudes for particles and antiparticles occur inside the dispersion integral. Suppose the difference in the total cross sections for particles and antiparticles for high s does not tend to zero, but is some value $\delta\sigma$. Then applying optical theorem, one can show that the difference in the imaginary parts of the forward scattering amplitudes is $\propto \sqrt{s}\delta\sigma$. Evaluating the dispersion integral over the difference of the imaginary parts, one gets a value for the difference of the real parts which is $\propto \sqrt{s}\delta\sigma \ln s$. The assumed behavior of the imaginary parts implies that for each individual process XY, the imaginary part of the forward scattering amplitude behaves like $\sqrt{s}\times$constant, while the real part behaves like $\sqrt{s}\ln s \times$ constant. The result implies that the real part of the elastic scattering amplitude dominates. This behavior is contrary to intuitive expectations. The imaginary part of the scattering amplitude gets positive contribution from every inelastic process, while the real part gets contributions from a large number of terms with random distribution of positive and negative signs. The expectation that the imaginary part must be dominant over the real part is supported by experience with many models that have been constructed to explore the behavior. The contradiction to experience is removed if we change the assumption on the difference in particle and antiparticle cross sections at high energies. We thus require $\delta\sigma = 0$, which implies that both the particle as well as the antiparticle cross sections tend to the *same* constant at high energies.

The theorem has been further generalized by Pomeranchuk and Okun who showed that at high energies, scattering amplitudes are dominated by elastic scatterings in the t-channel [336]. Only for elastic scattering, the exchanged object has the quantum numbers of vacuum, an object called the *pomeron*. This leads to exchange amplitudes which are not real for forward direction. The pomeron plays a role in peripheral inelastic processes.

Positron

The positron is the antiparticle to the electron. It was predicted as a necessary consequence of the Dirac theory for the electron. It has the same mass and spin as the electron, only its charge is opposite to that of the electron. Its magnetic moment and the value of $g-2$ have been measured and shown to be equal in magnitude to that of the electron. See further in sections under "Dirac Equation".

Positronium

This is a bound state of a positron and an electron. It resembles a hydrogen atom—the proton of the hydrogen atom is replaced by the positron. The reduced mass in this case is half the electron mass, and hence the ionization energy is half that of the hydrogen atom, namely, about 6.8 eV. The energy levels of this system are similar to those in the hydrogen atom; the separations of the energy levels are half those in the hydrogen atom because of the difference in the reduced mass. The states are characterized by the principal quantum number n and the total angular momentum J which is the vector sum of the orbital angular momentum l and the total spin s. Because this system is electrically neutral, it also possesses a charge conjugation parity quantum number C.

We can show that for such a fermion-antifermion system with orbital angular momentum l and total spin s, the eigenvalue of C is $(-1)^{l+s}$. For this purpose, consider the behavior of the wave function of the bound state under the interchange of the fermion and the antifermion, which is equivalent to inversion of the relative coordinate $\vec{r}$ into $-\vec{r}$. The total wave function of the bound state is a product of three factors belonging to the spatial wave function, the spin wave function, and the charge wave function. The spatial wave function is a product of the radial wave function and an angular function, which for angular momentum l, is the spherical harmonic $Y_{lm}(\theta, \phi)$. The spin wave function for total spin s is χ_s. The system has two possible values of total spin, $s = 0$ or $s = 1$, corresponding to the singlet and the triplet spin states, respectively. The particle interchange, equivalent to inversion of the relative coordinate $\vec{r}$ between the particles, is achieved by $r \to r$, $\theta \to \pi - \theta$, $\phi \to \pi + \phi$. Under this transformation the radial wave function is unchanged, while the spherical harmonic acquires a factor $(-1)^l$, so that the spatial wave function acquires the factor $(-1)^l$. The spin wave function χ_s, under the interchange of the spins, is antisymmetric for $s = 0$ (singlet), and symmetric for $s = 1$ (triplet). These symmetries are accommodated with the factor $(-1)^{s+1}$ multiplying the spin wave function due to the interchange. Thus the product of the space and spin wave functions acquires the factor $(-1)^{l+s+1}$ when the fermion and the antifermion are interchanged. Let the charge wave function acquire a factor C under the interchange. The total wave function thus acquires the factor $(-1)^{l+s+1}C$.

It is experimentally observed that $^1S_0$ state of positronium (J=0) decays into two 2γ's, while the triplet state $^3S_1$ (J=1) decays into 3γ's. The charge conjugation parity C for an n photon state is $(-1)^n$; thus for $n = 2$, C is $+1$ and for $n = 3$, C is -1. These charge conjugation parities are reproduced by choosing $C = (-1)^{l+s}$. With this choice, the total wave function acquires a factor -1 when the particles are interchanged (overall antisymmetry), even though the system does not consist of identical fermions but is made up from a fermion and an antifermion. Thus, we have the result $C = (-1)^{l+s}$ for the charge conjugation parity for a fermion-antifermion bound state of orbital angular momentum l and total spin s.

Proton

This is the nucleus of the simplest of atoms, namely, the hydrogen atom. It is positively charged and has a mass for which the modern value is 938.27231 ± 0.00028 MeV. Like the electron, it has spin 1/2 and is assigned a positive parity. The proton is a hadron and participates in strong interactions. Dirac equation applied to the proton would predict a magnetic moment for the proton of 1 μ_n, (μ_n= nuclear magneton=$e/(2M)$). The experimentally measured value for the magnetic moment is not one nuclear magneton but considerably deviates from it as shown by I. I. Rabi and collaborators [151] by developing the high precision molecular beam magnetic resonance method. The value obtained in this work for the proton was 2.785 ± 0.02 nuclear magnetons. The modern value for the magnetic moment of the proton is $2.79284739 \pm 0.00000006$ μ_n. These values suggest that the proton is not a structureless particle like the electron; its hadronic structure is responsible for the additional contribution to its magnetic moment. Theoretical calculation of the magnetic moment of the proton will only be possible when its hadronic dynamical structure is correctly described, a problem which is still awaiting solution.

The electromagnetic structure of the proton is unraveled in high energy electron scattering experiments. The elastic form factors of the proton, over a wide range of square of the momentum transfer, and the deep inelastic form factors, over a wide range of square of the momentum transfer and energy transfer, have been measured by electron scattering experiments. Through high energy neutrino scattering, the weak form factors have also been measured. The size of the electromagnetic radius of the proton have been determined to have a mean squared value 0.71 fermi^2.

The dipole polarizability of the proton has also been the subject of investigations. First theoretical estimates of this quantity were provided by Baldin [337] using data on photoproduction of π mesons and Compton effect on nucleons. He also pointed to the use of elastic photon scattering on deuterons to obtain information about the polarizability of the neutron. A measurement of this quantity for the proton was carried out by observations on the elastic scattering of photons on protons by Goldansky et al. [338]. From the cross section for γ-proton scattering at 90 deg, they obtained a value for the electric polarizability, $\alpha_E = (11 \pm 4) \times 10^{-43}$ cm^3. Further, using dispersion relations and data on pion photoproduction, they obtained a value for the sum of the electric and magnetic polarizabilities, $\alpha_E + \alpha_M = 11 \times 10^{-43}$ cm^3. Combining these results they obtained for the individual quantities, $\alpha_E = (9 \pm 2) \times 10^{-43}$ cm^3 and $\alpha_M = (2 \pm 2) \times 10^{-43}$ cm^3, respectively.

Proton Spin Crisis
See discussion under "Deep Inelastic Scattering with Polarized Particles".

ψ Particle (J/ψ)
This particle was discovered as a narrow resonance in (p, Be) collisions leading to massive e^+e^- pair in Brokhaven National Laboratory by Ting, as well as in e^+e^- annihilations at SLAC by Richter, at an energy of 3.1 GeV. This particle was soon interpreted as the bound state of a new quark, the charm quark c, and its antiquark $\bar{c}$ in a $^3S_1$ state. It has baryon number zero, and is hence a vector meson. The modern value of its mass is 3096.88 ± 0.04 MeV, and it has a full width of $\Gamma = 87 \pm 5$ keV; its leptonic width is $\Gamma_{ee} = 5.26 \pm 0.37$ keV. The dominant branching ratio of its decays is to hadrons: $(87.7 \pm 0.5)\%$. Many other decay modes are known with branching ratios of order 1% or less and details can be obtained from the "Review of Particle Physics" [62]. Excited states $2S$, $3S$, and $4S$ are also known for this system. For further details regarding charm mesons, see section under "Charm Particles".

QCD—Quantum Chromodynamics
Quantum chromodynamics is the theory of the interaction of colored quarks and colored gluons, the gluons being the quanta of the *chromodynamic* field. It is formulated in a manner similar to quantum electrodynamics (QED), which is the theory of interaction of electrically

charged particles through the exchange of photons, the quanta of the *electromagnetic* field. The earliest suggestions for such a formulation were made by Nambu [339] and by Greenberg and Zwanziger [340].

In the formulation of the constituent quark model, quarks come in different types, called *flavor*, *up* (*u*), *down* (*d*), *strange* (*s*), etc. For each flavor, the need for color as an additional degree of freedom for the quarks is covered elsewhere (see section under "Colored Quarks and Gluons"). One needs a minimum of three colors for the quarks, usually red, blue, and green. A simple model incorporating color would be to consider another *global* symmetry, SU_3 symmetry of color, just as for flavor. Considering the quarks to transform in the triplet (fundamental) "3" representation of SU_3 color symmetry, and the antiquarks in the antitriplet or "$\bar{3}$" representation, the requirement is imposed that all hadronic wave functions be color singlets; that is, they be invariant under SU_3-color transformations. Under this condition, the only simple combinations that are allowed are: $\bar{q}_i q_i$, $\epsilon_{ijk} q_i q_j q_k$, and $\epsilon_{ijk} \bar{q}_i \bar{q}_j \bar{q}_k$ (where ϵ_{ijk} is the totally antisymmetric tensor, with $\epsilon_{123} = 1$). These represent objects with baryon number 0, 1, and -1, respectively, and are suitable to describe, mesons, baryons, and antibaryons, respectively. The baryon color wave functions are clearly antisymmetric in color, and when combined with totally symmetric functions in space, spin, and flavor, there is no difficulty in satisfying Pauli exclusion principle.

The suggestion of color singlet wave functions was clearly successful in describing the phenomenology of low lying hadronic states. However, there is no answer as to why color singlet states should be the only allowable ones. For the answer to this and other questions, it was found necessary to go from *global* to *local* symmetry and consider the color SU_3 gauge group. This leads one to a non-Abelian gauge theory which is a generalization of the SU_2 theory of Yang and Mills (see section under "Gauge Theories") [341]. The Lagrangian, which is invariant under transformations of the SU_3 gauge group, requires the introduction of eight gauge fields, $A_{\mu i}, i = 1, \ldots, 8$, and the form of the interaction of the gauge fields with the quarks is automatically specified. The full Lagrangian for the non-Abelian QCD theory is

$$\mathcal{L} = [i\bar{\psi}\gamma^\mu \partial_\mu \psi - m\bar{\psi}\psi] - \frac{1}{4\pi} \sum_{i=1}^{8} F_i^{\mu\nu} F_{\mu\nu i} - g \sum_{i=1}^{8} \bar{\psi}\gamma^\mu \lambda_i \psi A_{\mu i},$$

where $A_{\mu i}, i = 1, \cdots, 8$ are the gauge potentials, the $F_{\mu\nu i}$ are the gauge field strengths, ψ is the quark fields, g is the coupling constant between the quark and the gauge fields, and m is the mass of the quark of a given flavor. A sum over all flavors has also to be carried out. The quantization of this gauge field theory is not simple. It has been done and a set of Feynman rules for perturbative calculations have been derived. These are similar to the rules in QED, only more intricate and elaborate. In addition to the quark-gluon coupling vertex, which is the analog of the electron-photon coupling in QED, color gauge invariance of the theory dictates that there must exist three-gluon and four-gluon vertices, depending on the same coupling constant as that involved in the quark-gluon coupling. The theory is renormalizable, just like QED, allowing one to calculate higher order contributions. The quanta of the gauge field are called *gluons*. They are massless, spin 1 particles and carry color.

Two very important properties of this gauge field theory were discovered. The theory has been found to exhibit properties called *Asymptotic Freedom* [342] and *Color Confinement* [115]. The first property refers to the fact that the coupling of the quark with the gauge field tends to vanish at high momentum transfers to the quark, so that in this limit, the hadron can be considered a collection of noninteracting quarks (partons) (see section under "Parton Model"). The second property refers to the fact that at momentum transfers tending to zero (the infrared regime), the effective quark-gluon coupling constant, $\alpha_s = g^2/(4\pi)$, becomes large, and it is conjectured that it requires infinite energy to separate colored objects into free particle states. In other words, although there is no rigorous proof yet, there is a strong suggestion that color is confined within hadrons. Hadrons, which are color singlets, are the only objects that can be found in asymptotically free particle states. The property of asymptotic freedom provides the basis for the explanation of Bjorken scaling with the parton model. In the perturbative regime, QCD provides methods by which to calculate corrections to the parton model and how Bjorken scaling will be broken. Such calculations have been put to test in applications to deep inelastic lepton-proton scattering and in e^+e^- annihilations leading to hadrons. Within the perturbative approximations used, QCD has been found to describe these phenomena at the level of a few percent. The region of energies dealing with hadrons as bound states of quarks (and antiquarks) belongs to the nonperturbative regime of QCD.

To learn the details of perturbative QCD calculations, the reader may wish to consult a book on quantum field theory, for example, reference [98].

Quantum Electrodynamics—QED

Quantum Electrodynamics (QED) is the quantum field theory of the interactions of charged particles with electromagnetic fields, in particular, interactions of electrons and photons. It is formulated as an Abelian gauge field theory. The basic equations of the theory are a set of coupled equations with manifest Lorentz symmetry, which determine the electromagnetic fields for given charge-current sources of fields and the dynamics of the charges for given electromagnetic fields. Maxwell's equations determine the electromagnetic fields from the charge-current distributions, and the dynamics of the electron is determined by the Dirac equation for the electron in the presence of electromagnetic fields, in accordance with the demand of gauge invariance. Wave-particle duality is incorporated through the process of quantization of the electromagnetic and the Dirac fields, the classical field functions becoming operators with appropriate commutation relations imposed on them. At low energies, or at large distances (low resolution), one needs to deal with states which involve at most a few quanta. As the energy increases, one is proceeding toward shorter distances and smaller time scales (higher resolution), and the states with larger numbers of quanta begin to participate. A dimensionless coupling constant, called the *fine structure constant*, equal to $\alpha = e^2/(4\pi) \simeq 1/137$, where e is the charge of the electron, determines the coupling of the electron to the photon. This being a small number, methods are developed to solve the coupled equations using perturbation theory. This involves expansions in powers of this small coupling constant.

The most elegant formulation of the perturbation method is due to Feynman who invented the diagrammatic technique (now called the *Feynman diagram technique*). With it one can carry out calculations systematically for any process involving electrons and photons to any finite order of perturbation theory [343]. At about the same time, Schwinger [344], in a series of papers, and Tomonaga [345], also in a series of papers, gave covariant formulations of electrodynamics. The equivalence of all these formulations was established in a work by Dyson [346].

The matrix element for a given process can be written by first drawing all possible Feynman diagrams which are relevant for the process and associating a factor with each element of the diagram, according to the given rules. Using this input into the standard formalism of relativistic quantum mechanical calculations, cross sections or decay rates are derived, which are in a form suitable for comparison with experimental data for the process under consideration.

The number of vertices in a Feynman diagram is determined by the order to which we are calculating the contribution to the process in perturbation theory. Thus, a process to order e^2 will contain two vertices, to order e^4 four vertices, etc. In general, to any order in perturbation theory, one will generate Feynman diagrams, some of which will contain no loops, while others will contain loops. The diagrams with no loops are called *tree diagrams*, while the ones with loops are called *loop diagrams*. To obtain the contribution from diagrams involving loops, the rules demand that one must carry out an integration over the infinite range of four-momentum associated with the loop. Such loop integrations in general give contributions which are infinite at the large momentum end (ultraviolet divergences); sometimes (for loop or tree diagrams) there are infinite contributions which come also from the low momentum end (infrared divergences). To make unambiguous calculations of such divergent contributions possible in a Lorentz invariant way, two steps have to be taken. The first is a process called *regularization*, according to which the integral in question is defined through some limiting process which respects all the symmetries of the theory including gauge invariance. Before the limit is taken, one has a well defined mathematical quantity on which one can perform operations in an unambiguous way and can separate out the contributions which will become infinite when the limiting process is carried out. The beauty of QED is that, to all orders of perturbation theory, all the ultraviolet divergences can be absorbed into redefintions of mass, charge, and the normalization of the wave function of the particles by a process called *renormalization*. For the renormalized values of the mass and charge, one uses the experimentally observed values for these quantities. The infrared divergences cancel among themselves and with contributions from processes in which real soft photon emissions occur. Since extremely soft photons cannot be detected by any real detector with a finite energy resolution, their residual effect in processes in which any number of soft photons is emitted can be represented in terms of an exponential of the form $\exp(-\alpha/2\pi)\ln(-q^2/m^2)\ln(-q^2/E^2)$, where q^2 is the (spacelike) momentum transfer in the scattering process in which soft quanta are emitted, m is the mass of the electron, and E is the energy of the electron.

The results of such calculations for many observable quantities, when confronted with experimental measurements, give exceedingly good agreement in many phenomena with distance scales ranging all the way from large (macroscopic) scales down to scales of the order of 10^{-15} to 10^{-16} cm. Quantum electrodynamics is one of the most successful and well-tested theories. It has also served as a model for the construction of other quantum field theories of interacting particles.

The subject of QED is a vast one; for learning the details of the subject, it is recommended that a book on the subject be consulted, for example [98].

Quantum Field Theory

In particle mechanics one has a system with a finite number N of degrees of freedom. The system is described by giving the generalized coordinates, $q_i, i = 1, \ldots, N$, and the generalized velocities, $\dot{q}_i, i = 1, \ldots, N$. The solution of the dynamical problem is obtained when one can give the development of each degree of freedom as a function of time. Toward this end, the Lagrangian formulation of particle mechanics starts with the construction of the Lagrangian, in terms of the kinetic and potential energies of the system, and then derives equations of motion for each degree of freedom by extremizing the *action*. *Action* is the time integral of the Lagrangian. The Euler Lagrange equations that result from extremizing the action by a variational principle are the equations of motion. These equations are *total* differential equations of second order in time, which when solved in terms of given initial conditions, give the solution of the dynamical problem. In general, the solutions will involve $2n$ constants of integration which are determined from the initial condition on the coordinates q_i and their time derivatives $\dot{q}_i$. The Lagrangian formulation also shows that there is an intimate connection between symmetry properties possessed by the Lagrangian and conservation laws that follow from it. Such formal considerations of symmetries, etc. help in getting a great deal of insight into the system behavior even without a complete solution of the equations of motion.

Extension of this method to systems, such as fluids, involving continuous degrees of freedom have also been formulated. The main difference in such situations is that the formalism has to be extended to deal with an infinite number of degrees of freedom. Action in this case involves integrals over all space and time of a Lagrangian density which is a function of the generalized coordinates and its space and time derivatives, describing the fluid. The resulting Euler-Lagrange equations instead of being total differential equations in time variable, now become *partial* differential equations involving both space and time variables. The solutions of these involve the solution of boundary value and initial value problems. Symmetry principles and conservation laws play as important a role here as in the case of particle mechanics.

Fields, such as the electromagnetic field, which satisfy Maxwell's equations for the electric and magnetic field vectors, are another example of a system with infinite number of degrees of freedom. The field vectors, being continuous functions of space and time, involve continuous (infinite number of) degrees of freedom when treated as dynamical quantities, just as the quantities in fluid mechanics. One can construct a Lagrangian formulation of Maxwell's equations of electromagnetic theory. A choice of the Lagrangian is made such that it incorporates various invariances, such as Lorentz invariance, and such that the resulting Euler-Lagrange equations give the Maxwell equations.

A customary procedure in classical electromagnetic theory is to introduce scalar and vector potentials in terms of which the electric and magnetic fields are expressed. It is found that the potentials belonging to the fields are not unique. These undergo transformations, called *gauge transformations*, which lead to different potentials all of which lead to the same physical fields. These different potentials are said to be in different gauges, and the fact that they lead to the same physical consequences is referred to as *gauge invariance* of the theory. Gauge invariance is a very important symmetry of electromagnetic theory. When treating interactions of charged particles with electromagnetic fields, gauge invariance plays a particularly significant role, especially in quantum theory, in restricting the form of the interaction between the charged particle and the electromagnetic field. The law of conservation of electric charge can be shown to be a direct consequence of invariance under gauge transformations.

The transition from classical to quantum mechanics in field theory follows a procedure similar to that followed in particle mechanics. In particle mechanics, one defines momenta $p_i, i = 1, \ldots, N$ conjugate to the coordinates $q_i, i = 1, \ldots, N$ by $p_i = \partial L/\partial \dot{q}_i$, and constructs a quantity called the *Hamiltonian*, $H = \sum_i p_i \dot{q}_i - L$, from the chosen Lagrangian. In quantum theory, the $q_i's$ and the $p_j's$ become operators between which commutation relations are introduced: $[q_i, p_j] = i\delta_{ij}, [q_i, q_j] = [p_i, p_j] = 0$ (the first of these would have a factor $\hbar$ multiplying δ_{ij} on the right-hand side, but we should remember that in the natural units in which we are working, $\hbar = c = 1$).

A corresponding procedure can be extended to quantize a field theory. For simplicity, consider a real scalar field $\phi(\vec{x}, t)$ which satisfies the Klein-Gordon equation: $\Box \phi + m^2 \phi = 0$, where $\Box = \partial^2/\partial t^2 - \nabla^2$ and m is a parameter which will turn out to be the rest mass associated with the quantum of the scalar field. One constructs a Lagrangian density $\mathcal{L}$, which is a function of the field ϕ and its space-time derivatives $\partial_\mu \phi$, whose integral over all space and time gives the *action*. It is chosen such that when we seek the extremum of the action we get the Klein-Gordon equation as the equation of motion for the scalar field. The "momentum" density conjugate to the field $\phi(\vec{x}, t)$, called $\pi(\vec{x}, t)$, is defined by $\pi = \partial \mathcal{L}/\partial \dot{\phi}$, where $\dot{\phi}$ represents the time derivative of the scalar field. The Hamiltonian density is obtained from $\mathcal{H} = \pi(\vec{x}, t)\dot{\phi}(\vec{x}, t) - \mathcal{L}$. In quantizing the theory, the quantities ϕ and π become operators and one has to introduce commutation relations between them. One introduces "equal time" commutation relations,

$$[\phi(\vec{x},t),\pi(\vec{y},t)] = i\delta^3(\vec{x}-\vec{y}),$$

$$[\phi(\vec{x},t),\phi(\vec{y},t)] = 0, \quad [\pi(\vec{x},t),\pi(\vec{y},t)] = 0.$$

The spectrum of the Hamiltonian is obtained by analogy with the treatment of the harmonic oscillator in quantum mechanics. A Fourier integral decomposition of the field operator ϕ is carried out, the different modes being characterized by a variable $\vec{p}$,

$$\phi(\vec{x},t) = \int \frac{d^3p}{(2\pi)^3} \frac{1}{2\omega_{\vec{p}}} (a_{\vec{p}}e^{i(\vec{p}\cdot\vec{x}-\omega_{\vec{p}}t)} + a_{\vec{p}}^{\dagger}e^{-i(\vec{p}\cdot\vec{x}-\omega_{\vec{p}}t)}),$$

where $\omega_{\vec{p}} = \sqrt{|\vec{p}|^2 + m^2}$, and $a_{\vec{p}}, a_{\vec{p}}^{\dagger}$ are called *annihilation* and *creation* operators in the mode $\vec{p}$ and play a role similar to the ladder operators of the harmonic oscillator. The Fourier integral for π, similar to that of the one for ϕ, follows. Using these, from the equal time commutation relations for ϕ and π stated previously, one can derive that

$$[a_{\vec{p}}, a_{\vec{p}'}^{\dagger}] = (2\pi)^3\delta^3(\vec{p}-\vec{p}'),$$

and $[a_{\vec{p}}, a_{\vec{p}'}] = 0 = [a_{\vec{p}}^{\dagger}, a_{\vec{p}'}^{\dagger}]$. Finally, the expression for the total Hamiltonian for the system, has the mode expansion,

$$H = \int \frac{d^3p}{(2\pi)^3} \omega_{\vec{p}}(a_{\vec{p}}^{\dagger}a_{\vec{p}} + \frac{1}{2}[a_{\vec{p}}, a_{\vec{p}}^{\dagger}]).$$

The second term inside the integral for H is $\delta^3(0)$ which is infinite and represents the sum over all modes of the zero point energies. This term is not detectable, as experiments only find differences between the energy of a state and that of the ground state, this infinite quantity cancels out.

Quantum field theories for other fields, such as the electromagnetic field, the electron field, and the massive vector boson field, have been likewise constructed. The quanta of these fields are the photon, the electron, W, Z bosons, etc. Theories of interacting fields have also been constructed. For electromagetic interaction of charged particles with photons, requirement of gauge invariance restricts the form of the interaction almost completely. Details on this may be found in the section under "Gauge Theories". The solution of the problem of interacting fields is possible with the use of perturbation theory, valid for those cases in which the interaction is characterized by a coupling constant which is numerically small, so that a solution involving an expansion in powers of the coupling constant is possible.

Quarkonia

Quarkonia are bound states of a quark and an antiquark with zero net flavor. Examples are charmonium ($c\bar{c}$ bound state) and bottomonium ($b\bar{b}$ bound state). (See also under "Charm Particles—Charmonium" and under "Bottomonium".)

Quarks

The SU_3 symmetry, which has been successful in explaining octet and decuplet families of baryons and octets of mesons, has among its representations, a fundamental triplet 3 representation and a distinct anti-triplet $\bar{3}$ representation. Gell-Mann [119] and, independently, Zweig [120] proposed that there could be particles having quantum numbers associated with these representations. The particles associated with the triplet representation are called *quarks*, specifically the u, d, and s quarks, while those associated with the anti-triplet representation are the *antiquarks*, $\bar{u}$, $\bar{d}$, and $\bar{s}$, respectively. It was further shown that meson states could be constructed as bound states of a quark and an antiquark, while the baryons could be constructed as a bound state of three quarks. Examples are the proton and the neutron, which could be made up of the combinations uud and ddu, respectively, if an electric charge $+(2/3)|e|$ is associated with the u quark, while a charge $-(1/3)|e|$ is associated with the d quark. The quarks must also carry a baryon number $(1/3)$. It has been shown that the entire SU_3 family of particles could be constructed out of these more fundamental objects, the quarks. Such a model of elementary particles goes under the name of *constituent quark model* and details may be seen in the section under that heading.

Subsequent work after those early suggestions has revealed the existence of three further quarks, denoted by c, b and t and called the *charm* quark for c, *bottom* (or *beauty*) quark for b, and *top* quark for t. There is experimental evidence to show that these quarks are much more massive than the u and d quarks. The six quarks are arranged in three distinct families of doublets: (u, d), (c, s), and (t, b), where the first member of each doublet has the electric charge $+(2/3)|e|$, while the second member of each doublet has the charge $-(1/3)|e|$. Some of the hadrons which contain c, s, b quarks have been found experimentally. The "top" quark has been found to have an incredibly high mass of about 175 GeV!

Among the quarks, the lightest of them should be stable, and if it exists as a free particle, it should be possible to find it in suitably designed experiments. The distinct signature it would carry would be its fractional electric charge. Despite many attempts to discover the quark, none have been found in the free state. Yet in describing the structure of many hadrons they play a role as constituents. They seem to be permanently confined within hadrons. The confinement within hadrons has been understood in terms of another property the quarks carry. This property has been named *color*. Each quark comes in three distinct colors, red, green, and blue. Colored quarks set up color fields in the space around them. The quanta of these color fields are the gluons. An SU_3 color gauge theory called *Quantum Chromodynamics* has been constructed to describe quarks and gluons and their mutual interactions. This theory strongly suggests that colored objects cannot exist in free particle states, and only combinations which are color singlets can occur freely in nature. For more details on this subject, see in section under "QCD—Quantum Chromodynamics".

Quark Jets

First evidence for the production of quark jets was obtained at the SLAC-SPEAR machine studying e^+e^- annihilations at center of mass energies of 6.2 and 7.4 GeV [31]. The jets that are seen are hadrons. They have their origin in the primary quark-antiquark pair that is produced as a result of the e^+e^- annihilation. The fundamental process is the production of a virtual photon as a result of the electron-positron annihilation. Following this, the virtual photon produces a back-to-back quark-antiquark pair from the vacuum with extremely small distance separating them. The property of asymptotic freedom of the quark color interactions (see section under "QCD—Quantum Chromodynamics") implies that the quark and the antiquark can be treated as free particles at production. However, as the color carrying quark-antiquark pair separate from the point of production, strong color confinement forces come into play. At some stage during their separation, it becomes energetically more favorable to create further quark-antiquark pairs from the vacuum than to separate further. These newly created colored quarks and antiquarks pair off with the originally created quark-antiquark pair into color singlets. This results in the back to back jets of hadrons that are seen in an electron-positron collider. The hadronization process of the initial quark and antiquark is clearly separated from the process in which they were produced but does occur along the direction of motion of the initial quark and antiquark. Thus the produced hadronic jets are aligned in the directions of the original quark and antiquark.

The quark jet axis angular distribution can be obtained from measurements on the pencil jet of the hadrons that are produced in cones around the initial quark and antiquark directions; all one has to do is to integrate over the azimuthal angle around the axis of each of the cones. Experimentally, the jet angular distribution was determined to be proportional to $1 + (0.78 \pm 0.12) \cos^2 \theta$. This can be compared with theoretical expectations as follows. A straightforward QED calculation of the angular distribution of $\mu^+\mu^-$ pair produced in e^+e^- annihilation shows that it has the form $(1 + \cos^2\theta)$ at high energies in the center of mass. At the very short quark-antiquark distances at which the pair is produced, the property of asymptotic freedom exhibited by the strong color interactions allows us to treat them as free, spin-half, point particles. They couple through their (fractional) electric charge to the photon just like the muon. Thus the angular distribution of the quark-antiquark pairs should be the same as that for the muon pairs, namely $(1 + \cos^2\theta)$ which is very close to what is observed.

The total cross section for $e^+e^- \rightarrow$ hadrons is also obtainable from that for $\mu^+\mu^-$ production. The total cross section for $e^+e^- \rightarrow \mu^+\mu^-$ reaction calculated in QED is $\sigma = 4\pi\alpha^2/(3E_{cm}^2)$, where $\alpha \simeq (1/137)$, and E_{cm} is the center of mass energy at which the e^+e^- annihilation is studied. The total cross section for hadron production is the same as that for quark-antiquark production, because the quark and the antiquark convert into hadrons with one hundred percent probability. It differs from that for μ-pair production in a couple of factors: first, the quarks carry fractional charges, and second, they come in three colors. If we write the charge carried by the quark as $f_q|e|$, where f_q is a fraction $+2/3$ for u quark, and $-1/3$ for d and s quarks, etc., the ratio R of the quark-antiquark production cross section to that of the muon pair production cross section is $3\sum_q f_q^2$, where the sum has to be carried out over all the quarks that have a mass less than $E_{cm}/2$, and 3 is the number of colors. When E_{cm} is such that only u, d, s quarks are produced, $R = 3((2/3)^2 + (1/3)^2 + (1/3)^2) = 2$. As the energy is increased through the c quark threshold, R takes a jump to $2 + 3(2/3)^2 = 10/3$. When the energy is increased further through the b quark threshold, $3(1/3)^2$ is added to R making it $11/3$. Finally, when the energy crosses the t quark threshold, R should increase by a further $3(2/3)^2 = 4/3$, that is, to $15/3$. Experimentally, the behavior of R is consistent with these calculations. At low energies (~ 2 GeV) R has value 2 and jumps to the values enumerated above as the additional quark thresholds are crossed with increase in energy.

This shows that quarks indeed possess three colors, and that asymptotic freedom is valid for color interactions. Without the color factor all the theoretical ratios would have been too low by a factor of three.

Quark-Parton Model

The parton model was put forward by Feynman to offer explanation for the scaling observed in deep inelastic scattering of electrons by protons. The deep inelastic form factors, which are called the *deep inelastic struc-*

ture functions, can be shown to be functions of two variables, the square of the four-momentum transfer q transferred by the electron, q^2, and another variable called $\nu = q \cdot .p/M$, where p is the four-momentum of the initial proton, and M its mass. In the rest frame of the proton, ν is the energy transferred by the electron during the inelastic scattering process. Experimentally it was observed that at high values of q^2 and ν, the structure functions became functions of the ratio, q^2/ν, instead of being functions of the two separate variables, q^2 and ν. This is called *Bjorken scaling*. Feynman showed that if the proton was considered as a collection of free partons, each carrying a fraction x of the momentum of the proton, the deep inelastic scattering process could be viewed, at high q^2 and ν, as the sum of the incoherent contributions to the scattering from all types of partons integrated over all values of x of the partons. The integrals contain parton distribution functions which are functions of x, and which represent the probability that a given parton carries a momentum fraction x of the proton. These parton distribution functions are parameters which describe the proton and are not calculated by the model but have to be obtained by fitting to experimental data. It turns out that the fraction x of the momentum of the proton carried by the parton is nothing but $q^2/(2M\nu)$, the scaling variable in Bjorken scaling. This success raises the question as to what the partons are.

It has been mentioned above that Gell-Mann and Zweig proposed the quark model of hadrons, according to which baryons are considered as bound states of three quarks and mesons as bound states of a quark-antiquark pair. In the interest of economy of concepts, it was considered desirable to identify the partons with the quarks. Thus was born the quark-parton model. The proton is described by quark distribution functions, which represent the probability that a given type of quark in the proton carries a momentum fraction x of the proton. These quarks of the quark-parton model must be distinguished from the quarks of the constituent quark model of the proton. The former are usually called *current quarks*, while the latter are called *constituent quarks*. At low resolving power (low energies), the proton consists of just the three constituent quarks. As the resolution increases (high energies), the proton is seen to be a sea of quark-antiquark pairs and gluons, each of which carries a fraction x of the momentum of the proton. The quarks which carry moderately large values of x (about $1/3$ or more) are called *valence quarks*, and the others with much smaller values of x are called *sea quarks*. The formulation of QCD makes the connection between quarks and partons even clearer. The parton model is obtained in the limit in which asymptotic freedom is realized in QCD, that is, at infinite q^2. For finite q^2, QCD predicts logarithmic violations of Bjorken scaling due to the possibility of emission of gluons by the quarks. These scaling violations of precisely the amounts predicted by QCD have been observed and put the quark-parton model on a more sound theoretical footing. Reference may be made to other sections under "Parton Model—Bjorken Scaling Violations" and "QCD—Quantum Chromodynamics" for additional information.

Regge Poles

Regge [347] undertook the challenge of demonstrating the validity of the Mandelstam representation (see section under "Mandelstam Representation") for the scattering of two particles which interact with one another through a potential. He took the unorthodox approach of studying the behavior of the scattering amplitude (which is usually expressed as a sum of contributions from *integer* angular momentum partial waves) as a function of the angular momentum continued analytically into the complex plane. He found an alternative expression for the scattering amplitude which gets its contributions from isolated poles in the complex angular momentum plane. These poles are now called *Regge poles*. The idea that Regge poles may contribute also in cases where the interactions between particles are not representable by potential functions has proven quite useful, for understanding certain features of scattering processes involving hadrons. We explain below briefly, starting from potential scattering problems, the origin of the idea of Regge poles and the role they play in determining the scattering amplitude. Then the extension of these ideas to the relativistic scattering of two particles will be presented briefly.

In potential scattering theory, the scattering amplitude for a two particle to two particle scattering process is a function of two variables, the energy and the scattering angle. It is convenient to work in the center of mass system, where convenient variables are, the Mandelstam variable s representing the square of the total center of mass energy, and $\cos\theta$, the cosine of the scattering angle in the center of mass sytem. The partial wave expansion for the scattering amplitude has the form

$$A(s, \cos\theta) = \sum_{l=0}^{\infty}(2l + 1)a(l, s)P_l(\cos\theta).$$

The domain of analyticity of this function in the variable $\cos\theta$ is rather limited; outside of the region $-1 \leq \cos\theta \leq +1$, it is limited to the region of the ellipse with foci at ±1 and extending up to the nearest singularity of $A(s, \cos\theta)$ in the complex $\cos\theta$ plane. Regge sought an extension of the domain of analyticity for the scattering amplitude, by extending l to take on complex values, as follows. First, the above sum is rewritten

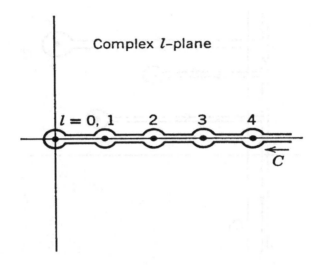

Figure 4.18: Contour C for integration in the complex l plane. (Reprinted with the kind permission of S. Gasiorowicz from his book, *Elementary Particle Physics*, John Wiley & Sons, New York, 1966.)

as a contour integral,

$$A(s, \cos\theta) = \frac{i}{2} \int_C dl (2l+1) a(l,s) \frac{P_l(-\cos\theta)}{\sin\pi l},$$

where the contour C is as shown in Figure 4.18. This contour integral when evaluated gives $-2\pi i$ times the sum of the residues of all the poles contained inside the contour. The poles are due to $\sin\pi l$ which occurs at integer l and the residues are $(-1)^l/\pi$. Thus, one recovers the previous expression.

Regge obtained an alternative expression by deforming the previous contour so that one integrates along a contour parallel to the imaginary axis with $\operatorname{Re}(l) \geq -1/2$. In this we make use of the properties of $P_l(\cos\theta)$ for complex argument and complex order. If $a(l,s)$ has isolated poles in the complex l plane at the locations shown, then these poles have to be taken into account during the deformation of the contour. One can take the contour past these poles so that the contour runs parallel to the imaginary axis (C_1), as shown in Figure 4.19 on the following page, but then one has to take the contribution coming from these poles. Thus

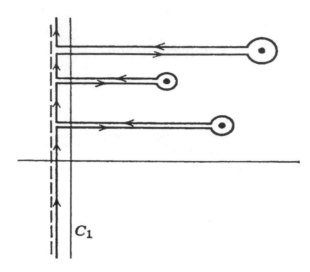

Figure 4.19: Contour C_1 for integration in the complex l plane. (Reprinted with the kind permission of S. Gasiorowicz from his book, *Elementary Particle Physics*, John Wiley & Sons, New York, 1966.)

the alternative expression to the above contour integral is,

$$
\begin{aligned}
A(s, \cos\theta) \; &= \; \frac{i}{2}\int_{C_1} dl \frac{2l+1}{\sin\pi l} a(l,s) P_l(-\cos\theta) \\
&\quad - \sum_n \frac{(2\alpha_n(s)+1)\beta_n(s)}{\sin\pi\alpha_n(s)} P_{\alpha_n}(s)(-\cos\theta),
\end{aligned}
$$

where $a(l,s)$ has been assumed to have poles at $l = \alpha_n(s), n = 1, 2, \ldots$ with residues $\beta_n(s), n = 1, 2, \ldots$. The first term is called the *background integral* and the contribution from the second term is due to the *Regge poles*. This representation for scattering amplitude is valid for complex $\cos\theta$ resembling the old form but with l interpolated to nonintegral values. It turns out that the contribution from the background integral is small in many cases. Neglecting it, Regge could establish the validity of the Mandelstam representation for a class of potentials in nonrelativistic potential scattering theory.

Studying the behavior of the Regge poles in the complex l plane, for the Yukawa potential, one gets a number of interesting insights. It can be shown that the rightmost Regge pole determines the asymptotic behavior of $A(s, \cos\theta)$ as $\cos\theta$ becomes large; this is because $P_n(z) \to z^n$ for large z. Further, the position of the poles $\alpha_n(s)$ as a function of s

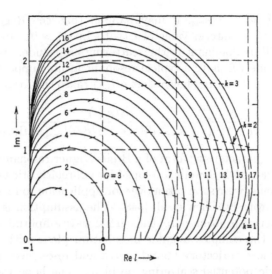

Figure 4.20: Typical trajectories of a Regge pole in the complex l plane. (Reprinted with the kind permission of S. Gasiorowicz from his book, *Elementary Particle Physics*, John Wiley & Sons, New York, 1966.)

in the complex l plane has the following behavior. $\alpha_n(s)$ for some n is found to be located on the real axis when $s < 0$ (unphysical region for this scattering channel). Initially, when s is negative and large, $\alpha_n(s)$ starts at values $l \leq -1$. As s increases, the pole moves along the real l axis to the right until s reaches the threshold value, equal to the square of the sum of the masses of the particles s_{th}, at which point the pole moves up into the complex l plane ($\mathrm{Im}\,(l) > 0$). Then as s increases further, beyond s_{th} to large positive values, the pole moves back to the line at $l = -1$. The location on the real l axis at which the pole starts moving into the upper half-plane is determined by the strength of the potential, and for larger strengths, the pole leaves the real l axis at larger and larger real l. Typical trajectories of a Regge pole in the complex l plane are shown in Figure 4.20 for increasing values, G, of the strength of the potential. It is clear from this figure that the potential needs to have a minimum strength for the trajectory to reach the point $\mathrm{Re}\,(l) = 0$. If it does so at some $s = s_0(< s_{th})$, the scattering amplitude will have the form $[\beta/(s - s_0)]P_0(\cos\theta)$. This form is characteristic of a bound state with zero angular momentum. If the potential is weak, there may be no bound states. Thus Regge poles are closely connected with bound states of a system.

For positive $s = s_{\text{res}}$, it may happen that the Regge trajectory passes close to an integer value of l (with $\text{Im}\,(l) > 0$). In such a case, this term will be dominant in the scattering amplitude, and the partial wave amplitude will have a typical resonant form. Suppose at $s = s_{\text{res}}$, $\text{Re}\,(\alpha_n(s)) = l$, it can be shown that $a(l,s)$ takes the form

$$a(l,s) = \frac{\beta_n(s_{\text{res}})}{(2l+1))[(s - s_{\text{res}})(d\text{Re}\,\alpha_n(s)/ds))|_{s_{\text{res}}} + i\text{Im}\,\alpha_n(s_{\text{res}})]}.$$

This has the characteristic form of a Breit-Wigner resonance at $s = s_{\text{res}}$.

Regge's ideas, which were developed for nonrelativistic potential scattering problems, were soon taken over for application to relativistic scattering of hadrons by Chew et al. [348]. The assumption is that in relativistic scattering problems also, the partial wave amplitudes are analytic functions in the complex angular momentum plane with Regge poles, $\alpha_n(s)$, on whose trajectory bound states and resonances lie, as in the nonrelativistic potential scattering problem. The hope was that these poles would give a representation of the amplitude and provide an elegant parameterization of hadronic scattering processes at high energies. There was intense activity in this field in the 1960's and early 1970's. It was soon found that hadronic interactions based on exchange of a single Regge pole were not sufficient to describe the experimental data well, and more complicated exchanges were included. This resulted in the addition of many more parameters and resulted in loss of predictive power. The discovery of Bjorken scaling and later the discovery of the J/ψ particle in 1974 led to the subsequent formulation of quantum chromodynamics with quarks and gluons as the fundamental constituents of hadrons. This shifted the activity away from studies involving Regge pole exchanges. However, one basic idea that survives from that period is that, instead of considering elementary particle exchanges of spin J, one has to exchange Regge poles. This amounts to interpolating the amplitude for nonintegral (or non-half-integral) values of J, a procedure which is called *Reggeization*.

Since the idea of Reggeization may well play a role in the further development of particle physics, we include here a brief look at Regge pole phenomenology. We start by briefly reviewing Mandelstam variables for describing two body processes, and the role of crossing symmetry in the present context.

Let us consider a reaction, $a + b \rightarrow c + d$, where a, b, c, d are all hadrons, with four-momenta p_a, p_b, p_c, p_d, respectively. The Mandelstam variables $s = (p_a + p_b)^2$ and $t = (p_a - p_c)^2$ can be used to describe the kinematics of this process. (The third variable $u = (p_a - p_d)^2$ is not an independent variable, as $s + t + u = \sum_{i=a}^{d} m_i^2$.) In the center of mass system of a and b, since the vector momenta of a and b are opposite, $s = W^2$, where W is the total center of mass energy, so s may be loosely called the *energy variable*. Bound states may occur for $W < (m_a + m_b)$, and scattering states occur for $W > (m_a + m_b)$. If $q = p_a - p_c$ is the momentum transfer four-vector, then $t = q^2 \simeq -2|\vec{p}_a||\vec{p}_c|(1 - \cos\theta)$, where θ is the scattering angle in the center of mass. From here one sees that $\cos\theta \simeq [1 + (2t/s)]$. The physical region for the collision process is $|\cos\theta| \leq 1$, so that s is positive and t is negative. This is called the *s-channel process*. The unphysical region in this channel is the region of large positive t ($|\cos\theta| \gg 1$).

Related to the s-channel process is another one which is obtained by switching particle b to the right-hand side and replacing it by its antiparticle $\bar{b}$, and particle c to the left-hand side and replacing it by its antiparticle $\bar{c}$. This gives the crossed-channel reaction $a + \bar{c} \rightarrow \bar{b} + d$. This is called the *t-channel process*. Since we are replacing particles b and c by their antiparticles, the four-momenta p_b, p_c of the s-channel process go over into $-p_b, -p_c$, respectively. In this channel, the variable t becomes positive, while s becomes the square of a momentum transfer and is negative. Large positive values of t for the s-channel reaction occur only in its unphysical region of $\cos\theta$, while large positive s can occur only in the unphysical region of the t-channel reaction. The *principle of crossing symmetry* states that there is one analytic function $A(s, t)$ which describes the amplitude for both processes. This means that the amplitude for the s-channel process, when continued analytically into its unphysical region, will give the amplitude for the t-channel process in its physical region and vice versa. The analytic continuation gives a representation for the amplitude in terms of sums over Regge poles in the complex angular momentum plane in the s- and t-channel, respectively.

As stated before, the Regge trajectory connects bound states and resonances which occur whenever $\mathrm{Re}\,\alpha(s)$ takes on integral (or half-integral) values, and determines the asymptotic behavior of the amplitude in the unphysical region when $\cos\theta($ or $2t/s) \rightarrow \infty$. If we use crossing symmetry, this means that, for $s \rightarrow \infty$ and $t < 0$, the amplitude will have the form (recalling that for z large, $P_\alpha(-z) \sim z^\alpha$)

$$A(s, t) \sim \sum_n s^{\alpha_n(t)} \beta_n(t),$$

where $\alpha_n(t)$ refers to the Regge poles in the crossed channel, that is the t-channel. Thus, the Regge poles in the crossed channel allow us to make some very useful statements about the asymptotic behavior of the amplitude for two particle high energy processes in the s-channel.

We may well ask, what can we say about the trajectories of the Regge poles? We can deduce some properties of $\alpha_n(t)$ through the following considerations. For forward scattering (elastic), $t = 0$, and we know that there, $A(s, 0) \sim s$ (recall that the total cross section, which is related to the imaginary part of the forward scattering amplitude, is nearly constant for large s). This fact allows us to conclude that there must be a leading Regge trajectory denoted by $\alpha_P(t)$, such that $\alpha_P(0) = 1$. This trajectory is called the *Pomeranchuk trajectory*. It is the Regge trajectory that is relevant for elastic scattering at high energies. All other Regge trajectories will have $\alpha(0) < 1$. Because in elastic diffraction scattering the quantum numbers and identities of the particles are the same before and after the reaction, the exchanged Regge pole must carry the quantum numbers of the vacuum. (The relevance of Pomeranchuk trajectory for the Pomeranchuk theorem on the relation between particle and antiparticle cross sections has been discussed in the section under "Pomeranchuk Theorem".)

We can also say something more about the behavior of the real part of $\alpha_P(t)$ in the neighbourhood of $t = 0$. Since the real part increases as t increases, we must have near $t = 0$, $\mathrm{Re}\,\alpha_P(t) \simeq 1 + \delta_P t$, where δ_P must be positive. Using this when we work out the elastic differential cross section as a function of t, we can write it as

$$\frac{d\sigma(\text{elastic})}{dt} = (\frac{d\sigma}{dt})_{t=0} \exp 2(\alpha_P(t) - 1) \ln(s/s_0),$$

where s_0 is some constant having the dimension of s to make the argument of the logarithm dimensionless. Regge theory does not have anything to say about the size of s_0. The theory predicts that the elastic cross section falls off exponentially for small t (remember $t < 0$). The above expression also predicts that the peak close to $t = 0$ shrinks logarithmically with s. Such a shrinkage is actually seen experimentally and establishes the usefullness of Regge theory for describing data in this region.

For an s-channel process, we are concerned with the particles and resonances which are associated with Regge trajectories in the t-channel, that is, whenever the $\mathrm{Re}\,\alpha(t)$ takes on integral or half-integral values. All the other quantum numbers, such as baryon number, strangeness, isospin, parity, G parity, must be the same for a given trajectory. A number of Regge trajectories are known which are nicely exhibited in *Chew-Frautschi Plots*. If one plots along the abscissa s (or what is equivalently the square of the mass of the bound state or the resonance), and along the ordinate $\mathrm{Re}\,\alpha(s)$, the "spin" of the particle or resonance, one finds that the known particles and resonances fall on straight lines in the plot (see Figure 4.8 under "Chew-Frautschi Plot"). We do not go into further details on this vast subject. A comprehensive list of other leading trajectories which are exchanged and which are relevant for particular processes is available in the book by Collins [349], which the reader interested in further details may consult.

Before we leave this section, we would like to conclude with the possibility that quantum chromodynamics may provide an explanation for the linearity of Regge trajectories. In quantum chromodynamics, the color interactions between quarks at large distances grow linearly with distance. If one replaces this interaction by a model string stretching between the quarks with a string tension k, one can calculate the relation between the angular momentum of the system and the mass of the hadron. A simple calculation shows that the angular momentum J can be expressed as $J = \alpha' M^2 + \text{const}$, where α' (the slope of the Regge trajectory) is related to the string tension parameter k. Fitting α' to a particular trajectory, for example, the ρ trajectory, one gets a value, $\alpha' = 0.9 \text{ GeV}^{-2}$. This corresponds to a string tension $k = 0.2 \text{ GeV}^2$.

Regularization

Amplitudes for physical processes are calculated in gauge field theories using Feynman diagrams in perturbation theory. When the Feynman diagrams contain closed loops, the integration over the four-momentum of the loops diverges. A prerequisite for handling divergent Feynman integrals and extracting parts from them which can be absorbed into the definition of the mass, charge, etc. and other parts which give rise to finite contributions (which have observable consequences) is to have unambiguous and consistent methods for achieving these ends respecting all symmetries, such as Lorentz invariance and gauge invariance. Thus methods must be found which make the divergent Feynman integral finite through some limiting process. Any such methods which render Feynman integrals finite are called methods of *regularization.*

A number of methods have been devised to give definition to divergent integrals through suitable limiting processes which respect all the symmetries in the problem. We will describe here two of the most commonly used methods. The first is called *Pauli-Villars Regularization* [350], and the second one is the *method of dimensional regularization* due to 't Hooft and Veltman [351]. In the first of these methods, the integral is rendered finite by introducing large mass fictitious particles. In the end, observable quantities should be independent of the fictitious large parameters. In the second method, the four-dimensional loop integral is analytically continued to complex dimensions n where it can be manifestly shown to be finite, and in the end the limit taken as the number of dimensions n approaches four. We illustrate these methods below by using a specific typical example.

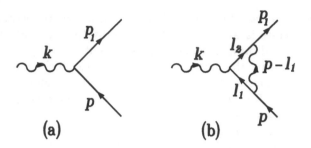

Figure 4.21: Feynman diagrams of the bare electron-photon vertex and its correction to order e^2.

Let us consider the correction to the bare lowest order vertex which represents the coupling of the photon to the electron in QED. If the four-momentum and the polarization four-vector of the photon incident at the vertex are k and $\epsilon^\mu(k)$, respectively, and the initial and final electrons have four-momenta p and $p_1 = p + k$, respectively, the vertex is $\bar{u}(p_1)(-ie)\gamma_\mu u(p)$, where the u's are the Dirac spinors for the electron. The bare vertex and the correction to the bare vertex to order e^2 are shown in Feynman diagrams of Figure 4.21. Following the Feynman rules, we can write the correction to the vertex as $\bar{u}(p_1)\Gamma_\mu(p_1, p)u(p)$, where $\Gamma_\mu(p_1, p)$ is given by the Feynman loop integral

$$
\begin{aligned}
\Gamma_\mu(p_1, p) &= \int \frac{d^4 l_1}{(2\pi)^4} \frac{-ig^{\rho\sigma}}{(p - l_1)^2 + i\epsilon} \bar{u}(p_1)(-ie\gamma_\rho) \frac{i(\slashed{l}_2 + m)}{l_2^2 - m^2 + i\epsilon} \gamma_\mu \\
&\qquad \frac{i(\slashed{l}_1 + m)}{l_1^2 - m^2 + i\epsilon}(-ie\gamma_\sigma)u(p) \\
&= 2ie^2 \int \frac{d^4 l_1}{(2\pi)^4} \frac{\bar{u}(p_1)[\slashed{l}_1\gamma_\mu\slashed{l}_2 + m^2\gamma_\mu - 2m(l_1 + l_2)_\mu]u(p)}{((p - l_1)^2 + i\epsilon)(l_2^2 - m^2 + i\epsilon)(l_1^2 - m^2 + i\epsilon)},
\end{aligned}
$$

where $l_2 = l_1 + k$. The evaluation of this integral is done by combining the denominators using Feynman parameter method,

$$
\frac{1}{abc} = \int_0^1 dx\,dy\,dz\,\delta(1 - x - y - z) \frac{2!}{(xa + yb + zc)^3}.
$$

Using here $a = ((p - l_1)^2 + i\epsilon)$, $b = (l_2^2 - m^2 + i\epsilon)$, $c = (l_1^2 - m^2 + i\epsilon)$, and calling the denominator Den we have, (using $l_2 = l_1 + k$ and $x + y + z = 1$),

$$
Den = [l_1^2 + 2l_1 \cdot (yk - zp) + yk^2 + zp^2 - (x + y)m^2 + i\epsilon]^3.
$$

We shift the variable of integration l_1 such that in the denominator, the linear term proportional to l_1 is removed. This is achieved if we define the new loop variable $q = l_1 + yk - zp$, and the denominator can be simplified to the form

$$Den = q^2 - A^2 + i\epsilon,$$

where $A^2 = -xyk^2 + (1 - z)^2 m^2$. The numerator, when expressed in terms of q and simplified, will have terms involving q which are quadratic, linear, and independent of q. The terms linear in the four-vector q^λ integrate to zero, because the rest of the integrand is a function of q^2 only. The term quadratic in q can be written in the form

$$\int \frac{d^4 q}{(2\pi)^4} \frac{q^\lambda q^\nu}{Den} = \int \frac{d^4 q}{(2\pi)^4} \frac{g^{\lambda\nu} q^2 / 4}{Den},$$

because of the symmetry of the integrand. It is easy to see that the quadratic term in the numerator gives a divergent contribution. Focussing attention on this divergent integral, we have to evaluate

$$\int \frac{d^4 q}{(2\pi)^4} \frac{q^2}{Den}.$$

Giving meaning to this integral involves the process of regularization and will be our sole concern here.

(a) **Pauli-Villars regularization** In this method, we make the integral finite, by replacing the photon propagator by an expression which is the difference between it and that of another propagator with a very large mass Λ, called the *regulator mass*,

$$\frac{1}{(p - l_1)^2 + i\epsilon} \rightarrow \frac{1}{(p - l_1)^2 + i\epsilon} - \frac{1}{(p - l_1)^2 - \Lambda^2 + i\epsilon}.$$

In the limit that $\Lambda \to \infty$, the subtracted term is zero and we have the old loop integral. For any finite Λ, however, the subtraction helps in giving a convergent integral. The prescription is to first perform the integral for finite Λ and afterwards take the limit Λ to infinity. The result of the modification of the photon propagator, by a repetition of the previous calculation for the subtracted term, leads to

$$\int \frac{d^4 q}{(2\pi)^4} \left(\frac{q^2}{Den} - \frac{q^2}{Den(\Lambda)} \right),$$

where $Den(\Lambda) = q^2 - A^2(\Lambda) + i\epsilon$, with $A^2(\Lambda) = A^2 + z\Lambda^2$. To perform the integral over $d^4 q$, one carries out a transformation $q^0 = iq^4$ (called a *Wick Rotation*), which makes the Minkowski four-vector q into the

Euclidean four-vector q, and $q^2 = (q^0)^2 - \vec{q}^2 = -[(q^4)^2 + \vec{q}^2] = -q_E^2$. The integral over $d^4 q_E$ to be done is

$$\int \frac{d^4 q}{(2\pi)^4} \left(\frac{q^2}{Den} - \frac{q^2}{Den(\Lambda)} \right),$$

which evaluates to

$$\frac{i}{(4\pi)^2} \ln(\frac{A^2(\Lambda)}{A^2}).$$

For large Λ we see that $A^2(\Lambda) \sim z\Lambda^2$, we see that the regularized integral shows it has a logarithmic dependence on the mass associated with the regulator. When the regulator mass is taken to infinity, we get a divergence, but now we know the exact value of the logarithmic divergence. The calculation can be carried out to further stages to extract the infinite and finite parts in the above expression for the correction to the vertex. Note that this regularization method has respected Lorentz symmetry and gauge symmetry.

(b) **'t Hooft-Veltman Dimensional regularization** In this method, we define an integral which represents an analytic continuation of the original four-dimensional integral into n-dimensional space time, where n can in general be complex. After we evaluate the integral, the original integral is recovered in the limit when $n \to 4$. We start with the integral in terms of the Euclidean four-momentum q_E,

$$\int \frac{d^4 q_E}{(2\pi)^4} \frac{q_E^2}{(q_E^2 + A^2)^3},$$

of which the part that diverges is

$$\int \frac{d^4 q_E}{(2\pi)^4} \frac{1}{(q_E^2 + A^2)^2}.$$

Analytically continuing this integral from 4 to sufficiently small n will lead to convergent result for any integral and will serve to define it. The analytically continued integral is

$$\int \frac{d^n q_E}{(2\pi)^n} \frac{1}{(q_E^2 + A^2)^2}.$$

This integral is easily evaluated by going into spherical coordinates in Euclidean n-dimensional space. If $d\Omega_n$ is the element of solid angle on the unit n-dimensional sphere, we can write $d^n q = q^{n-1} d\Omega_n$, and we have

$$\int \frac{d^n q_E}{(2\pi)^n} \frac{1}{(q_E^2 + A^2)^2} = \int \frac{d\Omega_n}{(2\pi)^n} \int_0^\infty dq_E \frac{q_E^{n-1}}{(q_E^2 + A^2)^2}.$$

In polar coordinates in n-dimensions, we have

$$d\Omega_n = \sin^{n-2}\theta_{n-1}\sin^{n-3}\theta_{n-2}\ldots\sin\theta_2 d\theta_{n-1}d\theta_{n-2}\ldots d\theta_1,$$

where $0 \leq \theta_{n-m} \leq \pi$, for $2 < m < n-1$, and $0 \leq \theta_1 \leq 2\pi$. The integral over the solid angle is easily performed, and we get

$$\int d\Omega_n = \frac{2\pi^{n/2}}{\Gamma(n/2)}.$$

The other integral is written as

$$\int_0^\infty dq_E \frac{q_E^{n-1}}{(q_E^2+A^2)^2} = \frac{1}{2}\left(\frac{1}{A^2}\right)^{(2-(n/2))}\int_0^1 dx\, x^{1-(n/2)}(1-x)^{(n/2)-1}.$$

In writing this, we have used the variable $x = A^2/(q_E^2+A^2)$. Now the integral over x can be performed using the definition of the beta function, and we get

$$\int_0^1 dx\, x^{1-(n/2)}(1-x)^{(n/2)-1} = \frac{\Gamma(2-n/2)\Gamma(n/2)}{\Gamma(2)}.$$

Thus we have

$$\int \frac{d^n q_E}{(2\pi)^n}\frac{1}{(q_E^2+A^2)^2} = \frac{1}{(4\pi)^{n/2}}\frac{\Gamma(2-(n/2))\Gamma(n/2)}{\Gamma(2)}\left(\frac{1}{A^2}\right)^{(2-(n/2))}.$$

The gamma function has poles at zero and all negative integers. When we take the limit as $n \to 4$, writing $n = 4-\epsilon$, $\Gamma(2-(n/2)) = \Gamma(\epsilon/2)$ develops a pole at $\epsilon = 0$. The other gamma function $\Gamma(n/2) = \Gamma(2-(\epsilon/2))$ and tends to $\Gamma(2)$ when $\epsilon \to 0$. From the property of the gamma function, we have $\Gamma(\epsilon/2) \sim \frac{2}{\epsilon}-\gamma+\cdots$, where $\gamma \approx 0.5772\ldots$ is called the *Euler-Mascheroni constant*. Thus the result for the divergent integral in the limit when $n \to 4$ is

$$\int \frac{d^n q_E}{(2\pi)^n}\frac{1}{(q_E^2+A^2)^2}\lim_{n\to 4}\frac{1}{(4\pi)^2}\left(\frac{2}{\epsilon}-\ln A^2-\gamma+O(\epsilon)\right).$$

The $1/\epsilon$ pole term signifies the divergence in the method of dimensional regularization, which, in the Pauli-Villars regularization, appeared as a logarithmic divergence. The correspondence this suggests is $1/\epsilon \sim \ln\Lambda^2$, between the two methods of regularization.

When working with dimensional regularization, we must remember to set $g_{\rho\sigma}g^{\rho\sigma} = n$ and not 4, and in an integral, replace $k^\rho k^\sigma$ by $(k^2/n)g^{\rho\sigma}$ when the rest of the integrand is a function of k^2 only. Also, the Dirac matrices need to be extended to n-dimensions. Various identities which involve Dirac matrices will get, in the limit that n approaches 4, ϵ dependent contributions. When these terms mulitply the $1/\epsilon$ terms, one will generate finite contributions which have to be taken into account. We do not go into details on this subject here which involves the technicalities of renormalization procedure and the extraction of finite parts; reference may be made to a book on quantum field theory [98]

Renormalization

Perturbative calculations in quantum field theories start by writing the Lagrangian for the interacting fields as a sum of the free field Lagrangians and the Lagrangians representing the interactions between the fields. In the case of QED, for example, the full Lagrangian can be written as a sum of the Lagrangian for the free Dirac electron, for the free electromagnetic field, and an interaction Lagrangian representing the interaction of the electromagnetic field with the electron. The solution of the equations for this coupled set of fields is possible perturbatively if the coupling between the electron and the electromagnetic field is small. Then one starts from the unperturbed basis provided by the eigenstates of the free Lagrangians and uses perturbation theory from there. The mass parameter and the charge parameter that appear in the free Lagrangian are called *bare parameters*, because they are not the quantities that are measured in the laboratory. (One cannot switch off the cloud of virtual photons that surrounds an electron or the cloud of virtual electron-positron pairs that accompany a photon.) The "dressing" of the particle from the "bare" state is accomplished order by order in perturbation theory by "renormalization" and one builds the dressed particle from the bare particle. The methods of renormalization were first formulated by Feynman, Schwinger, and Tomonaga. Many other persons, such as Dyson, Salam, Weinberg, Bogolyubov and Parasiuk, have also made important contributions to the field.

Calculations in quantum field theories using the Feynman diagram approach involve, in general, calculation of diagrams with closed loops in them. The four-momenta associated with the loops have to be integrated over infinite ranges and they give rise to divergent contributions. By the procedure of regularization, one gives meaning to these integrals through some limiting procedures, so that one can deal with the terms, which become divergent in the limit, in an unambiguous and consistent manner. After regularization, one can proceed to subtract the terms that become divergent. The procedure of renormalization involves absorbing these divergences into the redefinition of various parameters, such as the bare mass, the bare coupling constant, and wave function normalization.

The infinite renormalization constants renormalize the bare parameters by an infinite amount. These renormalized parameters still play the role of mass, charge, etc., of the particles for which we substitute the experimental values.

A theory is said to be renormalizable if to all orders of perturbation theory, one can absorb all the divergences into the redefinition of a finite number of parameters, which appear in the bare Lagrangian, by unambiguous subtraction procedures. For the redefined parameters, such as charge and mass, the experimentally measured values for these quantities are used. The finite contributions left over after the subtraction of the divergences represent finite corrections to physical quantities to various orders in perturbation theory and will have observable effects.

Gauge theories, such as QED, Yang-Mills theories, and QCD, provide classes of interacting field theories which are renormalizable. The renormalization corrections to the anomalous magnetic moment of the electron in QED have been calculated up to order α^4 and agree very well with the experimental measurements to eight significant figures. The standard model with spontaneous symmetry breaking has also been proven to be renormalizable; here, however, one needs experimental inputs for many more parameters than those required in QED. Here also there are many impressive successes.

Renormalization Group

It has been mentioned under "Regularization" that infinities arise in the perturbation theory evaluation of Feynman diagrams that contain loops. After regularization, the would-be divergences are removed by absorbing them into redefinitions of physical quantities through the procedure of renormalization. This procedure actually introduces an arbitrary scale into the problem. To see this, let us recall that in Pauli-Villars regularization (see section under "Regularization"), the regularized expression is of the form

$$\ln \frac{A^2(\Lambda)}{A^2}.$$

To extract the Λ^2 dependence, we rewrite this expression as

$$\ln \left(\frac{A^2(\Lambda)}{\mu^2} \right) + \ln \left(\frac{\mu^2}{A^2} \right),$$

where we have introduced an arbitrary parameter μ with the dimension of mass so that both logarithmic terms have dimensionless arguments. Renormalization removes the first of these terms; the second term represents the finite contribution. We have made the subtraction of the divergent term at a mass scale μ^2; μ is what is called the *renormalization scale* which is arbitrary. Different prescriptions for renormalization

call for different μ, and the corresponding finite parts will depend upon
what μ is chosen. Although we have shown the appearance of the renormalization scale in the Pauli-Villars regularization scheme, it is easy to
see that it also appears in the dimensional regularization scheme in a
subtle way. In this case the evaluation above led to

$$\frac{2}{\epsilon} - \ln A^2 - \gamma + O(\epsilon),$$

which is rewritten as

$$\frac{2}{\epsilon} - \ln \mu^2 + \ln \frac{\mu^2}{A^2} - \gamma + O(\epsilon).$$

Introduction of μ^2 makes the logarithm of the finite part have dimensionless argument; it is not obvious that the ϵ in the logarithm of the pole
term carries dimension but it does, and this is made explicit by adding
$-\ln \mu^2$ to it. Thus with any regularization scheme, the renormalization
prescription introduces a renormalization scale.

The renormalized quantities, such as the mass and coupling constant,
depend upon the renormalization scale, but physical quantities, which
are functions of these, cannot depend upon the renormalization scale.
Thus the renormalized quantities must transform in such a way that
physical quantities which depend upon the renormalized quantities remain invariant under change of renormalization scale. It turns out that
the transformation of the renormalized quantities, as the parameter μ
is changed, form a Lie group called the *Renormalization Group*. This
was noted a long time ago by Stückelberg and Petermann [352]. Renormalization group equations express the invariance of physical quantities
under changes of the renormalization scale.

Consider the contribution from a set of connected Feynman graphs
(those which cannot be split into two separate graphs by cutting any
single internal line) with external propagators removed. Let us denote this by G; it is called the *single particle irreducible (1PI) Green's
function*. Suppose that in regularizing it we introduce a cut-off in the
loop momentum Λ, then we obtain the unrenormalized Green's function
$G_U(p_i, g_0, \Lambda)$, where p_i are external particle momenta, and g_0 is the bare
vertex coupling constant. In a renormalizable theory, it turns out that
it is possible to define renormalized Green's function $G_R(p_i, g, \mu)$ by

$$G_R(p_i, g, \mu) = Z_G(g_0, \Lambda/\mu) G_U(p_i, g_0, \Lambda),$$

where the renormalized Green's function G_R is finite in the limit $\Lambda \to \infty$,
and depends upon the scale μ and the renormalized coupling g. Z_G is
a product of renormalization factors (such as wave function renormalization and vertex renormalization), one for each external particle i in

the Green's function G. Since G_U is independent of μ, the statement $\frac{dG_U}{d\mu} = 0$ leads to the renormalization group equation. The form of the equation is:

$$\left(\mu \frac{\partial}{\partial \mu} + \beta(g) \frac{\partial}{\partial g} + \gamma(g) \right) G_R(p_i, g, \mu) = 0.$$

Here the functions $\beta(g)$ and $\gamma(g)$ stand for

$$\beta(g) = \mu \frac{\partial g}{\partial \mu}, \quad \gamma(g) = \frac{\mu}{Z_G} \frac{\partial Z_G}{\partial \mu},$$

and are called the *beta-function* and *anomalous dimension* respectively. The anomalous dimension depends on the Green's function being considered, while the beta function is the same for all Green's functions. In finding the beta function and the anomalous dimension, the differentiations with respect to μ are carried out keeping Λ constant, and the limit $\Lambda \to \infty$ is taken afterwards.

To illustrate the use of renormalization group equations we consider a simple but useful example. If, in some problem, a single large momentum P is present, all the particle momenta p_i can be expressed as fractions x_i of P. Introducing P^2 the square of the four-momentum P and the change of variable $t = (1/2) \ln (P^2/\mu^2)$, the renormalization group equation becomes

$$\left(-\frac{\partial}{\partial t} + \beta(g) \frac{\partial}{\partial g} + \gamma(g) \right) G(x_i, g, t) = 0.$$

Defining a *running coupling constant* $g(t)$ by

$$t = \int_{g(0)}^{g(t)} dg' \frac{1}{\beta(g')},$$

the solution to the renormalization group equation can be written as

$$G(x_i, g(0), t) = G(x_i, g(t), 0) exp \left[\int_{g(0)}^{g(t)} dg' \frac{\gamma(g')}{\beta(g')} \right].$$

In problems with a single large momentum scale, all the P^2 dependence arises from the running of the coupling constant $g(t)$.

There are many applications of the renormalization group equations in particle physics and statistical mechanics, but we do not go into this vast subject here. Interested persons may seek additional information from the book by Peskin and Schroeder [98].

Resonance States

In elementary particle collision processes at high energies, a number of product particles are produced, many of which are unstable. The unstable ones decay and give rise to products which themselves may be stable or unstable. The chain of decays ends when all the products are stable. When examined, the unstable states which decay are found to have a unique mass, a definite lifetime, and a unique set of quantum numbers, for example, angular momentum, parity, isospin, and strangeness. Through the uncertainty principle, the existence of a finite lifetime implies an energy width (or mass width) associated with the state. Such unstable states are called *Resonance states* and are found in baryonic systems as well as in meson systems. Broadly speaking, the lifetime of the state gives information on the interaction responsible for the decay. Very short lifetimes, of the order of 10^{-23} s, give rise to very broad resonances, with widths an order of magnitude smaller than the mass, and decay occurs through strong interactions. Lifetimes in the range 10^{-17} to 10^{-19} s are characteristic of decays through electromagnetic interactions, and long lifetimes such as 10^{-10} s or longer are associated with weak interactions. In addition, in each case there are characteristic selection rules on the quantum numbers of the product particles; for example, strangeness is conserved in strong interaction processes, but violated in weak interactions.

To find the mass and quantum numbers associated with a resonant state, one has to identify the product particles and measure the energy and the three-momenta of sets of particles that are produced in the reaction. If the same resonant state is produced in various reactions, the invariant mass distribution for certain identified set of product particles will have a peak at a certain value of the invariant mass which is the same in the different reactions. Its width gives information about the decay rate (or lifetime). A specific example will clarify these considerations. Consider the reaction $\pi^- + p \rightarrow \Sigma^+ + \pi^- + K^0$, with π mesons of some energy E incident on a hydrogen target. One measures the energy and three-momenta of the Σ^+ and π^-, and constructs the square of the invariant mass

$$M_{\Sigma\pi}^2 = (E_\Sigma + E_\pi)^2 - (\vec{p}_\Sigma + \vec{p}_\pi)^2.$$

If one plots the square of the invariant mass distribution for incident pion energies of 2.17, 2.25, and 2.36 GeV, one clearly sees a peak distinguished well from the phase space distribution expected in the case of no resonances.

In cases where the final state consists of three particles, a method using Dalitz plots (see section under "Dalitz Plot") is very useful in identifying the existence of invariant mass peaks for combinations of identified particles in the final state. The spin and parity of resonances are also found using the plot as described under "Dalitz Plot". In this way, resonances have been found in $K-\pi$, $\Xi-\pi$, $\pi-\pi$, $\pi-n$, etc. systems. The existence of meson resonances ω and ρ was also established. The study of resonant states experimentally gives the spectrum of elementary particle states. The object of the theory is to find the fundamental interactions responsible for these states as bound states and resonances of the underlying theory. In the last four decades hundreds of resonant states have been found, studied, and classified. The constituent quark model, in which baryonic states are considered as excitations of a bound three quark system and mesonic states as excitations in a bound quark-antiquark system, is very successful in describing the data well with six flavors of quarks and three colors for each flavor. A comprehensive listing of the known resonances and their qauntum numbers may be found in Reference [62].

Rho Meson (ρ) Resonance

The ρ-meson resonance was found in 1961 by analyzing the invariant mass of pairs of pions in the reaction $\pi^- + p \rightarrow \pi^+ + \pi^- + n$ or $\pi^- + \pi^0 + p$. The invariant mass distribution for the combinations $\pi^- \pi^0$ and $\pi^+ \pi^-$ showed clear peaks at a mass value of 765 MeV, with a width of 125 MeV. In the reaction $\pi^+ + p \rightarrow \pi^+ + \pi^+ + n$ or $\pi^+ + \pi^0 + p$, no peak was found for the $\pi^+ \pi^+$ combination while for the $\pi^+ \pi^0$ combination a peak was found at the same mass value of 765 MeV. These peaks in the three charge states suggests that the resonance has isospin 1, corresponding to ρ^+, ρ^0, ρ^- states. The width being so large, we can conclude that the decay is through strong interactions. Thus we can use the decay distributions to gain information about spin, parity, and G-parity quantum numbers. G-parity conservation in the decay to two pions leads to the assignment of $G = +1$ to this resonance. This, together with the $I = 1$ assignment, immediately lets us conclude that the charge conjugation parity of ρ^0 must be odd. ρ^+ and ρ^- are charge conjugates of one another.

Information about spin and parity of the ρ resonance may be obtained as follows. Because the two pion $I = 1$ state is odd under the interchange of isospin variables, it must also be odd under the interchange of spatial variables, since there are two bosons. This means that the orbital angular momentum for the decay has to be odd, and the parity also has to be odd. This allows for the possibilities, $1^-, 3^-, \ldots$. To decide between these, one has to study the angular distribution of the two pions in the ρ rest frame. In the rest frame of the ρ, its state is described by $\sum_m C_m \psi_{Jm}$. The distribution of the pions with respect to the z axis will then have the form $\sum_m P(m)|Y_J^m(\theta, \phi)|^2$, where $P(m) = |C_m|^2$ and $\sum_m m P(m) = 0$. The larger the value of J the more complex the angular distribution is, and by such studies it is found that one can exclude spins higher than 1. Thus the spin and parity of the iso-triplet ρ is determined to be 1^-. The determination of the quantum numbers of this resonance is typical of the methods used.

Salam-Weinberg-Glashow Model

In this model, the synthesis of electromagnetic and weak interactions is achieved by inroducing the ingenious mechanism of spontaneous breaking of gauge symmetry. A model Lagrangian was proposed in which the photon, mediating electromagnetic interactions, and the intermediate vector boson, mediating weak interactions, are unified into one spin 1 multiplet of gauge fields [353]. However, a big important difference between these gauge particles had to be accommodated; namely, the photon has zero rest mass, while the intermediate vector bosons must have a large mass to provide short range weak interactions. The symmetry of the spin 1 gauge multiplet had to be somehow broken in such a way so as to leave the photon massless and yet allow the intermediate vector bosons to acquire nonzero masses. Salam and Weinberg proposed that the symmetry of the Lagrangian is spontaneously broken by the vacuum (see section under "Higgs Mechanism"), with the result that the would-be Goldstone bosons provide the longitudinal modes for the intermediate vector bosons, making them massive and still leaving the photons massless. In his paper Weinberg expressed the hope that a theory with spontaneous breaking of gauge symmetry would be renormalizable. This conjecture has been proven to be correct with 't Hooft's proof of the renormalizability of such theories in 1972 (see also under "Standard Electroweak Model" for details).

Scaling behavior

See section under "Parton Model".

Second Quantization

This is a procedure adopted in quantizing field theories. One starts with a classical field theory, where the field functions, which are functions of space and time, satisfy certain field equations arising from a Lagrangian. A quantum version of the field can be obtained by looking upon the field function as the wave function of a single particle in quantum mechanics, satisfying the field equation (procedure sometimes called *first quantiza-*

tion). However, with this procedure one runs into serious problems with interpretations of the theory (see section under "Causality Condition"). These difficulties are avoided if one does not identify the field function as the wave function for a single particle but instead quantizes the classical field directly. The procedure is *second quantization* only in a picture in which the *first quantization* involved the interpretation of the field function as a one particle wave function for a quantum mechanical system. Without that interpretation of the field function as a one particle wave function, there is just one "quantization" of the classical field theory [98]. The procedure to "quantize" a field theory involves a redefinition of the classical field functions as operators on which certain commutation relations are imposed. One can show that the Hamiltonian has eigenstates which are obtainable by analogy with the simple harmonic oscillator. The eigenstates represent not just one particle states but many particle states.

Selectron
This is a hypothetical particle of spin 0, which is the supersymmetric partner of the spin 1/2 electron. (See under "Supersymmetry".)

Σ Particles
Sigma particles come in three charge states, $\Sigma^+, \Sigma^0, \Sigma^-$, and are called *Sigma hyperons*. Their masses are measured to be: $M(\Sigma^+) = 1189.37 \pm 0.07$ MeV, $M(\Sigma^0) = 1192.642 \pm 0.024$ MeV, and $M(\Sigma^-) = 1197.449 \pm 0.030$ MeV. Their lifetimes are determined to be: $\tau(\Sigma^+) = (0.799 \pm 0.004) \times 10^{-10}$ s, $\tau(\Sigma^0) = (7.4 \pm 0.7) \times 10^{-20}$ s, and $\tau(\Sigma^-) = (1.479 \pm 0.011) \times 10^{-10}$ s. The decay modes for the charged Σ's are as follows: $\Sigma^+ \to p + \pi^0 (51.57\%)$, $\Sigma^+ \to n + \pi^+ (48.31\%)$, and $\Sigma^+ \to p + \gamma (1.23 \times 10^{-3})$. Decays to $n + \pi^+ + \gamma$ and to $\Lambda^0 + e^+ + \nu_e$ have also been observed with branching ratios of order 10^{-4} and 10^{-5}, respectively. For Σ^-, the dominant decay mode is $\Sigma^- \to n + \pi^- (99.848\%)$. Other modes are $n + \pi^- + \gamma (\sim 10^{-4})$, $n + e^- + \bar{\nu}_e (\sim 10^{-3})$, $n + \mu^- + \bar{\nu}_\mu (\sim 10^{-4})$, and $\Lambda^0 + e^- + \bar{\nu}_e (\sim 10^{-5})$. Σ^0 decays almost exclusively to $\Lambda^0 + \gamma$. It is an electromagnetic decay.

The fact that the Σ hyperon comes in three charge states suggests that they are members of an isospin triplet. The spin of Σ^+ was determined by studying its decay after it was produced in the reaction, $K^- + p \to \Sigma^+ + \pi^-$. The measurements are consistent with assigning a spin of 1/2 to it. Σ^- likewise has spin 1/2. The Σ's are assigned a strangeness $S = -1$. Through the study of the decay, $\Sigma^0 \to \Lambda^0 + \gamma$ or equivalently to $\Lambda^0 + e^+ + e^-$, the parity of Σ^0 has been determined to be even. As all these particles fit into the baryon octet, they are all assigned even parity and spin $1/2$. The quark structure of these particles is: $\Sigma^+ = (uus)$, $\Sigma^0 = (uds)$, and $\Sigma^- = (dds)$.

Spatial Parity
See under "Parity Conservation" and "Parity Nonconservation in Nuclear β Decay."

Spin Statistics Theorem
This is an important result first derived by Pauli [354]. It states that for local field theories, demands of Lorentz invariance, positive energies, positive norms, and causality mean that particles with integral spin obey Bose-Einstein statistics, and particles of half-odd integral spin obey Fermi-Dirac statistics. In quantizing such theories, for particles obeying Bose-Einstein statistics, *commutation* relations have to be imposed on the field operators, while for particles obeying Fermi-Dirac statistics, one has to impose *anticommutation* relations between the field operators. If this is not adhered to, one will run into trouble with positive energies, positive norms, or causality.

Spontaneous Breaking of Chiral Symmetry
The vacuum of QCD may have a special structure which may not respect chiral symmetry. This occurs because, for massless quarks, quark-antiquark pair condensates will be formed from the attractive interaction in QCD of quarks with antiquarks. In the vacuum state they will have zero total momentum and zero angular momentum; this requirement will necessarily lead to a net chirality for the vacuum state. The vacuum expectation value of the scalar operator $\bar{q}q$ will be nonzero, which will lead to a mixture of chirality projections. This will in turn allow the quarks to develop effective masses in the vacuum. The existence of net chirality for the operator $\bar{q}q$ signals the onset of spontaneous breaking of the full symmetry group in which the right and left chirality quarks suffer different transformations, down to one in which the right and left chirality quarks suffer the same transformations. Thus spontaneous symmetry breaking will lead to the lack of conservation of axial vector currents, while the vector currents will be conserved. Goldstone's theorem then leads to massless, spinless Goldstone bosons. (The pion is approximately this Goldstone boson.)

Spontaneously Broken Symmetry
See section under "Higgs Mechanism".

Squark

This is a hypothetical particle of spin 0, which is the supersymmetric partner of the spin 1/2 quark. (See under "Supersymmetry".)

Standard Electroweak Model

This is a gauge theory unifying electromagnetic and weak interactions due to Salam, Weinberg, and Glashow [353]. The general idea is to start with massless fermions and massless gauge bosons belonging to some gauge group. A Lagrangian is written down which possesses the gauge symmetry corresponding to the gauge group. Spontaneous breaking of gauge symmetry is achieved by introducing a Higgs field, which has a nonvanishing vacuum expectation value (abbreviated as *vev*). The Higgs field is so chosen that several of the gauge bosons as well as the fermions will acquire mass, but the photon will be left massless. The gauge symmetry that achieves all this is the gauge group $SU_{2,L} \times U_{1,Y}$. We outline this development below.

Let us start with the Lagrangian for massless Dirac fields f,

$$\begin{aligned}
\mathcal{L} &= \bar{f} i \gamma_\mu \partial^\mu f \\
&= \bar{f}_L i \gamma_\mu \partial^\mu f_L + \bar{f}_R i \gamma_\mu \partial^\mu f_R,
\end{aligned}$$

where $f_L = \frac{1-\gamma_5}{2} f$, $f_R = \frac{1+\gamma_5}{2} f$ are the left-handed and right-handed chiral components of the fermion f. In applying this to the electron and its neutrino in the lepton sector, the electron has both left-handed and right-handed components, while the electron neutrino has a left-handed component only. Taking this fact into account we may write for the (e, ν_e) sector,

$$\mathcal{L} = \bar{e}_R i \gamma_\mu \partial^\mu e_R + \bar{e}_L i \gamma_\mu \partial^\mu e_L + \bar{\nu}_e i \gamma_\mu \partial^\mu \nu_e.$$

This is for the first generation. We have to add a similar term for each of the other generation of leptons. The considerations which we develop below for the first generation only have to be duplicated for the other generations.

Now we have to decide what internal symmetry to impose on the fermions. We choose to put (ν_e, e_L) into an isospin doublet L and put e_R into an isospin singlet R. This isospin is called *weak isospin* to distinguish it from the isospin we use with strongly interacting particles. Thus ν_e has $T = 1/2, T_3 = +1/2$, e_L has $T = 1/2, T_3 = -1/2$, and e_R has $T = 0, T_3 = 0$. The above Lagrangian can be written in the compact form,

$$\mathcal{L} = \bar{L} i \gamma_\mu \partial^\mu L + \bar{R} i \gamma_\mu \partial^\mu R$$

This $\mathcal{L}$ is invariant under global SU_2 transformations $L \to L' = e^{\frac{-i}{2} \vec{\tau} \cdot \vec{\alpha}} L$, $R \to R' = R$. (Here the $\vec{\tau}$ has as its components, the three Pauli matrices.) We note that we can write the electric charge Qe in terms of the generator T_3 as $Q = (T_3 - 1/2)$ for the L state, and $Q = (T_3 - 1)$ for the R state. In this form the charge is different for each member of the isospin doublet. A related generator which gives the same quantum numbers to all members of the doublet is $Y = 2(Q - T_3)$. It is called the *weak hypercharge*. Its values for one generation of leptons and quarks are: $Y(l_L) = -1$, $Y(e_R) = -2$, $Y(q_L) = 1/3$, $Y(u_R) = 4/3$, and $Y(d_R) = -2/3$. We also note that Y commutes with all the generators $T_i, i = 1, 2, 3$ of SU_2. The generators $T_i, i = 1, 2, 3$ and Y belong to the direct product group $SU_{2,L} \times U_{1,Y}$.

We considered above a global transformation with space-time independent parameters $\vec{\alpha}$. Let us make this a local symmetry by making these parameters space-time dependent. Then to preserve the symmetry of the Lagrangian, we have to replace the partial derivatives by the covariant derivatives with three massless gauge fields. The photon should not be among these three gauge fields because e_R being an SU_2 singlet will not interact with the photon, and we need it to interact. The way to make this interaction possible is to enlarge the gauge group with a U_1 factor associated with the transformations of e_R. We choose this U_1 in such a way that the electric charge Q is a linear combination of the U_1 generator and T_3 of the isospin. The hypercharge Y is the natural choice, so that the direct product group will be $SU_{2,L} \times U_{1,Y}$, and the gauge invariant Lagrangian is

$$\mathcal{L} = \bar{f} i \gamma_\mu D^\mu f,$$

where

$$D^\mu f = (\partial^\mu + ig\frac{1}{2}\vec{\tau} \cdot \vec{A}^\mu + ig'\frac{Y}{2}B^\mu)f,$$

and where $\vec{A}^\mu$ is a triplet of gauge fields with coupling g, and B^μ is an SU_2 singlet.

Now we want to break this symmetry spontaneously to $U_{1,em}$. We do this by introducing a set of scalar fields Φ and let it have a nonzero vacuum expectation value $\langle\Phi\rangle_0$ (abbreviated as *vev*), which is $U_{1,em}$ symmetric. We choose the scalar to be a complex doublet with hypercharge $Y(\Phi) = +1$, and write it in the form

$$\Phi = \begin{pmatrix} \phi^+ \\ \phi^0 \end{pmatrix}.$$

The Lagrangian for the scalar field Φ is

$$\mathcal{L}_s = (D_\mu \Phi)^\dagger (D^\mu \Phi) - V(\Phi),$$

where

$$D^\mu \Phi = \left(\partial^\mu + ig\frac{1}{2}\vec{\tau} \cdot \vec{A}^\mu + ig'\frac{1}{2}B^\mu \right) \Phi,$$

and $V(\Phi) = -\mu^2(\Phi^\dagger\Phi) + \lambda(\Phi^\dagger\Phi)^2$. The scalars can couple to fermions, too. We introduce a coupling term with the Lagrangian which has a Yukawa form and respects $SU_2 \times U_1$ invariance,

$$\mathcal{L}_{\text{Yukawa}} = g^e \bar{l}_L \Phi e_R + g^u \bar{q}_L \tilde{\Phi} u_R + g^d \bar{q}_L \Phi d_R + \text{Hermitian Conjugate},$$

where the conjugate scalar field $\tilde{\Phi} = i\tau_2\Phi^*$ which has hypercharge $Y(\tilde{\Phi}) = -1$, and g^e, g^u, and g^d are the Yukawa coupling strengths of the electron, the u-type quark, and the d-type quark respectively.

The total Lagrangian is the sum of all three Lagrangians above. Now, to break the symmetry spontaneously, we let Φ have a nonzero vev,

$$\langle 0|\Phi|0\rangle = \begin{pmatrix} 0 \\ v/\sqrt{2} \end{pmatrix}, \qquad v = \sqrt{\mu^2/\lambda}.$$

To see what particle spectrum emerges, we have to shift the fields about their vev. We write

$$\Phi = U^{-1}(\xi) \begin{pmatrix} 0 \\ \frac{v+\eta(x)}{\sqrt{2}} \end{pmatrix}, \qquad U(\xi) = e^{i\vec{\xi}(x)\cdot\vec{\tau}/v},$$

where the original complex fields, ϕ^+ and ϕ^0 are parametrized in terms of 4 real fields, $\xi_i(x)$, $i = 1, 2, 3$, and $\eta(x)$. These shifted fields are the physical fields and have zero vev,

$$\langle 0|\xi_i|0\rangle = 0, \ i = 1, 2, 3; \ \langle 0|\eta|0\rangle = 0.$$

We carry out a further gauge transformation to go into what is called the *unitary gauge* (where only the physical fields are present), by writing

$$\Phi' = U(\xi)\Phi.$$

Let us write $\Phi'(x) = \frac{v+\eta(x)}{\sqrt{2}}\chi$, with $\chi = \begin{pmatrix} 0 \\ 1 \end{pmatrix}$. The fermion fields transform as $l'_L = U(\xi)l_L$, $e'_R = e_R$, $q'_L = U(\xi)q_L$, $u'_R = u_R$, and $d'_R = d_R$, and the gauge fields as

$$\frac{\vec{\tau} \cdot \vec{A}'_\mu}{2} = U(\xi)\frac{\vec{\tau} \cdot \vec{A}}{2}U^{-1}(\xi) + \frac{i}{g}[\partial_\mu U(\xi)]U^{-1}(\xi); \qquad B'_\mu = B_\mu.$$

Now we need to express the scalar Lagrangian in terms of the new fields. We get

$$\mathcal{L}_s = (D_\mu\Phi')^\dagger(D^\mu\Phi') - V(\Phi'),$$

where

$$D_\mu \Phi' = \left(\partial_\mu + ig\frac{1}{2}\vec{\tau} \cdot \vec{A}'_\mu + ig'\frac{1}{2}B'_\mu \right) \frac{(v+\eta(x))}{\sqrt{2}}\chi,$$

and

$$V(\Phi') = \mu^2 \eta^2 + \lambda v \eta^3 + \frac{\lambda}{4}\eta^4.$$

The Yukawa coupling Lagrangian becomes

$$\mathcal{L}_{\text{Yukawa}} \quad = \quad \frac{v}{\sqrt{2}}[g^e \bar{e}'_L e'_R + g^u \bar{u}'_L u'_R + g^d \bar{d}'_L d'_R]$$

$$+ \quad \frac{1}{\sqrt{2}}[g^e \bar{e}'_L e'_R + g^u \bar{u}'_L u'_R + g^d \bar{d}'_L d'_R] + \text{Hermitian Conjugate.}$$

The above work is sufficient to extract some of the results of the model. We can read off the mass terms which are coefficients of quadratic field terms. (1) The mass of the Higgs scalar, $m_\eta = \sqrt{2}\mu$; (2) the fermion masses, $m_e = g^e v/\sqrt{2}$, $m_u = g^u v/\sqrt{2}$, $m_d = g^d/\sqrt{2}$; (3) the vector boson masses which are obtained from the part of the $|D_\mu \Phi'|^2$ which depends quadratically on the gauge fields. The parts of interest are

$$\text{Vector Boson mass terms} \quad = \quad \frac{v^2}{2}\chi^\dagger \left(\frac{g}{2}\vec{\tau} \cdot \vec{A}'_\mu + \frac{g'}{2}B'_\mu \right)^2 \chi$$

$$= \quad \frac{v^2}{8}\left(g^2[(A'^1_\mu)^2 + (A'^2_\mu)^2] \right.$$

$$+ \quad \left. [gA'^3_\mu - g'B'_\mu]^2 \right).$$

Here the A'^1_μ and A'^2_μ appear quadratically, but A'^3_μ and B'_μ have cross terms. The terms involving these can also be expressed as quadratic terms by forming suitable linear combinations of A'^3_μ and B'_μ as follows. We write

$$\frac{v^2}{8}[gA'^3_\mu - g'B'_\mu]^2 =$$

$$\frac{v^2}{8}\begin{pmatrix} A'^3_\mu & B'_\mu \end{pmatrix} \begin{pmatrix} g^2 & -gg' \\ -gg' & g'^2 \end{pmatrix} \begin{pmatrix} A'^3_\mu \\ B'_\mu \end{pmatrix}.$$

The matrix in the middle can be diagonalized by the following orthogonal transformation,

$$Z_\mu = \cos\theta_w A'^3_\mu - \sin\theta_w B'_\mu$$

$$A_\mu = \sin\theta_w A'^3_\mu + \cos\theta_w B'_\mu,$$

the previous expression becomes

$$\begin{pmatrix} Z_\mu & A_\mu \end{pmatrix} \begin{pmatrix} g^2 + g'^2 & 0 \\ 0 & 0 \end{pmatrix} \begin{pmatrix} Z_\mu \\ A_\mu \end{pmatrix}.$$

From the diagonalization it is readily seen that the eigenvalues of the matrix in the Z_μ, A_μ basis are $\frac{v^2}{8}(g^2+g'^2)$ and 0. Thus one linear combination of A'^3_μ and B'_μ becomes the Z boson and the other combination is the massless photon, reflecting a residual symmetry $U_{1,\text{em}}$ present. The mixing angle is called the *electroweak mixing angle*. We also note that $\tan\theta_w = g'/g$. Finally, the vector boson mass terms can be combined into the form,

$$M_W^2 W_\mu^+ W^{-\mu} + \frac{1}{2}M_Z^2 Z^\mu Z_\mu,$$

with $W_\mu^\pm = \frac{A'^1_\mu \mp A'^2_\mu}{\sqrt{2}}$, $M_W^2 = \frac{g^2 v^2}{4}$, and $M_Z^2 = \frac{v^2}{4}(g^2+g'^2)$. M_W^2 is the square of the mass of the charged W bosons, while M_Z^2 is the square of the mass of the neutral Z boson.

Next we look at the fermion-gauge boson coupling terms from the above. These can be written in the form $\frac{g}{\sqrt{2}}(J_\mu^+ W^{+\mu} + J_\mu^- W^{-\mu})$, where $J_\mu^+ = \bar{\nu}'_e\gamma_\mu e'_L + \bar{u}'_L\gamma_\mu d'_L$, and J_μ^- is the Hermitian adjoint of J_μ^+. It is clear from the form that J_μ^+ is a charge changing current, changing a left-handed electron by adding one unit of $(+)$ charge to it and making it into a (left-handed) neutrino, or changing a left-handed d quark into a left-handed u quark. J_μ^- changes the charge by (-1) unit.

If we apply these coupling terms to calculate the effective four fermion interaction at low energies (i.e., low invariant momentum transfers), we get

$$\mathcal{L}_{\text{cc}}^{\text{eff}} = -\frac{g^2}{2M_W^2}J_\mu^+ J^{-\mu},$$

which is just the $V - A$ theory. This lets us identify

$$\frac{g^2}{8M_W^2} = \frac{G_F}{\sqrt{2}},$$

where G_F is the Fermi weak coupling constant. This relation allows us to estimate the size of the vev of the Higgs field. We have

$$v^2 = \frac{4M_W^2}{g^2} = \frac{1}{\sqrt{2}G_F}.$$

Putting in numerical values, we get $v \approx 246$ GeV.

The theory also predicts that there must exist neutral current interactions. Going back and picking out the appropriate terms in the fermion-gauge boson coupling, we can identify

$$\mathcal{L}_{\text{nc}} = gJ_\mu^3 A'^{3\mu} + \frac{1}{2}g'J_\mu^Y B'^\mu,$$

where J_μ^Y is the hypercharge current $J_\mu^Y = J_\mu^{em} - J_\mu^3$. Using the definition of $A_\mu'^3$ and of B_μ' in terms of Z_μ and A_μ, we can write this as

$$\mathcal{L}_{nc} = e J_\mu^{em} A^\mu + \frac{g}{\cos\theta_w}(J_\mu^3 - \sin^2\theta_w J_\mu^{em})Z_\mu,$$

where $e = g\sin\theta_w$.

A big triumph for this theory of electroweak unification came from the discovery of the neutral currents experimentally of magnitude comparable to charged current effects. These experiments help to determine $\sin^2\theta_w$. Precision experiments at LEP and at SLC have determined this parameter to high accuracy. Using the rough value $\sin^2\theta_w = 0.23$, the theory gives $M_W = 2^{-5/4}eG_F^{-1/2}/\sin\theta_w$, $M_Z = M_W/\cos\theta_w$. Putting in numerical values, we get $M_W \simeq 80$ GeV, $M_Z \simeq 90$ GeV. The W and Z bosons were discovered by UA1 and UA2 collaborations at CERN in 1983 at about these predicted masses.

These experimental discoveries clearly establish the electroweak theory on a sound footing. Because the theory has also been proven to be renormalizable, higher order calculations can be undertaken. These allow experimental tests of the model with higher and higher accuracy. Much work has been done along these lines at LEP, SLC, and other machines. The standard electroweak model, together with the QCD Lagrangian for strong interaction of quarks, constitutes what is called the *standard model*. Predictions of the standard model have been tested to very high accuracy in a number of experimental measurements. Some of the details of these comparisons may be found in other sections in this book.

Statistics—Bose-Einstein, Fermi-Dirac

In many problems a knowledge of the distribution of the number of particles in terms of different energies is important. A perfect gas is an assembly of free noninteracting point particles. When the perfect gas is treated as a quantum mechanical system, the fact that the particles are *indistinguishable* plays a very important role in determining the statistics the assembly obeys. The wave function for the assembly is given by the product of the wave functions of the individual particles. Under the interchange of all the coordinates (including spin, isospin, etc.,) of any two identical particles, the wave function of the system is either *symmetric* or *antisymmetric*. An assembly, described by wave functions which are symmetric under the interchange of coordinates of any two particles, obeys Bose-Einstein statistics. An assembly, described by wave functions which are antisymmetric with respect to exchange of coordinates of any two particles, obeys Fermi-Dirac statistics. In the latter case, the antisymmetry of the wave function under the exchange of any two coordinates automatically limits the occupancy of the quantum state to just one particle; no two particles can be in the same quantum state with the same coordinates, which is *Pauli's exclusion principle*. The distribution formula for the number of particles n_r with energy E_r is derived to be of the form,

$$n_r = \frac{1}{\exp(\alpha + \beta E_r) \pm 1},$$

where α is a constant determined by the condition that the total number of particles in the system is N, and $\beta = (1/kT)$, T being the absolute temperature, and k the Boltzmann constant. Here the minus sign in the denominator, refers to the Bose-Einstein statistics, and the plus sign to Fermi-Dirac statistics. In the case of Fermi-Dirac statistics, α is negative and can be written in the form $\alpha = -\frac{E_F}{kT}$, where E_F is called the *Fermi-energy*. In the case of Bose-Einstein statistics, α cannot be negative but can be positive or zero. The highest possible $n_{r=0}$ corresponds to $\alpha = 0$, in which case infinite number is associated with the lowest energy level $E_{r=0} = 0$. This has relevance to the phenomenon of Bose-Einstein condensation.

Stochastic Cooling

This method of increasing the luminosity of particle beams in accelerators has been used at the $p\bar{p}$ collider at CERN with which the intermediate vector W and Z bosons were found. The aim is to reduce the phase volume associated with the beam particles. In attempting to do this, one has to respect Liouville's theorem, which states that one cannot reduce the overall phase volume; one can only change its shape when conservative forces act on the system. The total phase volume is made up of the phase volume of particles in the beam and the phase volume associated with the empty space surrounding the beam. What may be attempted without conflict with Liouville's theorem, for a given total phase volume, is to decrease the phase volume associated with the beam particles, by moving the particles individually toward the center of the distribution, while at the same time increasing the phase volume associated with empty space. However, to achieve this, one has to locate the individual particles in the beam, sense their characteristics, act on them individually, and make them go into the desired location of phase volume. The procedure reminds one of Maxwell's demon. The only reason that it works here is that in moving the particles around one is exchanging energy with them in such a way as not to come into conflict with the second law of thermodynamics.

Particles circulating in a beam undergo, in general, betatron oscillations (transverse to the beam's circulation) and longitudinal oscillations. Considering transverse oscillations first, if there were a single particle going around in a circular orbit, and one wanted to reduce the amplitude of the horizontal betatron oscillation about this orbit, one could pick up information about the transverse position with a transverse pickup where its tranverse position is nonzero, and give it just the right kick at a point where its transverse deviation from the orbit is zero, so that it reaches the proper orbit, within a single turn of circulation. This can likewise be done for the vertical betatron oscillations.

If we consider more than one particle, the same considerations can be extended to the other particles if the response of the electronic system is fast enough and resolves the individual particles. The total cooling time will then be equal to the system response time multiplied by the number of particles. If there are too many particles and the electronics is not able to resolve them separately, then this estimate for the cooling time is only a lower limit. In such a case, the system gain should be reduced so that in each turn only a small fraction of the needed correction is given. This is because, along with the signal from each particle, there are also signals from all the other particles which are not resolved by the system. Every particle is cooled by its own signal, but it is also simultaneously heated by the uncorrelated signals from all the other unresolved particles. The cooling rate is proportional to the system gain, while the effect of the random heating gives rise to Brownian motion in phase space, the mean square displacements of which vary linearly with time. Thus the heating rate varies as the square of the system gain. It is, therefore, possible to find an optimum system gain where one gets overall cooling. For example, one might choose the system gain such that the heating rate is about half the cooling rate.

To influence longitudinal oscillations, one requires "momentum cooling". Here one needs to develop a signal proportional to the momentum of the particle. In the Palmer method, a pickup is used that measures the dipole moment of the particles in a region of nonzero dispersion among the particles. This produces a (root mean squared) voltage proportional to the fluctuations in momentum in the beam. This voltage is then used in a kicker to accelerate or decelerate the particles to reduce the momentum spread. It cools the beam. Another method, the Thorndahl method, uses a notch filter. A notch filter is one whose response changes sign depending upon whether the particle circulation frequency is above or below the desired one. Evidence for stochastic cooling was obtained by Carron et al. in 1978 [355].

Storage Ring—Colliding Beam Machines

In the last couple of decades, much of the knowledge we have gained about elementary particles has come from studies in colliding beam machines. Colliding beams of electrons with positrons, protons with protons, protons with antiprotons, and electrons with protons have all been constructed. The main advantage in using colliding beams is in the enormously large center of mass energies available for the production of new particles as compared to machines in which an accelerated beam of particles strikes a fixed target, called *fixed target machines*. All these colliding beam machines have one common characteristic; the beams, which are to be used in the collision, have first to be produced and then stored in "storage rings" where they circulate for a long time. Clearly one of the factors that affects the performance of the storage ring is the length of time over which the beams can be held in the ring. Another factor has to do with the ability to focus the beams to as small a region as possible and with as little momentum spread as possible.

Once one has beams in storage rings with the desired characteristics, two counter-rotating beams of particles are made to meet one another at several points where collisions occurring between the beams and the reaction products can be studied. The quantity that determines the rate of reactions is called *luminosity*. If σ is the cross section for the interaction of the beams, and L is the luminosity, then the number of collision events per unit time (the reaction rate) is given by the product σL. The expression for L in the case of two oppositely directed relativistic beams is $L = fnn_1n_2/A$, where f is the revolution frequency of the beam, n is the number of bunches of particles in either beam, n_1, n_2 are the numbers of particles per bunch in the two beams, and A is the area of cross section of the beams. To obtain large luminosity, one has to diminish A as much as possible and increase all the other quantities as much as possible. Also, the bunches of particles should have a small momentum spread. Typical values for the luminosities are about 10^{31} cm^{-2}s^{-1} for electron-positron colliders and about 10^{32} cm^{-2}s^{-1} for pp colliders. To reduce the momentum spread in the bunches, electron cooling and stochasitc cooling methods are used. These also include methods to reduce A and make the beams last long in the storage rings.

An interesting property of relativistic beams in storage rings is the fact that they are polarized. This arises as a result of the emission of synchrotron radiation by the circulating particles and was pointed out by Ternov et al. [356].

The entire field of colliding beam accelerator physics has a very rich literature and much exiciting work is being done. We have presented only a brief summary here.

Strange Particles
See sections under "Kaons: The τ-θ Puzzle" and under "Hyperons—Decay Selection Rules".

Strangeness Conservation
See sections under "Associated Production" and "Gell-Mann, Nishijima Formula".

String Theory
This is a theory of elementary particles in which particles, instead of being considered points, are pictured as the quantum modes of vibrating strings. The first developments of this theory arose from efforts to understand the spectra of the enormous numbers of hadronic resonances, produced in strong interactions, with rather high values of spin. The mass squared of a particle of spin J was found to be given roughly by $M^2 = J/\alpha'$, where α' is a constant called the *slope of the Regge trajectory*, and has the value $\alpha' \sim 1 (\text{GeV})^{-2}$. Although this relation has been found to be true up to values of $J \sim 13/2$, it is quite possible that this relation continues on indefinitely.

A behavior of the sort described in the previous paragraph is in sharp contrast to expectations from point-particle quantum field theories. In quantum field theory, at high energies, the two-particle elastic scattering amplitude is constructed, as the sum of exchanges in the s- and t-channel tree diagrams (that is, diagrams involving no closed loops), where s and t are the Mandelstam variables. Tree diagrams in which particles of spin J are exchanged are found to give rise to bad high energy behavior, violating the Froissart bound. For example, if all the external particles are scalars, the t-channel amplitude, due to the exchange of a particle of spin J, has the form, $A_t(s,t) = \frac{g^2(s)^J}{t - M^2}$, and becomes worse and worse for larger and larger J. If there are several t-channel contributions, due to various values J, with masses M_J and couplings g_J, the contribution will be

$$A_t(s,t) = \sum_J \frac{g_J^2(s)^J}{t - M_J^2}.$$

According to this, the high energy behavior (at large s) will be determined by the exchange of the hadron of the highest J, a behavior which is completely different from the much softer variation that is is found experimentally. In other words, there does not seem to be a hadron of highest spin. The experimentally observed behavior can be understood, if, in the above expression, the sum over J extends to infinity and the resulting sum is finite, as happens, for example, in the power series expansion of $e^{-x} = \sum_{J=0}^{\infty} (-x)^J/J!$.

There is also the contribution from the s-channel exchanges. In a development similar to the discussion of the t-channel amplitude, the s-channel amplitude $A_s(s,t)$ is then

$$A_s(s,t) = \sum_J \frac{g_J^2(t)^J}{s - M_J^2}.$$

Symmetry under cyclic permutations of the momenta of the external particles demands that the same masses and couplings must appear in the t-channel and s-channel amplitudes. Again, we will have bad behavior for large t if the sum is over a finite number of terms, and to get a reasonable behavior, the sum will have to be extended over infinite range.

Now, suppose the couplings and masses are such that the s- and t-channel amplitudes are equal, then it seems that the amplitude should be written *either* as an infinite s-channel sum *or* as an infinite t-channel sum, but not as a sum over both s- and t-channel contributions. In fact, analyzing the experimental data available at that time, Dolen, Horn, and Schmid [357] showed that the two were very nearly equal. If this equality is taken to be exact, the equality of the s- and t-channel amplitudes was called *duality*, because these provide alternative descriptions of the same process. This presented a challenge to theorists to construct a theory which exhibits this duality.

Veneziano met the challenge and generated a "dual-resonance" amplitude in 1968 [358]. He wrote the amplitude for two body scattering in terms of the beta function $B(a,b)$, where $B(a,b) = \Gamma(a)\Gamma(b)/(\Gamma(a) + \Gamma(b))$: $A(s,t) = B(-\alpha(s), -\alpha(t))$, where the $\alpha(s)$ and $\alpha(t)$ are Regge trajectories in the s- and t-channels, respectively. Such a function is symmetric under $s \leftrightarrow t$, and exhibits Regge behavior $s^{\alpha(t)}$ for large s and fixed t and $t^{\alpha(s)}$ for large t and fixed s, and has poles at all integer values of the Gamma functions. The behavior of $B(a,b)$, near $b = -J$ (J a non-negative integer), is

$$B(a,b) \sim \frac{(-1)^J}{J!}\frac{1}{b+J}(a-1)(a-2)\cdots(a-J).$$

This expression exhibits all the singularities of the function B as a function of b for fixed values of a. It can be shown that the right-hand side of the expression for B, when summed over all J from 0 to ∞, converges for Re$(a) > 0$, that is,

$$B(a,b) = \sum_{J=0}^{\infty} \frac{(-1)^J}{J!}\frac{1}{b+J}(a-1)(a-2)\cdots(a-J).$$

A simlar result holds for B interchanging a and b, and in this case the infinite series converges for $\text{Re}\,(b) > 0$.

Veneziano chose to write linear Regge trajectory functions, $\alpha(t) = \alpha(0) + \alpha't$. The function $B(-\alpha(s), -\alpha(t))$ has singularities corrsponding to the exchange of particles in the t-channel of mass $M^2 = (J - \alpha(0))/\alpha'$, $J = 0, 1, 2, 3, \ldots$. The parameters $\alpha(0)$ and α' in the Regge trajectory function are called the *intercept* and *slope* of the Regge trajectory, respectively. If α_0 and α' are positive, the pole with $J = 0$ has negative mass squared, and such a particle is called a *tachyon*. If α' were not restricted to be positive, all the higher J values would also lead to tachyons. We also note that the residue of the pole, $\alpha(t) = J$, is a polynomial of Jth order in s, which we want to be positive (so as not to have negative norm states, called *ghosts*). The appearance of tachyons and ghosts were two unappealing features of the dual resonance models, and a lot of effort was put into curing these diseases. It was found that to get no ghosts, one had to work in 26 space-time dimensions and α_0 had to have the value 1. If α_0 is chosen to be 1, then the $J = 1$ state has a mass which vanishes. Clearly, the dual resonance model predicts a massless $J = 1$ hadron, in direct contradiction to observations. In fact, in another form of the dual resonance model, consistency demanded the existence of a massless $J = 2$ hadron, also. Another serious deficiency of the dual resonance models was soon found when it was applied to other strong interaction processes, such as deep inelastic scattering. In the region where $s \to \infty$, $t \to -\infty$, but s/t is held fixed (the Bjorken limit), the Veneziano amplitude falls off exponentially and fails to describe Bjorken scaling and parton-like behavior of hadrons. These results led to a waning of interest in dual resonance models for hadronic interactions.

At about this time, it was realized that the Veneziano model could be understood in terms of the relativistic model of a closed string. An action integral for the string was constructed, and the equations of motion were derived. Quantization of the theory was carried out using the canonical formalism. For the closed string, periodicity properties allow one to write the solution of the equations of motion in the form of Fourier series. The Fourier coefficients become creation and annihilation operators, and a vacuum state can be defined on which the action of the annihilation operators will give zero. Creation operators acting on the vacuum state will give excited states. One obtains a corresponding mass spectrum. The elementary particles are identified with the different rotational and vibrational states built on the vacuum state. The square of the mass of the particle is proportional to the frequency of vibration of the string and turns out to be an integer multiple of some basic parameter, agreeing with the linear Regge trajectory of the dual resonance model. One feature of the bosonic string theory was that it was free of ghosts only in 26 space-time dimensions, and its ground state was a tachyon. It is interesting to note that these features are shared by the dual resonance model.

In an earlier paragraph we mentioned the existence of massless spin 1 and spin 2 particles. An examination of the couplings of these spin 1 and spin 2 objects revealed that they resembled those of the Yang-Mills fields and gravitational fields, respectively. These features motivated Scherk and Schwarz [359] to suggest abandoning such a string theory for describing hadrons, and instead use it as a method of unifying all the fundamental forces including gravity. In this way, string theory could provide an approach to quantum gravity. The hadronic mass scale is of the order of 1 GeV (length scale of order 1 fermi), while that of gravity is the Planck mass 10^{19} GeV (length scale $\sim 10^{-33}$ cm), the change of viewpoint takes us nineteen orders of magnitude up in the mass scale or down in the length scale. Based on the changed viewpoint, all the excited states will be of the order of Planck mass, and the ground state is essentially massless on this scale. There is a large amount of degeneracy in the mass states of string theories giving the hope that many of the observed particles could be accommodated in the ground state of the theory.

The observed particles consist of leptons and quarks in addition to the gauge bosons. Leptons and quarks being fermions, the problem was to construct a spinning string theory. This was accomplished by Ramond [360] who proposed the string equivalent of the Dirac equation for the fermion. Soon thereafter, Neveu and Schwarz [361] constructed a string theory of interacting bosons and fermions. They found that such a theory could be consistently formulated only in 10 space-time dimensions. They found a further bonus in this theory—the ground state was not a tachyon.

A supersymmetric string theory was successfully constructed by Green and Schwarz in 1982 [362]. This theory was also found to have no tachyon in its ground state. Such string theories are called *super-string theories*. Even superstring theories, while they do not suffer from problems with ghosts and tachyons, still have *anomalies* (such as the triangle Adler-Jackiw anomaly). In 1984, a major advancement was made with the discovery that an anomaly-free superstring theory could be constructed in 10 space-time dimensions. Since we want a theory applicable in the four-dimensional world, attention was focused on what to do with the six extra dimensions.

To deal with the problem of extra dimensions, an old theory of Kaluza and Klein [363], in which they tried to unify electromagnetism and gravity by going to five-dimensional space-time, looked like an avenue that could be explored in this context. The ground state of the five-dimensional general relativity was assumed to be the direct product of four-dimensional Minkowski space-time with a circle S_1. If the radius of this circle is extremely small ($\sim 10^{-33}$ cm), all observed phenomena, at even the highest energies so far explored, will be averages over the circle S_1. Such ideas could be extended to hold for the six extra dimensions.

Attempts to construct a superstring theory with all the desirable low energy phenomenological characteristics have been successful. These should have (1) four-dimensional space-time, (2) simplest form of supersymmetry at the Planck scale, (3) a gauge group which is large enough to contain the standard model gauge group, $SU_3 \times SU_2 \times U_1$, and (4) the standard model particles, the leptons and quarks with their associated quantum numbers. The most successful superstring theory incorporating many of these features is the so-called *heterotic string theory*, based on the exceptional gauge group $E_8 \times E_8$. To get to 4 dimensions from the 10 dimensions, different methods of compactifications of the extra dimensions have been found. Such a theory has a number of acceptable phenomenological features. There are only three generations of light fermions (consistent with the observation of the width of the Z^0). Space-time supersymmetry is broken in a clever way, so that symmetry breaking in the observable sector is reduced by a factor $1/M_{\text{Planck}}^2$, so that the weak gauge bosons have small enough masses compared to the Planck scale. There are only singlet and triplet representations of SU_3 present. Of course, there are a number of additional particles predicted, such as extra Z^0 boson.

Based on these modest successes, there is a lot of exciting activity taking place with superstring theories. However, the goal of achieving a "theory of everything (TOE)", is still probably a long way off.

Strong Focusing
See "Alternating Gradient Strong Focusing Machines" in Chapter 2.

SU_3—Model for Hadron Structure
The search for an underlying symmetry which would explain the similarity of patterns in the eight baryons and the eight pseudoscalar mesons was the major preoccupation of workers in the 1950's. The earliest such attempt was by Sakata. Sakata constructed a model [364] which laid the ground work for the full-fledged development of unitary symmetry. He pointed out that in order to construct particles with strangeness, it was necessary to add at least one strange particle to the proton and

the neutron and consider a symmetry enlarged from isospin. (Isospin symmetry, based on the observation of charge independence of nuclear particle interactions, was by then already well established.) The only known baryon having strangeness was the Λ^0, so Sakata's proposal was to build all the then-known mesons and baryons with the triplet, proton (p), neutron (n) and Λ^0, and their antiparticles. For example, in the Sakata model, K^+ is the composite $(\bar{\Lambda}^0 p)$, K^0 is $(\bar{\Lambda}^0 n)$, Σ^+ is $(\Lambda^0 p\bar{n})$, Ξ^- is $(\Lambda^0 \Lambda^0 \bar{p})$, etc. Other baryons can similarly be written in composite form in this model. In the limit of perfect symmetry, he set the masses of Λ^0, p, n, equal to one another and considered unitary transformations among three particles (p, n, Λ^0).

In isospin symmetry, the transformations of the $\psi = (p, n)$ doublet are of the form

$$\psi \to \psi' = U\psi,$$

where the transformations U form all unitary, unimodular 2×2 matrices. U is of the form $\exp i\vec{\alpha} \cdot \vec{T}$, where $\vec{\alpha}$ are the parameters of the transformation, and $\vec{T}$ are the generators of the transformations. The generators $\vec{T}$ can be expressed in terms of 2×2 Pauli matrices $\vec{\tau}$ as $\vec{T} = (1/2)\vec{\tau}$. The group corresponding to these transformations is the group SU_2.

The Sakata model considers transformations of the triplet, $\psi = (p, n, \Lambda^0)$, and may be written as

$$\psi \to \psi' = U\psi,$$

where the transformations U form all unitary, unimodular 3×3 matrices. Here U may be written in the form $\exp i \sum_{i=1}^{8} \alpha_i T_i$, where $\alpha_i, i = 1, 2, \ldots, 8$ represent the parameters of the transformation, and $T_i, i = 1, 2, \ldots, 8$ are the generators of the transformation. The $T_i, i = 1, 2, \ldots, 8$ can be expressed in terms of eight 3×3 matrices λ_i, $i = 1, 2, \ldots, 8$ as $T_i = (1/2)\lambda_i$. A standard form for the matrices for λ_i, $i = 1, 2, \ldots, 8$ was introduced by Gell-Mann and plays here a role similar to that played by the Pauli matrices for SU_2. The form given by Gell-Mann is quoted here for easy reference.

$$\lambda_1 = \begin{pmatrix} 0 & 1 & 0 \\ 1 & 0 & 0 \\ 0 & 0 & 0 \end{pmatrix}, \quad \lambda_2 = \begin{pmatrix} 0 & -i & 0 \\ i & 0 & 0 \\ 0 & 0 & 0 \end{pmatrix}, \quad \lambda_3 = \begin{pmatrix} 1 & 0 & 0 \\ 0 & -1 & 0 \\ 0 & 0 & 0 \end{pmatrix},$$

$$\lambda_4 = \begin{pmatrix} 0 & 0 & 1 \\ 0 & 0 & 0 \\ 1 & 0 & 0 \end{pmatrix}, \quad \lambda_5 = \begin{pmatrix} 0 & 0 & -i \\ 0 & 0 & 0 \\ i & 0 & 0 \end{pmatrix}, \quad \lambda_6 = \begin{pmatrix} 0 & 0 & 0 \\ 0 & 0 & 1 \\ 0 & 1 & 0 \end{pmatrix},$$

$$\lambda_7 = \begin{pmatrix} 0 & 0 & 0 \\ 0 & 0 & -i \\ 0 & i & 0 \end{pmatrix}, \quad \lambda_8 = \frac{1}{\sqrt{3}} \begin{pmatrix} 1 & 0 & 0 \\ 0 & 1 & 0 \\ 0 & 0 & -2 \end{pmatrix}.$$

The group corresponding to these transformations is SU_3. The fundamental representations in the Sakata model are the triplet (3) to which p, n, Λ^0 belong, and the ($\bar{3}$) to which the antiparticles $\bar{p}, \bar{n}, \bar{\Lambda}^0$ belong. The mesons which are formed from particle-antiparticle combinations would belong to the product representation $3 \times \bar{3}$, which reduces to an octet and a singlet. The seven pseudoscalar mesons, $\pi^+, \pi^0, \pi^-, K^+, K^0$, $K^-, \bar{K}^0$, can be fit into the octet; the discovery of the η^0 meson would complete the octet. On the baryon side, however, the Sakata model did not do as well. In the examples above, the baryons in this model were composites of two particles and one antiparticle (recall $\Sigma^+ = \Lambda^0 p \bar{n}$ etc.). This implies that the baryon multiplet would be found in the reduction of $3 \times 3 \times \bar{3}$. Working out the reduction, one finds many more particles in the multiplet than in the octet, and the Sakata model was in trouble.

The solution to this problem was proposed by Gell-Mann and independently by Ne'eman that the baryons also be accommodated in the eight-dimensional regular representation of SU_3 (see section under "Eightfold Way"). With the proposal of the quark model, with the quarks p, n, λ belonging to the fundamental triplet representation of SU_3, with their fractional charges $(+2/3, -1/3, -1/3)|e|$, respectively, and with fractional baryon numbers $1/3$ (now renamed u, d, s quarks), it was possible to accommodate the octet baryon multiplet in the reduction of $3 \times 3 \times 3$, and the mesons in $3 \times \bar{3}$. The Sakata model clearly provided the inspiration to investigate the relevance of the higher symmetry group SU_3 to particle physics, and this avenue has proved to be very successful.

Sudbury Neutrino Observatory (SNO)

This is a new facility that has been constructed in the INCO Creighton mine in Sudbury, Ontario, Canada, by a group of scientists from Canada, U. S., and U. K. This facility is a neutrino observatory and will primarily measure the intensities, energies, and directions of origin of electron-neutrinos from the sun. The detector relies on the use of heavy water (of which Canada has plenty), rather than ordinary water, for neutrino detection. Deuterium in the heavy water gives the further capability to detect the muon-neutrinos and tau-neutrinos, also. Taken together, the data are expected to advance our understanding of the properties of neutrinos and also provide a check on solar models which are based on our current knowledge of the energy generation and energy transformation processes in the sun.

SNO is a heavy water neutrino detector, and as such, it is unique in the world. It uses 1 kilo-tonne of heavy water contained in an acrylic sphere of diameter 12m. The neutrinos interact with the deuterium in the heavy water, resulting in the emission of Cherenkov radiation from the heavy water. This radiation is detected by an array of 9600 photomultiplier tubes (PMT's) mounted on a geodesic dome surrounding the sphere. The entire heavy water sphere is immersed inside a sphere containing ordinary, but highly pure, light water. The entire spherical assembly is contained in a 30m spherical cavity excavated within the rock of the mine, located at a depth of 6800 feet. This depth under the rock provides excellent shield from cosmic rays. A very large amount of effort has been devoted to making the laboratory extremely clean to reduce radioactive signals which might otherwise mask the signal from the neutrinos.

The SNO facility can make independent measurements of the fluxes of the different flavor neutrinos by the following methods:

- First, there is the charged current reaction, $\nu_e + d \to p + p + e^-$. The ejected electron carries most of the energy of the neutrino and, in its passage through the heavy water, gives rise to Cherenkov radiation, which is detected. The pattern of the PMT's that are hit by the Cherenkov radiation is used to determine the neutrino energies. Such measurements of the neutrino energy spectrum are expected to show a distortion, if neutrino flavor oscillations are occurring. About 30 events of this type are expected per day according to the standard solar model.

- Second, there is the neutral current reaction, $\nu_x + d \to p + n + \nu_x$. This deuteron breakup reaction can be initiated by all flavors of neutrinos, x. This feature of the detector is what gives SNO its uniqueness. The neutron produced from the breakup of the deuteron slows down in the heavy water and, after a characteristic slowing down time, will eventually be captured on a nucleus. The capture will lead to the emission of a cascade of gamma rays. These gamma rays scatter against electrons in the medium, and the electrons will generate Cherenkov light, which is detected in the PMT's. Clearly, the efficiency of this method of detecting neutrinos of any flavor will depend upon how efficiently the released neutrons are captured and the resulting cascade of gamma rays detected. Since the capture of neutron on deuterium is not all that efficient, SNO is proposing to use two different systems to improve neutral current detection. One of these will use ^{3}He ultra-clean proportional tubes suspended in a grid within the heavy water. Because ^{3}He has a large capture cross section for neutrons, the proton and triton products following the capture give rise to pulses in the counter wire. The other method will dissolve two tonnes of ultra-clean $MgCl_2$ in the heavy water. The chlorine has a high capture cross section for thermal neutrons. The capture leads to a cascade of gamma rays peaked at about 8 MeV of energy, which will be detected. About 30 neutrons per day are expected according to the standard solar model.

- Third, there is the scattering reaction, $\nu_x + e \rightarrow \nu_x + e$. This is the reaction which is used by other light water detectors, such as the SuperKamiokande. In principle, this reaction occurs for all flavor of neutrinos, but the electron neutrinos give signals which are about six times larger than for other flavors. Although good directional information on the neutrino can be obtained, this method is not very good for determining spectral information on the neutrino, since in the final state the energy is shared between the electron and the neutrino. The rate of these reactions in SNO is expected to be about 3 events per day.

The water system in SNO has to be capable of handling large volumes of extremely pure water. Products from uranium and thorium radioactive chains have to be reduced to extremely low concentrations. The purpose of the light water surrounding the heavy water is to provide for absorption of gamma rays and neutrons which originate from activity in the rock. For the 1 kilo-tonne of heavy water, impurities in the ordinary water must be less than 10^{-14} g/g of water. Anything that comes into contact with the water has to meet the stringent radio-impurity levels set. The water purification system must also make sure that no biological activity occurs to prevent the deterioration of the ultraviolet transmittance of the water. All these tremendous engineering challenges have been met in putting together the SNO system.

To handle the data, SNO has an extensive electronics and data acquisition system. Signals from each of the 9600 PMT's are processed individually. Each PMT channel has a large dynamic range and timing precision of better than one nanosecond. Power consumption must be kept at a minimum to avoid excessive heat load in the mine.

In the first phase of operation, SNO is expected to take data from the charged current mode. In the next phase, neutral current data are expected to be taken. In Figure 4.22, we have reproduced an event display from SNO, which illustrates the ring of hits on the PMT's from the Cherenkov cone. In the display, on the right-hand part, two rings of hits on PMT's can be seen. One of the rings is quite well defined, while the other is more diffuse. The event display represents a two particle event originating inside the detector. High energy muons usually give a clear ring, and so one of the particles in the display is probably a muon. Event displays like these will be analyzed for neutrino initiated events with software that has been developed.

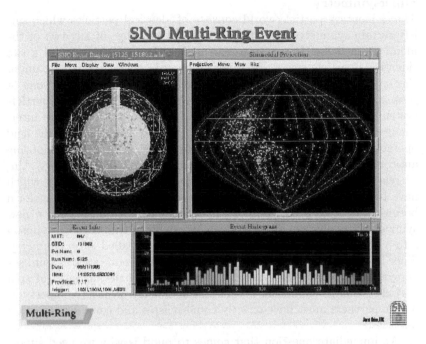

Figure 4.22: An event display from SNO showing the ring of PMT's that have received hits from the Cherenkov radiation originating in the heavy water. (Courtesy SNO collaboration.)

It will be very exciting should results from SNO confirm that neutrino flavor oscillations are indeed occurring. The unambiguous discovery of such oscillations would clearly establish that the neutrinos are not massless. The mass of the neutrino plays a role in determining the eventual fate of the universe, whether it will expand forever or fall back in a "big crunch".

SuperKamiokande Experiment
This is a large light water neutrino detector which has been operating in the Kamioka mine in Japan and has provided high statistics data on solar neutrinos. This detector has been able to map the solar neutrino spectrum by measuring the recoil electron energy. It has also been used to detect atmospheric neutrinos. Its data show that neutrino flavor oscillations may be occurring. (See further under "Neutrino Oscillations" and under "Atmospheric Neutrinos".)

Supersymmetry
The elementary particle world consists of identical particles which satisfy certain symmetry properties under the exchange of any two of the identical particles. In quantum mechanics, the wave function of a two (identical) particle system is either symmetric or antisymmetric under the interchange of the particles. Particles are classified as *bosons* or *fermions* depending on whether the wave function of the two particle system is *symmetric* or *antisymmetric*, respectively, under the interchange of the two identical particles. In quantum field theories, the field operators which describe fermions (bosons) are quantized by imposing anticommutation (commutation) relations between the operators. Until relatively recently, all the space-time and other symmetries which have been considered have had the property that they only transform bosons into bosons and fermions into fermions. Around 1970, a new symmetry was discovered [365] which is capable of transforming a boson into a fermion and vice versa. This symmetry is called *supersymmetry*. Since bosons carry integral spin and fermions half-odd integral spin, supersymmetry relates integral spin particles with half-odd integral spin particles. If supersymmetry is an exact symmetry of nature, supersymmetric partners must have exactly equal masses and couplings to other particles.

An immediate question that comes to mind is why we need supersymmetry. One answer is purely theoretical—the new mathematical structure provided by this extension of existing theories of symmetries deserves study in its own right. But more significantly from the point of view of particle physics, the existence of supersymmetry may provide an answer to a puzzling question of the standard model for which there is no satisfactory answer otherwise. Current theories put the energy scale at which the unification of all the fundamental forces of nature, including gravity occurs, at the Planck scale, about 10^{19} GeV. The electroweak scale is characterized by the standard model Higgs vacuum expectation value of about 246 GeV. These scales differ by some seventeen orders of magnitude. The mass of the Higgs particle in the electroweak theory gets radiative corrections which naturally tend to pull the particle masses to the highest available mass scale, the Planck scale. In other words, there is instability to radiative corrections. One way that this might be avoided is if the intermediate states which contribute to the radiative corrections, give contributions of opposite sign such that they cancel when all the effects are added up. This is exactly what happens in supersymmetric theories, where in the intermediate states, for every particle, there is a supersymmetric partner which also occurs. These supersymmetric partners give contributions, which are equal but opposite in sign to that from the particle, and cancel one another. In other words, the existence of supersymmetry will provide a natural explanation for the wide difference between the electroweak scale and the Planck scale. This is in line with what happens in other examples. The photon remains massless due to gauge symmetry, despite radiative corrections. The pion would be massless if chiral symmetry was exact.

It is clear that supersymmetry cannot be exact. The mass degeneracy expected between a particle and its supersymmetric partner (with spins differing by $1/2$) is not seen in nature. Hence supersymmetry is a broken symmetry. We recall that the gauge symmetry of electroweak interactions is broken by the Higgs field in such a way that the photon can remain massless while the W and Z bosons acquire a mass related to the electroweak scale. Suppose the supersymmetry breaking proceeds by the analogs of the Higgs particle. Since it is spontaneously broken, supersymmetry becomes a hidden symmetry in much the same way as gauge symmetry is a hidden symmetry of the electroweak theory. The scale at which the breaking occurs is assumed to be related to the electroweak scale and to give masses to W's and Z's and their superpartners which are comparable to the electroweak scale. In this scenario, one would expect to see a whole spectrum of new particles.

With this picture, the supersymmetric partners of the quarks and gluons, of leptons and photons, all will acquire masses which could be in the range of several hundred GeV to a TeV. It is possible to look for them at the Tevatron of Fermilab, the LEP collider at CERN, and the Large Hadron Collider (LHC) that is being constructed at CERN.

Supersymmetric theories with spontaneous symmetry breaking have been constructed and can be shown to be renormalizable. Conservation of $B - L$ is assumed, where B is the baryon number and L is the lepton number. The invariance corresponding to this conservation law is called *R-parity invariance* with $R = (-1)^{3(B-L)+2S}$, where S is the spin of the particle. A consequence of this invariance is that supersymmetric particles must be produced in pairs if the initial state has R-parity zero (corresponding to non-supersymmetric particles). Higher mass supersymmetric particles are generally unstable and decay into lower mass supersymmetric particles, conserving R parity; hence the least massive supersymmetric particle must be stable as it cannot get rid of its R-parity quantum number. Such a particle is called the *lightest supersymmetric particle (LSP)*.

The nomenclature for the particles which are paired as supersymmetric partners of quarks, gluons, leptons, and photons are as follows. The quarks have their spin zero partners, the squarks; the gluons, the gluinos (spin half partners); the leptons, the sleptons (spin zero partners); and the photon, the photino (spin half partner). In addition the W and Z bosons have their spin half partners, the w-inos and the z-inos. The graviton (spin 2) has a supersymmetric partner, the gravitino with spin (3/2). The Higgs particle structure contains two doublets of Higgs fields, which is the minimum required for the generation of masses of both the u-type and d-type quarks (and the charged leptons). Their supersymmetric partners are called the *Higgs-inos*, and both charged and neutral varieties are present. There is mixing among the charged particles by themselves and among the neutral particles by themselves, as the interaction eigenstates and the mass eigenstates may not coincide. Thus the terms *neutralinos* and *charginos* are used to describe the neutral mixture and charged mixture, respectively.

So far none of these new particles is seen in any of the Tevatron or LEP experiments. These experiments have served to rule out masses for these particles below certain values; for the lightest supersymmetric neutral particle the lower limit is about 11 GeV, while for the others, lower limits range from tens of GeV to 200 GeV. For details, reference may be made to the "Review of Particle Physics" [62]. The future LHC machine is expected to be a fertile place to look for these (and other) particles.

τ Lepton

First indications that there might be another lepton in addition to the electron and the muon was obtained by Perl et al [366] in 1975 by studying e^+e^- collisions using the SPEAR 8.4 GeV electron-positron ring. They found events at a center of mass energy around 4 GeV, in which the final state contained $e^+\mu^-$+*missing energy* or $e^-\mu^+$+*missing energy* and no other charged particles or photons were detected. The threshold for the production of these particles (3.56 GeV) was not far from the threshold for the production of the J/ψ (3.74 GeV). Hence a big challenge was faced at the time, in unraveling the existence of this particle, from the other observed hadronic background effects. Persistent and careful search helped establish the existence of the heavy lepton distinct from the hidden charm meson, the J/ψ.

The missing energy and momentum spectra appeared to suggest that at least two additional neutral particles were being produced in each event. In further studies [246] it was more definitely established that the data were completely compatible with the hypothesis that these anomalous events arise from the production, in the e^+e^- annihilation reaction, of a pair of oppositely charged heavy leptons, each of which decays into an electron or a muon and appropriate neutrinos. These heavy leptons were given the name—$\tau^\pm$ *leptons*. Thus what was observed was interpreted as $\tau^+ \to e^+ + \nu_e + \bar{\nu}_\tau$ and $\tau^- \to \mu^- + \bar{\nu}_\mu + \nu_\tau$ [367]. The cross section for the process $e^+e^- \to \tau^+\tau^-$, its threshold behavior, and its magnitude were found to be consistent with what is expected if $\tau^\pm$ were point-like, spin $1/2$, Dirac particles. Other spin possibilities have been explicitly ruled out [368,369].

From the energy spectra of the decay electrons and muons, the mass of the parent leptons could be deduced. The value found initially was 1.90 ± 0.10 GeV for the τ leptons. The experiment also determined that the mass of the neutrino associated with the τ lepton, ν_τ, was less than 600 MeV with 95% confidence. The lepton charged current involving (ν_τ, τ) was found to be $V - A$ rather than $V + A$, just like in the decay of the muon. The leptonic branching ratios, for decay to electrons and muons, was consistent with the assumption of e/μ universality and the strength of the decay coupling was found to be G_F as in muon decay. More up to date measurements place the mass at $m_\tau = 1,777.05^{+0.29}_{-0.26}$ MeV. Its mean lifetime has been measured to be $(290 \pm 1.2) \times 10^{-15}$ s.

Many decay modes of the tau leptons, both leptonic and semileptonic, have been found and their branching ratios have been determined; details may be found in the "Review of Particle Physics" [62]. Decay modes involving one charged particle account for 84% to 85% of the branching ratio. Of these, the leptonic decays of τ^- involving either $e^- + \bar{\nu}_e + \nu_\tau$ or $\mu^- + \bar{\nu}_\mu + \nu_\tau$ have branching ratios $(17.81 \pm 0.07)\%$ and $(17.37 \pm 0.09)\%$, respectively.

Technicolor Hadrons

We recall that supersymmetry was proposed to stabilize the radiative corrections to the Higgs mass so that it does not naturally become of the order of the Planck mass. Technicolor is an alternative method to supersymmetry to protect the low mass Higgs particle from acquiring a large mass of the order of the Planck mass [370]. This method considers the possibility that, in addition to the usual stongly interacting color sector, there exists, at a mass scale of the order of 1 TeV, another strongly interacting sector which has come to be called the *technicolor sector*. Specifically, the fermion sector has (1) leptons, which are flavor doublets (as in the standard model) and color and technicolor singlets; (2) quarks, which are flavor doublets, color triplets, and technicolor singlets; and (3) technicolor quarks, which are flavor doublets, color singlets, and technicolor n-plets. The coupling of the n-plet g_n is such that at a mass scale of about 1 TeV, it becomes strong.

One of the consequences of having the technicolor quarks is that there must exist a rich spectrum of technicolor hadrons with masses starting at around 3 TeV. Another main difference from the standard model approach concerns processes which produce longitudinal Z's and W's above an energy of the order of a TeV. In the standard model approach, e^+e^- annihilation forms a virtual photon or Z which can produce a pair of transverse W's, which by virtue of their weak coupling give a non-resonant contribution to R (the ratio for producing μ pairs to hadrons), which can be calculated perturbatively. The virtual photon or Z can also materialize into a pair of charged Higgs, which are essentially the longitudinal W's. With technicolor quarks, the virtual photons and Z's can also decay into a pair of technicolor quarks. The resulting contribution to R will have characteristic resonances, just like the J/ψ, Υ, etc., which may be called the *technicolor rho, technicolor omega*, etc. at mass of the order of a few TeV.

Technicolor theory, just like supersymmetry, requires the existence of a different spectrum of new particles in the several TeV range. If they are there, the Large Hadron Collider (LHC) has prospects of finding them.

Third Quark Family
First indications that there was a third quark family matching the third lepton family (ν_τ, τ), came from the observation of the dimuon resonance at 9.5 GeV produced in 400 GeV proton-nucleus collisions using the Fermilab proton synchrotron [371]. The process occurring was interpreted as the production of a new meson $\Upsilon(1S)$: $p + \text{nucleus} \rightarrow \Upsilon(1S) + X$ with subsequent decay of $\Upsilon(1S) \rightarrow \mu^- + \mu^+$. The $\Upsilon(1S)$ meson was interpreted as the bound state of a new quark, called the *b quark*, with its antiparticle $\bar{b}$. The $b\bar{b}$ bound state is also called *bottomonium*. The b quark also goes under the name of *beauty quark* or *bottom quark*. More details may be found in the section under "Bottomonium". The b quark carries charge $-(1/3)|e|$, like the d and the s quarks of the first and second generations of quark families. Its mass is in the range of 4.1 GeV to 4.4 GeV. The other quark of the third family, called the *t quark* (also called the *top* or *truth quark*), which carries a charge $+(2/3)|e|$, is necessary to complete the family of the third generation. It took a long time to find it. Persistent searches for it finally paid off when it was discovered at an incredibly high mass of about 175 GeV.

Three jets in e^+e^- annihilations
According to QCD, in addition to two-jet production in e^+e^- annihilations corresponding to the process $e^+e^- \rightarrow q\bar{q}$, there should also be processes in which the outgoing quark or the antiquark emits a gluon. The gluon would manifest itself as a further hadronic jet, thus resulting

in a final state with three jets. Two jet and three jet events have been clearly seen in e^+e^- annihilations. From a study of the three jet events, it has been possible to deduce that the spin of the gluon is 1. For more details see sections under "Gluon".

Three prong Kaon decay
The first example of the decay of a charged kaon into three charged particles was already obtained by Powell and collaborators in 1949 while studying particles in cosmic radiation using nuclear emulsion plates [372]. The mass of the initial particle, determined by grain counting in the nuclear emulsion, was about 1000 electron masses. The heavy particle came to rest in the emulsion and decayed emitting three charged particles all of which left tracks in the emulsion. This is the first example of the decay of K^+ into $\pi^+ + \pi^+ + \pi^-$.

Three-triplet Model
In an effort to overcome some of the kinematical and dynamical difficulties associated with the single-triplet quark model, a model was proposed in which the low lying baryons and mesons were built out of three triplets of quarks with integral charges [373]. Although the word *color* was not introduced to specify the quarks in this paper, the double SU_3 symmetry scheme proposed there can be looked upon as one in which, one SU_3 refers to color symmetry, while the other SU_3 refers to the usual flavor symmetry. In this picture, the low lying baryons and mesons are singlets in color SU_3. Particle states belonging to higher representations of the color SU_3 are considered to be separated from those in the singlet representation by considerably larger mass differences than those between particles existing within the singlet representation. The three triplet model is thus considered the first introduction of an additional symmetry for the quarks, now called *color symmetry*.

Time reversal transformation
The idea of time inversion transformation was introduced into quantum mechanics for the first time by Wigner in 1932 [374]. In many problems, the Hamiltonians of physical systems are invariant under a change in sign of the time variable. This amounts to interchanging the past and the future. The invariance of the Hamiltonian under time reversal leads to certain very general relations between transition probabilities and cross sections for direct reaction $A + B \rightarrow C + D$, and its inverse reaction $C + D \rightarrow A + B$. Such relations are obviously very useful.

Physical quantities can be classified according to their behavior under the time reversal transformation $t \rightarrow -t$. There are some, such as the coordiantes of points, kinetic energy, and total energy, which do not change under the transformation. There are others, such as velocity, linear momentum, angular momentum, and spin, which do change sign under the transformation. The former quantities depend upon even powers of t and, hence, are said to be even under the transformation, while the latter depend upon odd powers of t and, hence, may be said to be odd under the transformation.

We bring out the implications of time reversal transformation for a physical system by considerations based on the Schrödinger equation. The time rate of change of some state with wave function ψ is given by the time dependent Schrödinger equation

$$i\frac{\partial \psi}{\partial t} = H\psi.$$

Let us denote the state obtained by the reversal of time by ψ'. In this state all quantities which are even under the transformation have the same value as in the original state, while the quantities which are odd change sign but retain their magnitude. We now want to find the time reversal operator T which changes the state ψ into ψ'. By defintion ψ' satisfies

$$i\frac{\partial \psi'}{\partial t'} = -i\frac{\partial \psi'}{\partial t} = H\psi',$$

as $t' = -t$ and the Hamiltonian has been assumed to be unchanged under the transformation. Now let us take the complex conjugate of the equation given above for ψ, and we have

$$-i\frac{\partial \psi^*}{\partial t} = H^*\psi^*.$$

If there exists a unitary operator U satisfying the conditions

$$UH^* = HU, \qquad U^\dagger U = I,$$

operating with this operator U from the left on the equation above for ψ^*, we have

$$-i\frac{\partial U\psi^*}{\partial t} = HU\psi^*.$$

Comparing this equation with the one for ψ' above, we see that

$$\psi' = U\psi^*.$$

Defining an operator K which carries out complex conjugation on what it operates, we may write

$$\psi' = U\psi^* = UK\psi.$$

Thus the operator T which relates ψ' with ψ is $T = UK$; that is, it is a product of a unitary operator and an operator for complex conjugation.

It should be noted that the complex conjugation operator K is an *antilinear operator*. This is because, when K operates on the linear combination of wave functions, we have

$$K \sum_i a_i \psi_i = \sum_i a_i^* K \psi_i.$$

K also has the property that the absolute value of the scalar product of two arbitrary functions is unchanged, and the normalization of the wave functions is also unchanged:

$$|\langle K\psi | K\phi \rangle| = |\langle \psi^* | \phi^* \rangle| = |\langle \psi | \phi \rangle|.$$

Further, according to the general rule, the transformation of the wave function by T must be accompanied by the the transformation of any operator F in the form: $F' = TFT^{-1}$.

We consider a few simple examples to illustrate the operation of time reversal. First, we consider the Hamiltonian for a spin zero particle in the coordinate representation:

$$H = (\vec{p}^2/(2m)) + V(\vec{r}), \qquad \vec{p} = -i\vec{\nabla}.$$

In this case $U = 1$, because $H = H^*$, so that $T = K$. The transformed position and momentum operators are: $\vec{r}' = K\vec{r}K^{-1} = \vec{r}$ and $\vec{p}' = K(-i\vec{\nabla})K^{-1} = i\vec{\nabla} = -\vec{p}$, respectively.

Second, we consider the Hamiltonian for a spin zero particle interacting with an electromagnetic field having a vector potential $\vec{A}$, in the coordinate representation:

$$H = (1/2m)(\vec{p} - e\vec{A})^2 + V(\vec{r}),$$
$$\vec{p} = -i\vec{\nabla}.$$

Here from the defining equation for U, $UH^* = HU$, we see that U must have the property that it changes $\vec{A}$ into $-\vec{A}$. Then $T = UK$. We can easily verify that $\vec{r}' = T\vec{r}T^{-1} = \vec{r}$ and $\vec{p}' = T\vec{p}T^{-1} = -\vec{p}$.

Third, we consider a spin 1/2 particle in interaction with electromagnetic field specified by a vector potential $\vec{A}$:

$$H = (1/2m)(\vec{p} - e\vec{A})^2 - (e/2m)(\vec{\sigma} \cdot \vec{B}) + V(\vec{r}),$$
$$\vec{B} = \vec{\nabla} \times \vec{A},$$

where the components of $\vec{\sigma}$ are the Pauli matrices. In this case, in order to satisfy $UH^* = HU$, U has to be written as the product $U = U_1 U_{\vec{\sigma}}$, where U_1 is the operator which changes $\vec{A}$ into $-\vec{A}$, and $U_{\vec{\sigma}}$ is such that $U_{\vec{\sigma}}\vec{\sigma}^* = -\vec{\sigma}U_{\vec{\sigma}}$. If we choose $U_{\vec{\sigma}} = i\sigma_y$, this is satisfied. Thus in this case, $T = U_1(i\sigma_y)K$. This shows that in the case of a spin $1/2$ particle, the time reversed state, by virtue of the matrix-operator $i\sigma_y$, has the z-component of the spin flipped to a value opposite to that of the original state.

We now derive the relation between the matrix elements for direct and reverse transitions. Let us consider the matrix element between time reversed states with wave functions ψ'_a and ψ'_b: $M'_{ab} = \langle\psi'_a|M'|\psi'_b\rangle$. Since $\psi'_a = T\psi_a = U\psi^*_a, \psi'_b = U\psi^*_b$, we have

$$M'_{ab} = \langle U\psi^*_a|M|U\psi^*_b\rangle = \langle\psi^*_a|U^\dagger MU|\psi^*_b\rangle.$$

Making use of the hermiticity properties of the operator M, we deduce that $U^\dagger MU = M^{\dagger*} = M^T$, where M^T is the transpose of M. Thus we have

$$\langle\psi^*_a|U^\dagger MU|\psi^*_b\rangle = \langle\psi_b|M|\psi_a\rangle$$

and the relation $M'_{ab} = M_{ba}$ between the matrix elements of the direct and reverse transitions.

Although we have based the above discussion on the nonrelativistic Schrödinger equation, similar discussions can be carried out for the relativistic wave equations and in quantum field theory. We do not present these details here—the interested reader may wish to consult a book on quantum field theory [98].

Top Quark

The top quark is a member of the third quark family of which the other member is the b quark. In the standard model they are arranged as a weak isospin doublet, with the top quark carrying charge $+(2/3)|e|$ and $I_3 = +1/2$, while the b quark has charge $-(1/3)|e|$ and $I_3 = -1/2$. It was discovered by the CDF and D0 collaborations working at the Tevatron in Fermilab [375,376]. Both groups searched for the top quark in proton-antiproton collisions at a center of mass energy of 1.8 TeV. The dominant mechanism for the production involves the production of top quark-top antiquark pairs by a quark from the proton annihilating with an antiquark from the antiproton, $q\bar{q} \to t\bar{t}$. There is also a small contribution due to a gluon from the proton fusing with a gluon from the antiproton, $gg \to t\bar{t}$. The estimated production cross section through these channels, for a mass of the top quark of 175 GeV, is about 5 pb.

Top quarks can also be produced singly with a lower cross section than with the previous mechanism. Such production involves $q_1\bar{q}_2 \to W^* \to t\bar{b}$, where quarks of different flavor from the proton and the antiproton annihilate producing an intermediate virtual W which sub-sequently decays into the final $t\bar{b}$ pair, or quark-gluon fusion, $q_1 g \rightarrow q_2 t \bar{b}$. Diminished experimental acceptances for the top quarks produced through these channels make the contribution from these channels much smaller than the dominant contribution from the other channels via $t\bar{t}$ pairs. So one may ignore the contribution from single production of top quarks.

If the mass of the top quark is above the (W, b) threshold, the top quark decays predominantly into $W + b$. Decays to $W + d$ or $W + s$ are suppressed relative to $W + b$, because the Kobayashi-Maskawa matrix elements V_{td} and V_{ts} are so much smaller than V_{tb}. Expressions for the decay width can be worked out using the standard model, and it is found that the width varies from about 1 GeV for a top quark mass of 160 GeV to about 1.5 GeV for a top quark mass of 180 GeV. Such large widths imply very short lifetime for the top quark, which means that the top quarks decay long before they can form top flavored hadrons or $t\bar{t}$ bound states.

If the t quark decays in the (W, b) mode and the W itself decays to quarks or leptons, one might expect to see the following decay products from the decay of a t quark: $q_1 \bar{q}_2 b$ or $l \nu_l b$. For the $t\bar{t}$ then, the possible final decay products could belong to one of three possible sets: (a) $q\bar{q}_1 b q_2 \bar{q}_3 \bar{b}$; (b) $q\bar{q}_1 b l \nu_l \bar{b} + q\bar{q}_1 \bar{b} l \nu_l b$; or (c) $\bar{l}\nu_l b l' \bar{\nu}_{l'} \bar{b}$. The first set gives rise to final states which are hadronic jets only. The second set gives rise to a lepton and jets, while the last set gives rise to two lepton signal. To identify the signal for the top quark, one has to reconstruct the invariant mass from measurements on the decay products event by event. The characteristics of the emitted neutrinos are obtained by the large transverse momentum imbalance in the event (missing p_T).

Top quark-top antiquark pairs have been observed in all of the decay modes listed above by the CDF and D0 collaborations. The extraction of the properties of the top quark from measurements on the decay products requires the understanding of the mechanism of production of the top quark and also an understanding of the QCD backgrounds involved. The ratio of signal to background is improved much by selecting events in which a b quark is present, and the jets are very energetic, corresponding to their origin from the decay of a very massive particle. The mass of the top quark has been measured by both CDF and D0 collaborations. The best determination of the mass was from the lepton plus jets channel. The result obtained by CDF was $(175.9 \pm 4.8 \pm 4.9)$ GeV [377], and that obtained by D0 was $(173.3 \pm 5.6 \pm 5.5)$ GeV [378], in each of which the first uncertainty is due to statistical error, and the second one is due to systematic error arising mainly from uncertainties in the jet energy scale and in the Monte Carlo simulations. Putting together all the measurements, the Particle Data Group, which lists all the particle properties, gives $(173.8 \pm 3.5 \pm 3.9)$ GeV for the mass of the top quark.

These groups have also measured the production cross section for $(t, \bar{t})$ pairs in $(p, \bar{p})$ collisions. The measured values from the lepton plus jets modes are: (4.1 ± 2.0) pb (D0 collaboration, assuming top mass = 173.3 GeV) and $(6.7^{+2.0}_{-1.7})$ pb (CDF collaboration, assuming top mass = 175 GeV). Theoretical estimates of the production cross section are $(5.0–5.8)$ pb, and $(4.75–5.5)$ pb, respectively.

Clearly these measurements need to be improved as the Tevatron work provides the basis for the extrapolation for future measurements at LHC. The much higher cross sections expected at LHC will allow one to carry out further extensive studies on the properties of the top quark. Discrepancies between theoretical and experimental values for the $(t, \bar{t})$ production would signal the existence of new physics beyond the standard model. On the decay side, the study of angular distributions from t decays will allow one to see if the decay $t \to Wb$ is mediated by $(V - A)$ interactions. The fraction of longitudinal and transverse W's in the decay is fixed once the coupling is determined. Any deviations from such expectations would be very interesting because they would call into question the Higgs mechanism for symmetry breaking. A lot of interesting physics lies ahead to be explored at the LHC.

Two-Component Theory for Neutrino
The Dirac equation for a massive spin $1/2$ particle has in it four matrices, α_i, $i = 1, 2, 3$ and β, with the requirements

$$\alpha_i \alpha_j + \alpha_j \alpha_i = 2\delta_{ij}, \ i, j = 1, 2, 3,$$

$$\alpha_i \beta + \beta \alpha_i = 0, \ \beta^2 = 1.$$

The term containing the β matrix is multiplied by the rest mass of the particle and is necessarily present when the rest mass is not zero. In this case the minimum dimension of the matrices, α_i, $i = 1, 2, 3$ and β, has to be 4×4 and the wave function for the particle generally has four components.

When the rest mass of the particle is zero, as for example in the case of the neutrino, the term containing the β matrix is not necessary and we only have three matrices, α_i, $i = 1, 2, 3$, satisfying

$$\alpha_i \alpha_j + \alpha_j \alpha_i = 2\delta_{ij}, \ i, j = 1, 2, 3,$$

and the Dirac equation for the particle has the form

$$i\frac{\partial \psi}{\partial t} = -i\vec{\alpha} \cdot \vec{\nabla}\psi.$$

In this case it is possible to satisfy the anticommutations relation for α's by choosing them equal to the Pauli two-component matrices, $\alpha_i = \sigma_i$ $i = 1, 2, 3$ or $\vec{\alpha} = \vec{\sigma}$. The corresponding wave function for the massless spin $1/2$ particle will have only two components. If one writes a plane wave solution of the form,

$$\psi(\vec{x}, t) = u(p, s)e^{-i(Et - \vec{p} \cdot \vec{x})}, \quad E = |\vec{p}|,$$

the amplitudes $u(p, s)$ satisfy

$$|\vec{p}|u(p, s) = (\vec{\sigma} \cdot \vec{p})u(p, s).$$

The solution of this equation with the usual form of Pauli matrices, $\sigma_1 = \left(\begin{smallmatrix} 0 & 1 \\ 1 & 0 \end{smallmatrix}\right)$, $\sigma_2 = \left(\begin{smallmatrix} 0 & -i \\ i & 0 \end{smallmatrix}\right)$, and $\sigma_3 = \left(\begin{smallmatrix} 1 & 0 \\ 0 & -1 \end{smallmatrix}\right)$, with the direction of z axis chosen along $\vec{p}$, is of the form

$$u(p, +) = \begin{bmatrix} 1 \\ 0 \end{bmatrix}.$$

In this case, we have

$$\frac{\vec{\sigma} \cdot \vec{p}}{|\vec{p}|}u(p, +) = +1u(p, +).$$

The operator on the left $(\vec{\sigma} \cdot \vec{p})/|\vec{p}|$ is the helicity operator, which represents the projection of the spin along the direction of motion of the particle, and has eigenvalue $+1$ in the state $u(p, +)$. It represents a right-handed particle. To obtain a left-handed particle, we would have chosen $\vec{\alpha} = -\vec{\sigma}$ and have ended up with the solution

$$u(p, -) = \begin{bmatrix} 0 \\ 1 \end{bmatrix},$$

which would satisfy

$$\frac{\vec{\sigma} \cdot \vec{p}}{|\vec{p}|}u(p, -) = -1u(p, -).$$

Thus, in this state, the spin is oriented antiparallel to the momentum and, hence, represents a left-handed particle. It is suitable for describing left-handed neutrinos.

Such properties of the Dirac equation in the massless limit were discussed first by Weyl [70], but since the consequences led to violation of parity, the theory did not achieve acceptance. Now, however, parity violation in weak interactions has been experimentally established, and the neutrinos of the theory are describable by two component wave functions [379]. Because in weak interactions charge conjugation symmetry is also lost, the combined operation of CP for the two-component neutrino leads to the conclusion that, along with left-handed neutrinos, only right-handed antineutrinos will exist. Processes in which neutrinos appear can be treated in the two-component formalism.

Ultraviolet Divergences in Field Theories

The evaluation of the probabilities of occurrence of various physical processes in quantum field theories is achieved by calculating the S-matrix elements for the processes. When the interactions between the particles involved are weak, the calculation of these elements is carried out using perturbation theory, the total Hamiltonian for the interacting fields being written as the sum of the Hamiltonian for the free fields and an interaction Hamiltonian which represents the interactions between the free fields. In the lowest order of perturbation theory in which the physical process can occur, the relevant S-matrix elements lead to finite results. One can obtain corrections to the lowest order of perturbation theory by going to higher orders in perturbation theory. These correction terms in general involve intermediate states in which particles of arbitrarily large momenta are produced and reabsorbed. The contribution from such intermediate states is calculated by integrating over the infinite range for the arbitrary momentum of the particles in the intermediate states. Such integrals diverge in general. These are called *ultraviolet divergences* as the contribution to the integrals tends to diverge logarithmically when the upper limit of the integrals tends to infinity.

For quantum electrodynamics (QED), Tomonaga, Schwinger, and Feynman (see more in the sections under "Quantum Electrodynamics" and "Renormalization") made a systematic study of these divergent contributions and gave prescriptions for the separation of the infinite parts in an unambiguous and Lorentz invariant manner. They showed that these infinite parts could be absorbed in a redefinition of a finite number of parameters, such as the mass and the charge occurring in the theory, and the renormalization of the wave functions through a process called *renormalization*. The remaining finite parts represent higher order corrections to the physical process under consideration. Using these methods, higher order corrections for many physical processes have been evaluated and confronted with experimental measurements. The improvements in the experimental measurements necessitate the calculation of further higher order corrections which presents a great theoretical challenge. In the end, when both the experimental and theoretical challenges have been met, the results of experiment and theory agree remarkably well for many processes. Such agreements between experiment and theory inspire confidence in accepting QED as the fundamental quantum theory for the interactions of point-like charged particles and photons and validates the method of renormalization.

Another renormalizable field theory is quantum chromodynamics (QCD). In contrast with QED, QCD is a non-Abelian gauge theory, with the quanta of the chromodynamic fields, called the *gluons*, carrying a charge called *color*. As a result of the color assignment to quarks and gluons, the color interactions have a property called *asymptotic freedom*. The effective coupling between quarks and gluons tends to vanish at very high energies (in the ultraviolet region) and grows to large values as the energy tends to small values (in the infrared region). Thus at high energies a perturbative treatment of quark-gluon interactions is possible using renormalization methods just as in QED. Of course, such calculations are technically more complicated on account of the non-Abelian nature of the quark-gluon interactions. Perturbative QCD corrections have been worked out for a number of processes and the results are in reasonable accord with experimental findings. The degree to which theory and experiment agree in QCD is less than in QED because of the difficulties in carrying out calculations of QCD corrections and also because in many processes non-perturbative corrections which are not calculable with any accuracy at present are important (see further in the sections under "Asymptotic Freedom" and "QCD—Quantum Chromodynamics"). Properties of hadrons, which are color singlet-bound states of quarks, or of quarks with antiquarks are hard to calculate because they fall in the non-perturbative regime of QCD. Special methods (notably lattice gauge theories among them) have to be devised to handle the region of large couplings.

Universe—Baryon Asymmetry

The present state of our universe is predominantly made of matter (baryons) as opposed to antimatter (antibaryons). This is certainly observed in the local region around our galaxy, and there are further observational and theoretical reasons which lead one to conclude that this preponderance of baryons pervades the whole universe. One can ask, within the standard big-bang cosmology, how this situation has evolved from the beginning. One possibility is that the universe got created with a nonzero baryon number initially. In that case, no further explanations may be necessary if that nonzero baryon number is carried forward.

An alternative scenario in which equal numbers of baryons and antibaryons were created at the big bang (and the initial baryon number is zero) is, however, more likely, because the conservation law of baryon number does not follow from any fundamental symmetry principles of physics. It is then necessary to see how the universe, which was initially symmetric between baryons and antibaryons, has evolved to its present state of preponderance of baryons over antibaryons. In fact, grand unified gauge theories (GUTS) of weak, electromagnetic, and strong interactions have as their consequence the violation of baryon number as well as that of CP invariance [380]. This is because in these theories, quarks and leptons enter the model on the same footing, thus allowing for the possibility of transformations between quarks and leptons. Such transformations will violate baryon number as well as lepton number. The interesting question that arises concerns the mechanism for the generation of excess baryons over antibaryons in these theories.

Already in 1966, Sakharov [381] showed how violation of CP invariance could lead to a baryon asymmetry in the universe. Following developments of grand unified theories, a number of people [382] have calculated the expected baryon asymmetry and the expected ratio of the number of baryons to photons at present. Three important requirements are necessary for the evolution of baryon asymmetry from the original symmetric state. They are (a) baryon number non-conservation at a microscopic level, (b) CP non-conservation, and (c) departure from thermal equilibrium. Baryon number nonconserving interactions are mediated in GUTS by superheavy bosons carrying color, called X *bosons*. These could be gauge bosons or bosons of the Higgs type, coupling to quark-lepton combination or quark-quark combinations, having typically a mass of the order 10^{16} GeV. CP violation in GUTS depends on the particular model chosen and several example models have been worked out. The departure from thermal equilibrium occurs due to the decay of the X boson when the temperature falls below their rest mass energy M_X.

The sequence of steps that takes place may be described as follows. Initially one starts from an extremely high temperature T, the thermal energy corresponding to which is of the order of the Planck mass energy ($M_P = 10^{19}$ GeV). In this regime, the processes mediated by the X bosons proceed at a rate Γ, which is faster than the Hubble expansion rate H. Any baryon asymmetry that was present at the big bang will be wiped out as equilibrium is established. In the range $M_P > T > M_X$, the production and decay rates of the X bosons are small compared to the Hubble expansion rate. When the temperature starts dropping, both H and Γ start decreasing, relative to the decay rate of X bosons. When the Hubble expansion rate drops below the decay rate of the X bosons, one passes to a regime where X decay becomes very important. For M_X of order 10^{15} GeV and $T < M_X$, the inverse process to decay will proceed at a smaller rate than the expansion rate, and departure from thermal equilibrium occurs. Hereafter, any baryon asymmetries generated by X-boson decays will be maintained in later stages of the evolution.

The asymmetry can be related to the fundamental parameters occurring in the grand unified theory as follows. Suppose the X boson has two decay channels with branching ratios γ and $1-\gamma$, generating baryon numbers B_1 and B_2, respectively. With CP nonconservation, the $\bar{X}$ boson decays generating baryon numbers $-B_1$ and $-B_2$ with branching ratios $\bar{\gamma}$ and $1-\bar{\gamma}$ (and $\gamma \neq \bar{\gamma}$). Then, in the decays of X and $\bar{X}$, the average baryon number produced will be given by

$$
\begin{aligned}
\Delta B &= \frac{1}{2}(\gamma B_1 + (1-\gamma)B_2 - \bar{\gamma}B_1 - (1-\bar{\gamma})B_2) \\
&= \frac{1}{2}(\gamma - \bar{\gamma})(B_1 - B_2).
\end{aligned}
$$

The ratio of the number of baryons n_B to the number of photons n_γ is then found to be

$$
\frac{n_B}{n_\gamma} = \frac{N_X}{N}\Delta B,
$$

where N_X is the number of X bosons and N is the total number of helicity states. The experimental baryon to photon number ratio is 10^{-9}. Agreement with this number is possible in GUT models through the mechanism explained above.

Upsilon Particles

The upsilon particles (Υ) were found as a dimuon resonance in proton nucleus collisions as well as in electron-positron collisions. They are interpreted as bound states of a third generation quark b with its antiquark $\bar{b}$. In the proton-nucleus collisions, the reaction studied was $p+$nucleus $\to \mu^+\mu^- +X$, at a laboratory proton energy of 400 GeV from the Fermilab proton synchrotron [94,371]. A strong peak in the invariant mass spectrum of the two muons was found at 9.5 GeV. This state has been called the $\Upsilon(1S)$. This state has also been seen in electron-positron annihilations at DESY-Doris-II storage ring [383,384]. The mass as determined in the e^+e^- annihilation reactions was 9.46 ± 0.01 GeV, consistent with identifying this state as the $\Upsilon(1S)$.

Further investigations in the mass region for the dimuon pairs in the range 9.5 GeV to 10.5 GeV, at the electron storage rings, have revealed the existence of four states in this region of which the first is $\Upsilon(1S)$. The others have been called $\Upsilon(2S)$, $\Upsilon(3S)$, and $\Upsilon(4S)$. This nomenclature evolved from a potential model construction of the bound states of the $b\bar{b}$ system. Inspired from QCD, the interaction potential between the quark and the antiquark in a color singlet state was taken as consisting of an attractive Coulombic part (effective at short distances) and a linear or logarithmic confining potential at large distances. At extremely small distances the interaction potential is assumed to vanish to be in accord with the asymptotic freedom of QCD.

Generating the spectrum of states arising in such a potential for a two body problem is a standard exercise in quantum mechanics and is qualitatively similar to generating that of the hydrogen atom or positronium. The states can be labeled by the principal quantum number n, orbital angular momentum quantum number l, and the total angular momentum quantum number j. When the spins of the quark and the antiquark are combined, one can have spin states that are singlets or triplets in total spin. Thus in standard spectroscopic notation, we have 1^3S_1, 2^3S_1, 3^3S_1, 4^3S_1, ..., etc., S states for the triplet spin and 1^3S_0, 2^3S_0, 3^3S_0, 4^3S_0, ..., etc., and S states for the singlet spin, with principal quantum numbers $n = 1, 2, 3, 4, \ldots$. There are also triplet and singlet P states: $2^3P_j, j = 0, 1, 2$, $3^3P_j, j = 0, 1, 2, \ldots$; 2^1P_1, 3^1P_1, ..., etc. The actual positions of the energy levels can be calculated numerically by solving the Schrödinger equation once the parameters in the potential and the mass of the b quark is chosen. These parameters are fit so that the upsilon(1S), ..., upsilon(4S) particles, can be identified with the spin triplet S-wave bound states of the $b\bar{b}$ system. The existence of the ^3P-wave bound states implies that there must be radiative transitions between these and the ^3S states; measurements of the energy of these photons give the positions of the ^3P levels, which can be compared with the theoretical values. In this manner, the rich spectroscopy of the $b\bar{b}$ system has been studied in detail (see also section under "Bottomonium"). The mass of the b quark from these fits ranges between 4.1 and 4.4 GeV. The upsilon(4S) has a mass which is higher than the sum total of the mass of the b quarks; it decays into b and $\bar{b}$ quarks, each of which picks up light quarks (u, d, s) or $(\bar{u}, \bar{d}, \bar{s})$ from the vacuum to form color singlet B (and $\bar{B}$) mesons. Just as the triplet S-bound states are collectively labeled Υ, the singlet S-bound states are labeled η_b states, and the triplet P-bound states are labeled χ_b states, the subscript b refering to the fact that these are $b\bar{b}$-bound states.

The masses widths and branching ratios for prominent decay channels are now given.

(a) $\Upsilon(1S)$:

Mass $= (9460.37 \pm 0.21)$ GeV; Full width $= (52.5 \pm 1.8)$ keV; Branching ratios: $(e^+e^-) = (2.52 \pm 0.17)\%$, $(\mu^+\mu^-) = (2.48 \pm 0.07)\%$, $(\tau^+\tau^-) = (2.67^{+0.14}_{-0.16})\%$; other channels have very small branching ratios.

(b) $\Upsilon(2S)$:

Mass $= (10.02330 \pm 0.00031)$ GeV; Full width $= (44 \pm 7)$ keV; Branching ratios: $(\Upsilon(1S)\pi^+\pi^-) = (18.5 \pm 0.8)\%$, $(\Upsilon(1S)\pi^0\pi^0) = (8.8 \pm 1.1)\%$, $(e^+e^-) = (1.18 \pm 0.20)\%$, $(\mu^+\mu^-) = (1.31 \pm 0.21)\%$, $(\tau^+\tau^-) = (1.7\pm1.6)\%$, $(\gamma\chi_b(2^3P_0)) = (4.3\pm1.0)\%$, $(\gamma\chi_b(2^3P_1)) = (6.7\pm0.9)\%$, $(\gamma\chi_b(2^3P_2)) = (6.6\pm0.9)\%$; other channels have very small branching ratios.

(c) $\Upsilon(3S)$:

Mass $= (10.3553 \pm 0.0005)$ GeV; Full width $= (26.3 \pm 3.5)$ keV; Branching ratios: $(\Upsilon(2S) + X) = (10.6 \pm 0.8)\%$, $(\Upsilon(1S)\pi^+\pi^-) = (4.48 \pm 0.21)\%$, $(\Upsilon(1S)\pi^0\pi^0) = (2.06 \pm 0.28)\%$, $(\gamma\chi_b(3^3P_0)) = (5.4\pm0.6)\%$, $(\gamma\chi_b(3^3P_1)) = (11.3 \pm 0.6)\%$, $(\gamma\chi_b(3^3P_2)) = (11.4 \pm 0.8)\%$, where $X =$ anything; other channels have very small branching ratios.

(d) $\Upsilon(4S)$:

Mass $= (10.5800\pm0.0035)$ GeV; Full width $= (10\pm4)$ MeV; Branching ratios: $(B\bar{B}) > 96\%$, (non BB) $< 4\%$; other channels have very small ratios.

The spin-parity assignment J^P for the Υ states from the experimental measurements is 1^-; the spin-parity assignments for the χ states is based on theory (experimental confirmation pending).

For more detailed information on the upsilon particles, reference may be made to "Review of Particle Properties" [62].

V Particles
See discussion in sections under "Hyperons—Decay Selection Rules" and "Kaons: The τ-θ Puzzle".

$V - A$ Interaction
See discussion under "Beta Decay—Theory".

Vector Coupling Constant
See discussion under "Beta Decay—Strong Interaction Corrections".

Ward Identity
This is an identity that was derived by Ward [385] in QED. It proved that a conjecture by Dyson, on the finiteness of the renormalized charge, is indeed valid. The renormalization constants called Z_1 and Z_2 in QED, which are each infinite and refer respectively to vertex renormalization and self-energy renormalization, are equal to one another ($Z_1 = Z_2$), to

all orders of perturbation theory. It can be shown that this identity is intimately connected with charge conservation and gauge invariance of QED.

Weak Interactions

This is a fundamental interaction, which is responsible not only for the beta decay of nuclei but also for the decays of many elementary particles. It has been unified with electromagnetic interactions successfully under the scheme of "electroweak synthesis". The mediators of the weak interactions are the W and Z fields, whose quanta are the $W^\pm$ and Z^0 bosons. (See further discussion under "Gauge Theories" and "Standard Electroweak Model".)

$W^\pm$ Boson

These particles are the carriers of the weak interactions between elementary particles and are responsible for the weak decays of the particles. Their mass was predicted from theoretical considerations to be around 80 GeV. They were found at the SPS proton-antiproton collider at CERN in 1983 by the UA1 and UA2 collaborations. They looked for isolated large transverse energy electrons with large missing transverse energy. These searches converged on the same events. The events fitted the hypothesis that they originated from the two-body decay of a single parent particle of mass about 80 GeV [386,387]. The particles were identified to be the $W^\pm$ boson postulated in the unification of electromagnetic and weak interactions.

Since their early discovery, the detailed properties of these particles have been established by precision work done at the LEP collider in its latter phase, called *LEPII*, which produced these particles in pairs. For details regarding their properties, it is suggested that "Review of Particle Physics" [62] be consulted.

Wimps

The name *wimp* stands for weakly interacting massive particle. The neutralinos which could have mass in the range of tens to hundreds of GeV could be classified as wimps. They could be part of the cold dark matter in the universe. (See also under "Dark Matter, Machos, Wimps".)

W-inos

This hypothetical particle of spin 1/2 is the supersymmetric partner of the spin 1 charged W bosons. (See under "Supersymmetry".)

Ξ Particles
See discussion under "Hyperons". Ξ particles (and their antiparticles) containing a charm quark (antiquark) have also been found.

Yang-Mills Fields
See discussion under "Gauge Theories".

Yukawa Model
Originally proposed by Yukawa [113] to explain the origin of short-range nuclear forces, it involves the exchange of scalar or pseudoscalar quanta of mass between the electron and the proton. More recently, this notion has been generalized to exchange of Higgs scalars and other particles between fermions.

Z^0 Particles
These particles are the neutral counterparts of the $W^{\pm}$ bosons responsible for the weak interactions of elementary particles. The $W^{\pm}$ bosons are responsible for mediating charged current weak interactions. The Z^0, likewise, mediate neutral current weak interactions. The first observations of these particles were made by the UA1 and UA2 collaborations working at the CERN SPS proton-antiproton collider in 1983 [388,389]. They observed electron and muon pairs which appeared to have originated from the decay of a parent particle with mass about 92 GeV. This particle was called the Z^0 and identified to be the same as the particle postulated in the unified theory of electroweak interactions. Since the original discovery, much precision work has been done on measuring the properties of this particle with the LEP collider at CERN and at the SLC at SLAC. Details of these properties can be found in "Review of Particle Physics" [62].

Z-inos
This hypothetical particle of spin 1/2 is the supersymmetric partner of the spin 1 neutral Z^0 bosons. (See under "Supersymmetry".)

References

[1] J. D. Cockcroft, E. T. S. Walton, Nature, 129, 649, 1932.

[2] R. J. Van de Graaff, Phys. Rev. 38, 1919, 1931.

[3] E. O. Lawrence, M. S. Livingston, Phys. Rev. 37, 1707, 1931.

[4] D. W. Kerst, Phys. Rev. 60, 47, 1941.

[5] E. M. McMillan, Phys. Rev. 70, 800, 1946.

[6] N. C. Christofilos, Unpublished Report, 1950; US Patent No. 2,736,799, 1956.

[7] E. D. Courant, M. S. Livingston, H. S. Snyder, Phys. Rev. 88, 1190, 1952; E. D. Courant, H. S. Snyder, Ann. Phys. 3, 1, 1958.

[8] D. W. Kerst et al., Phys. Rev. 102, 590, 1956.

[9] Experimental Techniques in High Energy Physics, T. Ferbel (ed.), Addison-Wesley, Menlo Park, CA, 1987.

[10] C. Grupen, Particle Detectors, Cambridge University Press, New York, N. Y., 1996.

[11] H. A. Bethe, Annalen d. Physik, 5, 325, 1930; Zeits. f. Physik, 76, 293, 1932.

[12] F. Bloch, Zeits. f. Physik, 81, 363, 1933.

[13] E. Fermi, Phys. Rev. 56, 1242, 1939; Phys. Rev. 57, 485, 1940.

[14] R. M. Sternheimer, Phys. Rev. 88, 851, 1952.

[15] S. M. Seltzer, M. J. Berger, Int. J. Applied. Rad. 33, 1189, 1982; 35, 665, 1984.

[16] H. A. Bethe, W. Heitler, Proc. Roy. Soc. A146, 83, 1934.

[17] O. Klein, Y. Nishina, Zeits. f. Physik, 52, 853, 1929.

[18] D. A. Glaser, Phys. Rev. 91, 762, 1953.

[19] CERN-Courier, Vol 27, No.6, 25, 1987.

[20] P. A. Cherenkov, Phys. Rev. 52, 378, 1937.

[21] G. Charpak et al., Nucl. Inst. Meth. 62, 262, 1968.

[22] D. R. Nygren, Internal Report, Lawrence Berkeley Laboratories, 1974; Phys. Scripta. 23, 584, 1981.

[23] C. K. Hargrove et al., Phys. Scripta. 23, 668, 1981.

[24] S. Adler, Phys. Rev. 177, 2426, 1969.

[25] J. S. Bell, R. Jackiw, Nuovo Cimento. 51, 47, 1969.

[26] E. Rutherford, Phil. Mag. 21, 669, 1911.

[27] H. Geiger, Proc. Roy. Soc. (London), 83A,492, 1910.

[28] H. Geiger and E. Marsden, Proc. Roy. Soc. (London), 82A, 495, 1909.

[29] H. Geiger and E. Marsden, Phil. Mag., 25, 604, 1913.

[30] E. Rutherford, Phil. Mag. 47, 109, 1899.

[31] G. Hanson et al., Phys. Rev. Lett. 35, 1609, 1975.

[32] Tasso collaboration, R. Brandelik et al., Phys. Lett. 86B, 243, 1979.

[33] Pluto collaboration, C. Berger et al., Phys. Lett. 86B, 418, 1979.

[34] Jade collaboration, W. Bartel et al., Phys. Lett 91B,142, 1980.

[35] D. Dorfan et al., Phys. Rev. Lett. 14, 1003, 1965.

[36] F. Binon et al., Phys. Lett. 30B, 510, 1969.

[37] L. Agnew et al., Phys. Rev. 110, 994, 1958.

[38] O. Chamberlain et al., Phys. Rev. 100, 947, 1955.

[39] E. Hayward, Phys. Rev. 72,937, 1947.

[40] E. W. Cowan, Phys. Rev. 94,161, 1954.

[41] H. S. Bridge, H. Courant, H. C. de Staebler, B. Rossi, Phys. Rev. 95, 1101, 1954.

[42] E. Amaldi et al., Nuovo Cimento, 1, 492, 1955.

[43] O. Chamberlain et al., Phys. Rev. 102, 921, 1956.

[44] A. Pais, Phys. Rev. 86, 663, 1952.

[45] G. D. Rochester and C. C. Butler, Nature 160, 855, 1947; A. J. Seriff et al., Phys. Rev. 78, 290, 1950; G. McCusker and D. H. Millar, Nuovo Cimento 8, 289, 1951; V. D. Hopper and S. Biswas, Phys. Rev. 80, 1099, 1950; R. Armenteros et al., Nature 167, 501, 1951; Phil. Mag. 42, 1113, 1951; H. S. Bridge and M. Annis, Phys. Rev. 82, 445, 1951; R. W. Thompson, H. O. Cohn, and R. S. Flum, Phys. Rev. 83, 175, 1951; R. B. Leighton, S. D. Wanlass, and W. L. Alford, Phys. Rev. 83, 843, 1951; W. B. Fretter, Phys. Rev. 83, 1053, 1951 .

[46] D. D. Ivanenko, Nature 129, 798, 1932.

[47] W. Heisenberg, Z. Phys. 77, 1, 1932.

[48] E. Rutherford, Proc. Roy. Soc. A97, 324, 1920.

[49] C. J. Barkla, Phil. Mag. 21, 648, 1911.

[50] H. G. J. Moseley, Phil. Mag. 26, 1024, 1913; Phil. Mag. 27, 703, 1914.

[51] N. Bohr, Phil. Mag. 26, 1, 1913; Phil. Mag. 26, 476, 1913.

[52] H. Lehmann, K. Symanzik, W. Zimmermann, Nuovo Cimento 1, 205, 1955.

[53] A. S. Wightman, Phys. Rev. 101, 860, 1956.

[54] G. G. Raffelt, Stars as Laboratories for Fundamental Physics, University of Chicago Press, Chicago, 1996.

[55] J. Ellis, K. A. Olive, Phys. Lett. B193, 525, 1987; G. G. Raffelt, D. Dearborn, Phys. Rev. Lett.60, 1793, 1988.

[56] M. Basile et al., Lett. Nuovo Cimento. 31, 97, 1981.

[57] C. Albajar et al., Phys. Lett. 273B, 540, 1991.

[58] C. Bebek et al., Phys. Rev. Lett. 46, 84, 1981; K. Chadwick et al.,
      Phys. Rev. Lett. 46, 88, 1981.

[59] D. Decamp et al., Phys. Lett. 255B, 236, 1991; D. P. Acton et al.,
      Phys. Lett. 276B, 379, 1992.

[60] S. W. Herb et al., Phys. Rev. Lett. 39,252, 1977; C. Berger et al.,
      Phys. Lett. 76B, 243, 1978; C. W. Darden et al., Phys. Lett. 76B,
      246, 1978.

[61] J. K. Bienlein et al., Phys. Lett. 78B, 360, 1978; C. W. Darden et
      al., Phys. Lett. 78B, 364, 1978; K. Ueno et al., Phys. Rev. Lett.
      42, 486, 1979; T. Bohringer et al., Phys. Rev. Lett. 44, 1111, 1980;
      D. Andrews et al., Phys. Rev. Lett. 45, 219, 1980; G. Finocchiaro
      et al., Phys. Rev. Lett. 45, 222, 1980.

[62] Review of Particle Physics, Eur. Phys. J. C15, 1 - 878, 2000,
      Springer Verlag.

[63] C. Albajar et al., Phys. Lett. 186B, 247, (1987); Phys. Lett. 197B,
      565, 1987.

[64] H. Albrecht et al., Phys. Lett. 192B, 245, 1987.

[65] M. Artuso et al., Phys. Rev. Lett. 62, 2233, 1989.

[66] B. Adeva et al., Phys. Lett. 252B, 703, 1990.

[67] D. Decamp et al., Phys. Lett. 258B, 236, 1991.

[68] E. C. G. Stückelberg, Helv. Phys. Acta 11, 299, 1938.

[69] E. P. Wigner, Proc. Am. Phil. Soc. 93, 521, 1949.

[70] H. Weyl, Z. Phys. 56, 330, 1929.

[71] S. J. Lindenbaum, L. C. L. Yuan, Phys. Rev. 92, 1578, 1953.

[72] W. Pauli, translation into English, published in Physics Today, 31,
      27, 1978.

[73] R. Daudell et al., J. Phys. Radium 8, 238, 1947; J. N. Bahcall,
      Phys. Rev. 124, 495, 1961.

[74] S. Jung et al., Phys. Rev. Lett. 69, 2164, 1992.

[75] F. S. Crawford et al., Phys. Rev. Lett. 1, 377, 1958.

[76] P. Nordin et al., Phys. Rev. Lett. 1, 380, 1958.

[77] A. H. Snell and L. C. Miller, Phys. Rev. 74, 1217, 1948

[78] J. M. Robson, Phys. Rev. 77, 747(A), 1950 and Phys. Rev. 78, 311, 1950; A. H. Snell et al., Phys. Rev. 78, 310, 1950.

[79] S. S. Gershtein and Y. B. Zeldovich, Zh. Eksp. Teor. Fiz. 29, 698,1955; JETP 2, 576, 1956.

[80] R. P. Feynman and M. Gell-Mann, Phys. Rev. 109, 193, 1958; E. C. G. Sudarshan and R. E. Marshak, Phys. Rev. 109, 1860, 1958.

[81] A. F. Dunaitsev et al., Phys. Lett. 1, 138, 1962.

[82] E. Fermi, Z. Phys. 88, 161, 1934 and Nuovo Cimento. 11, 1, 1934.

[83] E. Segrè, Nuclei and Particles, W. A. Benjamin, Inc. New York, 1964.

[84] H. J. Bhabha, Proc. Roy. Soc. (London), A154, 195, 1935.

[85] A. A. Penzias, R. Wilson, Astrophys. J. 142, 419, 1965.

[86] J. D. Bjorken, Phys. Rev. 179, 1547, 1969.

[87] I. J. R. Aitchison and A. G. J. Hey, Gauge Theories in Particle Physics, Adam Hilger Ltd., Bristol, U. K. 1984.

[88] E. D. Bloom et al., Phys. Rev. Lett.23, 930, 1969.

[89] M.Breidenbach et al., Phys. Rev. Lett. 23, 935, 1969.

[90] G. Miller et al., Phys. Rev. D5, 528, 1972.

[91] J. I. Friedman and H. W. Kendall, Ann. Rev. Nucl. Sci. 22, 203, 1972.

[92] R. P. Feynman, Phys. Rev. Lett. 23, 1415, 1969.

[93] C. G. Callan and D. Gross, Phys. Rev. Lett. 22, 156, 1969.

[94] W. R. Innes et al., Phys. Rev. Lett. 39, 1240, 1977; Phys. Rev. Lett. 39, 1640, 1977.

[95] N. Cabibbo, Phys. Rev. Lett. 10, 531, 1963.

[96] M. Kobayashi, K. Maskawa, Prog. Theor. Phys. 49, 652, 1973.

[97] J. J. Thomson, Phil. Mag. 44, 293, 1897; Nature 55, 453, 1897.

[98] M. E. Peskin and D. V. Schroeder, Introduction to Quantum Field Theory, Addison-Wesley, Reading, Mass., 1995.

[99] M. Gell-Mann, M. L. Goldberger, W. E. Thirring, Phys. Rev. 95, 1612, 1954.

[100] M. Gell-Mann, A. Pais, Phys. Rev. 97, 1387, 1955.

[101] J. J. Aubert et al., Phys. Rev. Lett. 33, 1404, 1974; J. E. Augustin et al., Phys. Rev. Lett. 33, 1406, 1974.

[102] J. D. Bjorken, S. L. Glashow, Phys. Lett. 11, 255, 1964.

[103] E. G. Cazzoli et al., Phys. Rev. Lett. 34, 1125, 1975.

[104] C. Baltay et al., Phys. Rev. Lett. 42, 1721, 1979.

[105] S. F. Biagi et al., Z. Phys. C28, 175, 1985.

[106] S. F. Biagi et al., Phys. Lett. 122B, 455, 1983.

[107] P. Avery et al., Phys. Rev. Lett. 62, 863, 1989.

[108] M. S. Alam et al., Phys. Lett. 226B, 401, 1989.

[109] B. Knapp et al., Phys. Rev. Lett. 37, 882, 1976.

[110] I. Peruzzi et al., Phys. Rev. Lett. 37, 569, 1976.

[111] R. Brandelik et al., Phys. Lett. 70B, 132, 1977.

[112] E. Fermi, C. N. Yang, Phys. Rev. 76, 1739, 1949.

[113] H. Yukawa, Proc. Phys. Math. Soc. Jap. 17, 48, 1935.

[114] A. H. Compton, Phys. Rev. 22, 409, 1923.

[115] K. G. Wilson, Phys. Rev. D10, 2445, 1974.

[116] C. S. Wu et al., Phys. Rev. 105, 1413, 1957.

[117] M. Goldhaber, L. Grodzins, A. Sunyar, Phys. Rev. 109, 1015, 1958.

[118] A. Sakharov, Pisma Zh. Eksp. Teor. Fiz. 5, 32, 1967 (English translation: JETP Lett. 49,594, 1979)

[119] M. Gell-Mann, Phys. Lett.8, 214, 1964.

[120] G. Zweig, CERN-8419-TH-412, 1964.

[121] G. I. Budker, Atomic Energy. USSR. 22, 346, 1967.

[122] D. Möhl, G. Petrucci, L. Thorndahl, S. van der Meer, CERN/PS/AA 79-23, 1979; Physics Reports 58, 73-119, 1980.

[123] G. I. Budker et al., PAAC 7, 197, 1976.

[124] R. A. Alpher, R. C. Herman, Physics Today 41, No.8, 24, 1988.

[125] A. A. Penzias, R. Wilson, Astrophys. J. 142, 419, 1965.

[126] G. F. Smoot et al., Astrophys. J. 396, L1, 1992.

[127] P. deBernardis et al., Nature 404, 955, 2000.

[128] S. S. Gershtein, Y. B. Zeldovich, Pisma Zh. Eksp. Teor. Fiz. 4, 174, 1966; (English translation in Unity of Forces in the Universe, volume II, A. Zee, World Scientific, p 748, 1982).

[129] R. Cowsik, J. McLelland, Phys. Rev. Lett. 29, 669, 1972.

[130] G. Marx and A. S. Szalay, Proceedings of the "Neutrino 72" Conference; Acta Phys. Hung. 35, 113, 1974.

[131] J. H. Christenson et al., Phys. Rev. Lett. 13, 138, 1964.

[132] S. Bennett et al., Phys. Rev. Lett. 19, 993, 1967.

[133] L. Wolfenstein, Phys. Lett. 13, 562, 1964.

[134] W. Pauli, Niels Bohr and the development of physics, ed. W. Pauli, (pp 30-51), Pergamon Press, New York, 1955.

[135] R. S. Van Dyck, P. B. Schwinberg, H. G. Dehmelt, Phys. Phys. Rev. D34, 722, 1986; Phys. Rev. Lett. 59, 26, 1987.

[136] S. L. Adler, R. Dashen, Current Algebras and Applications to Particle Physics, W. A. Benjamin, N. Y., (1968); L. K. Pandit and V. S. Mathur in Advances in Physics, ed. R. L. Cool, R. E. Marshak, John Wiley, N. Y., (1968); B. Renner, Current Algebra, Pergamon Press, Oxford, (1968).

[137] S. L. Adler, Phys. Rev. Lett. 14, 1051, 1965.

[138] W. I. Weissberger, Phys. Rev. Lett. 14, 1047, 1965.

[139] R. H. Dalitz, Phi. Mag. 44, 1068, 1953; R. H. Dalitz, Phys. Rev. 94, 1046, 1954.

[140] J. Orear et al., Phys. Rev. 102, 1675, 1956.

[141] T. D. Lee, C. N. Yang, Phys. Rev. 104, 254, 1956.

[142] EMC collaboration; J. G. Ashman et al., Phys. Lett. 206B, 364, 1988.

[143] J. D. Bjorken, Phys. Rev. 148, 1467, 1966; Phys. Rev. D1, 1367, 1970.

[144] J. Ellis, R. L. Jaffe, Phys. Rev. D9, 1444, 1974; erratum D10, 1669, 1974.

[145] H. L. Anderson et al., Phys. Rev. 85, 934, 1952.

[146] H. C. Urey, F. G. Brickwedde, G. M. Murphy, Phys. Rev. 39, 164, 1932.

[147] H. C. Urey, F. G. Brickwedde, G. M. Murphy, Phys. Rev. 40, 1, 1932.

[148] J. Chadwick, M. Goldhaber, Nature, 134, 237, 1934.

[149] G. M. Murphy, H. Johnston, Phys. Rev. 45, 761(A108), 1934.

[150] I. Esterman, O. Stern, Phys. Rev. 45, 761(A109), 1934.

[151] J. M. B. Kellogg, I. I. Rabi, N. F. Ramsey, J. R. Zacharias, Phys. Rev. 56, 728, 1939.

[152] P. A. M. Dirac, Proc. Roy. Soc. A117, 610, 1928; Proc. Roy. Soc. A118, 351, 1928.

[153] H. A. Kramers, Atti. Congr. Intern. Fisici, Como, 2, 1927.

[154] R. Kronig, J. Opt. Soc. Amer. 12, 547, 1926.

[155] M. Gell-Mann, M. L. Goldberger, W. Thirring, Phys. Rev. 95, 1612, 1954.

[156] J. C. Christenson, Phys. Rev. Lett. 25, 1523, 1970.

[157] S. D. Drell, T. M. Yan, Phys. Rev. Lett. 24, 181,1970; Ann. Phys. (N. Y.) 66, 595, 1971.

[158] M. Gell-Mann, "The eightfold way: A theory of strong interaction symmetry," California Institute of Technology Synchrotron Laboratory Report CTSL-20, 1961, unpublished.

[159] Y. Ne'eman, Nucl. Phys. 26, 222, 1961.

[160] S. Okubo, Prog. Theoret. Phys. (Kyoto), 27, 949, 1962.

[161] D. R. Yennie, M. M. Levy, D. G. Ravenhall, Rev. of Mod. Phys. 29, 144, 1957.

[162] J. Schwinger, Phys. Rev. 73, 416, 1948.

[163] H. M. Foley, P. Kusch, Phys. Rev. 73, 412, 1948.

[164] R. A. Millikan, Phys. Rev. XXXII, 349, 1911.

[165] G. I. Budker, Atomic Energy USSR 22, 346, 1967.

[166] G. I. Budker et al., PAAC 7, 197, 1976.

[167] M. E. Rose, Phys. Rev. 73, 279, 1948.

[168] N. F. Mott, Proc. Roy. Soc. A124, 425, 1929.

[169] R. W. McAllister, R. Hofstadter, Phys. Rev. 102, 851, 1956.

[170] C. D. Anderson, Phys. Rev. 43, 491, 1933.

[171] P. M. S. Blackett, G. P. S. Occhialini, Proc. Roy. Soc. A139, 699, 1933.

[172] W. Pauli, Z. Phys. 31, 373, 1925; Z. Phys. 31, 765, 1925.

[173] G. E. Uhlenbeck, S. Goudsmit, Naturw. 13, 953, 1925.

[174] A. Salam, J. C. Ward, Phys. Lett. 13, 168, 1964.

[175] S. Weinberg, Phys. Rev. Lett. 19, 1264, 1967.

[176] J. Goldstone, Nuovo Cimento, 19, 154, 1961.

[177] J. Goldstone, A. Salam, S. Weinberg, Phys. Rev. 127, 1965, 1962.

[178] S. L. Glashow, Nucl. Phys. 22, 579, 1961.

[179] G. 't Hooft, Nucl. Phys. B33, 173, 1971; Nucl. Phys. B35, 167, 1971.

[180] UA1 collaboration, G. Arnison et al., Phys. Lett. 134B, 469, 1984.

[181] UA2 collaboration, P. Bagnaia et al., Phys. Lett. 129B, 130, 1983.

[182] A. Einstein, Annalen der Physik. Leipzig 17, 132, 1905.

[183] R. A. Millikan, Phys. Rev. 7, 18, 1916.

[184] C. V. Raman, S. Bhagavantam, Nature 129, 22, 1932.

[185] A. Pevsner et al., Phys. Rev. Lett. 7, 421, 1961.

[186] Alff et al., Phys. Rev. Lett. 9, 325, 1962.

[187] G. R. Kalbfleisch et al., Phys. Rev. Lett. 12, 527, 1964.

[188] M. Goldberg et al., Phys. Rev. Lett. 12, 546, 1964.

[189] B. Cassen, E. U. Condon, Phys. Rev. 50, 846, 1936.

[190] T. D. Lee, M. N. Rosenbluth, C. N. Yang, Phys. Rev. 75, 905, 1949.

[191] E. Fermi, Rend. Lincei 3, 145, 1926.

[192] P. A. M. Dirac, Proc. Roy. Soc. A112, 661, 1926.

[193] R. P. Feynman, Phys. Rev. 76, 749, 1950; Phys. Rev. 76, 769, 1950; Phys. Rev. 80, 440, 1950.

[194] V. A. Fock, Z. Phys. 75, 622, 1932.

[195] M. Froissart, Phys. Rev. 123, 1053, 1961.

[196] S. Mandelstam, Phys. Rev. 112, 1344, 1958.

[197] T. D. Lee, C. N. Yang, Nuovo Cimento, 3, 749, 1956.

[198] C. N. Yang, R. L. Mills, Phys. Rev.96, 191, 1954.

[199] J. Schwinger, Phys. Rev. 74, 1439, 1948; Phys. Rev. 75, 651, 1948.

[200] S. Tomonaga, Prog. Theor. Phys. 1, 27, 1946; Z. Koba, T. Tati, S. Tomonaga, Prog. Theor. Phys. 2, 101, 1947 and Prog. Theor. Phys. 2,198, 1947; T. Tati, S. Tomonaga, Prog. Theor. Phys. 3, 391, 1948.

[201] G. 't Hooft, M. Veltman, Nucl. Phys. B44, 189, 1972; Nucl. Phys. B50, 318, 1972.

[202] M. Gell-Mann, Phys. Rev. 92, 833, 1953.

[203] K. Nishijima, Prog. Theor. Phys. 13, 285, 1955.

[204] S. L. Glashow, J. Iliopoulos, I. Maiani, Phys. Rev. D2, 1285, 1970.

[205] D. P. Barber et al., Phys. Rev. Lett. 43, 830, 1979.

[206] R. Brandelik et al., Phys. Lett. 97B, 453, 1980.

[207] C. Berger et al., Phys. Lett. 97B, 459, 1980.

[208] M. L. Goldberger, S. B. Treiman, Phys. Rev. 110, 1178, 1958.

[209] H. Georgi, S. L. Glashow, Phys. Rev. Lett. 32, 438, 1974.

[210] J. Pati, A. Salam, Phys. Rev. D10, 275, 1974.

[211] H. Fritzsch, P. Minkowski, Ann. Phys. 93, 193, 1975.

[212] M. Gell-Mann, P. Ramond, R. Slansky, Rev. Mod. Phys. 50, 721, 1978.

[213] G. A. Milekhin, I. L. Rozental, Zh. Eksp. Teor. Fiz. 33, 187, 1957; B. Edwards et al., Phil. Mag. (7)3, 237, 1958; O. Minakawa et al., Japanese emulsion collaboration, Nuovo Cim. Suppl. 11, 125, 1959.

[214] E. Fermi, Prog. of Theor. Phys. 5, 570, 1950; L. D. Landau, Izv. Akad. Nauk. SSSR, Fiz. 31, 278, 1958.

[215] R. Hagedorn, Suppl. Nuovo Cimento 3, 147, 1965; R. Hagedorn, J. Ranft, Nuovo Cimento 6, 169, 1968; R. Hagedorn, Suppl. Nuovo Cimento 6, 311, 1968.

[216] F. Gürsey, L. A. Radicati, Phys. Rev. Lett. 13, 173, 1964.

[217] B. Sakita, Phys. Rev. 136B, 1756, 1964.

[218] H1: S. Aid et al., Nucl. Phys. B470, 3, 1996; C. Adloff et al., Nucl. Phys. B497, 3, 1997; Zeus: M. Derrick et al., Z. Phys. C72, 399, 1996; J. Breitweg et al., Phys. Lett. B407, 432, 1997.

[219] P. W. Higgs, Phys. Rev. Lett. 13, 508, 1964; Phys. Rev. 145, 1156, 1966; F. Englert, R. Brout, Phys. Rev. Lett. 13, 321, 1964; G. S. Guralnick, C. R. Hagen, T. W. B. Kibble, Phys. Rev. Lett. 13, 585, 1964.

[220] T. W. B. Kibble, Phys. Rev. 155, 1554, 1967.

[221] M. Danysz, J. Pniewski, Phil. Mag. 44, 348, 1953.

[222] W. F. Fry, Schneps, M. S. Swami, Phys. Rev. 101, 1526, 1956.

[223] M. Danysz et al., Phys. Rev. Lett. 11, 29, 1963.

[224] Y. B. Bushnin et al., Phys. Lett. 29B, 48, 1969; Sov. J. Nucl. Phys. 10, 337, 1969.

[225] F. Binon et al., Sov. J. Nucl. Phys. 11, 357, 1970.

[226] F. Bloch, A. Nordsieck, Phys. Rev. 52, 54, 1937.

[227] G. 't Hooft, Nucl. Phys. B79, 276, 1974.

[228] A. M. Polyakov, JETP Lett. 20, 194, 1974.

[229] J. Schwinger, Ann. Phys., 2, 407, 1957.

[230] R. Armenteros et al., Nature 167, 501, 1951; R. W. Thompson et al., Phys. Rev. 90, 329, 1953.

[231] L. W. Alvarez et al., Phys. Rev. 101, 503, 1956; V. L. Fitch, R. Motley, Phys. Rev. 101, 496, 1956.

[232] K. Lande et al., Phys. Rev. 103, 1901, 1956.

[233] E. Boldt, D. O. Caldwell, Y. Pal, Phys. Rev. Lett. 1, 150, 1958.

[234] U. Camerini et al., Phys. Rev. 128, 362, 1962.

[235] R. H. Good et al., Phys. Rev. 124, 1223, 1961.

[236] M. H. Alston et al., Phys. Rev. Lett. 6, 300, 1961.

[237] G. Smith et al., Phys. Rev. Lett. 10, 138, 1963.

[238] W. E. Lamb, R. C. Retherford, Phys. Rev. 72, 241, 1947.

[239] H. A. Bethe, Phys. Rev. 72, 339, 1947.

[240] T. D. Lee and C. N. Yang, Phys. Rev. 109, 1755, 1958.

[241] F. S. Crawford et al., Phys. Rev. Lett. 2, 114, 1959.

[242] M. H. Alston et al., Phys. Rev. Lett. 6, 698, 1961.

[243] S. H. Neddermeyer, C. D. Anderson, Phys. Rev. 51, 884, 1937.

[244] J. C. Street, E. G. Stevenson, Phys. Rev. 51, 1005, 1937.

[245] G. P. S. Occhialini, C. F. Powell, Nature 159, 186, 1947; D. H. Perkins, Nature 159, 126, 1947.

[246] M. L. Perl et al., Phys. Lett. 63B, 466, 1976.

[247] F. Reines, C. L. Cowan, Nature 178, 446, 1956.

[248] R. Davies, Phys. Rev. 97, 766, 1955.

[249] E. J. Konopinski, H. M. Mahmoud, Phys. Rev. 92, 1045, 1953.

[250] E. P. Hincks, B. Pontecorvo, Phys. Rev. 73, 257, 1948.

[251] K. Nishijima, Phys. Rev. 108, 907, 1957.

[252] G. Danby et al., Phys. Rev. Lett. 9, 36, 1962.

[253] P. A. M. Dirac, Rev. Mod. Phys., 21, 392, 1949.

[254] K. G. Wilson et al., Phys. Rev. D49, 6720, 1994.

[255] P. A. M. Dirac, Proc. Roy. Soc. A133, 60, 1931.

[256] E. Majorana, Nuovo Cim. 14, 171, 1937.

[257] B. Kayser, Comments Nucl. Part. Phys. 14, 69, 1985.

[258] B. Maier, Nucl. Phys. (Proc. Suppl.) B35, 358, 1994.

[259] J. L. Vuilleumier et al., Phys. Rev. D48, 1009, 1993.

[260] K. Inoue et al., Prog. Theor. Phys. 68, 927, 1982 [E:70, 330, 1983]
       and 71, 413, 1984; R. Flores, M. Sher, Ann. Phys. (N.Y.) 148,
       95, 1983; J. F. Gunion, H. E. Haber, Nucl. Phys. B272, 1, 1986
       [E:B402, 567, 1993]; J. F. Gunion et al., The Higgs Hunter's Guide,
       Addison-Wesley Publishing Company, Redwood City, CA, 1990.

[261] S. H. Neddermeyer, C. D. Anderson, Phys. Rev. 51, 884, 1937.

[262] B. Rossi, H. Van Norman Hilbery, J. B. Hoag, Phys. Rev 56, 837,
       1939.

[263] E. J. Williams, G. E. Roberts, Nature 145, 102, 1940.

[264] F. Rasetti, Phys. Rev 60, 198, 1941.

[265] B. Rossi, N. Nereson, Phys. Rev 62, 417, 1942.

[266] E. P. Hincks, B. Pontecorvo, Phys. Rev. 75, 698, 1949.

[267] R. B. Leighton, C. D. Anderson, A. J. Seriff, Phys. Rev. 75, 1432,
       1949; Phys. Rev. 76, 159, 1949.

[268] M. Conversi, E. Pancini, O. Piccioni, Phys. Rev. 71, 209, 1947.

[269] R. E. Marshak, H. A. Bethe, Phys. Rev. 72, 506, 1947; also S.
       Sakata, T. Inoue, Prog. Theor. Phys. 1, 143, 1946.

[270] B. Pontecorvo, Phys. Rev 72, 246, 1947.

[271] Y. Nambu, G. Jona-Lasinio, Phys. Rev. 122, 345, 1961; Phys. Rev. 124, 246, 1961.

[272] S. L. Glashow, Nucl. Phys. 22, 579, 1961.

[273] F. J. Hasert et al., Phys. Lett. 46B, 138, 1973; Nucl. Phys. B73, 1, 1974.

[274] L. M. Barkov, M. S. Zolotorev, Phys. Lett. 85B, 308, 1979; Zh. Eksp. Teor. Fiz. 79, 713, 1980; JETP 52, 360, 1980.

[275] C. S. Wood et al., Science 275, 1759, 1997; N. H. Edwards et al., Phys. Rev. Lett. 74, 2654, 1995; P. A. Vetter et al., Phys. Rev. Lett. 74, 2658, 1995; D. M. Meekhov et al., Phys. Rev. Lett. 71,3442, 1993; M. J. D. MacPherson et al., Phys. Rev. Lett. 67, 2784, 1991.

[276] C. Y. Prescott et al., Phys. Lett. 77B, 347, 1978; Phys. Lett. 84B, 524, 1979.

[277] B. Pontecorvo, Zh. Eksp. Teor. Fiz. 37, 1751, 1959; JETP 10, 1236, 1960.

[278] M. Schwartz, Phys. Rev. Lett. 4, 306, 1960.

[279] V. A. Lyubimov et al., Phys. Lett. 94B, 266, 1980.

[280] Belesev et al., Phys. Lett. B350, 263, 1995; Stoefel et al., Phys. Rev. Lett. 75, 3237, 1995; Ch. Weinheimer et al., Phys. Lett. B300, 210, 1993; H. Robertson et al., Phys. Rev. Lett. 67, 957, 1991.

[281] A. Picard et al., Nucl. Inst. Meth. in Phys. Res. B63, 345, 1992.

[282] B. Pontecorvo, Zh. Eksp. Teor. Fiz. 34, 247, 1958.

[283] L. Wolfenstein, Phys. Rev. D17, 2369, 1978; Nucl. Phys. B186, 147, 1981.

[284] S. P. Mikheyev, Yu. A. Smirnov, Sov. J. Nucl. Phys. 42, 913, 1985.

[285] S. J. Parke, Phys. Rev. Lett. 57, 1275, 1986; also Proceedings of the Fourteenth SLAC Summer Institute on Particle Physics, ed. E. C. Brennan, Stanford.

[286] R. Davis, Jr., D. S. Harmer, K. C. Hoffman, Phys. Rev. Lett. 20, 1205, 1968.

[287] J. N. Bahcall, M. H. Pinsonneault, Rev. Mod. Phys. 67, 781, 1995; J. N. Bahcall, S. Basu, M. H. Pinsonneault, Phys. Lett. B433, 1, 1998.

[288] J. N. Bahcall, Neutrino Astrophysics, Cambridge University Press, New York, N. Y., 1989, p 13.

[289] SAGE Collaboration, V. Gavrin et al., Neutrino 96. Proceedings of the XVII International Conference on Neutrino Physics and Astrophysics, Helsinki, Eds. K. Huitu, K. Enqvist, J. Maalampi (World Scientific, Singapore, 1997) p. 14; J. N. Abdurashitov et al., Neutrino 98. Proceedings of the XVIII International Conference on Neutrino Physics and Astrophysics, Takayama, Japan, 1998, Eds. Y. Suzuki, Y. Totsuka [Nucl. Phys. B (Proc. Suppl.) 77, 20, 1999].

[290] GALLEX Collaboration, P. Anselmann et al., Phys. Lett. B342, 440, 1995; W. Hampel et al., Phys. Lett. B388, 364, 1996.

[291] Kamiokande Collaboration, Y. Fukuda et al., Phys. Rev. Lett.77, 1683, 1996.

[292] SuperKamiokande Collaboration, Y. Fukuda et al., Phys. Rev. Lett. 81, 1158, 1998; also Y. Suzuki, Neutrino 98 [289].

[293] J. N. Bahcall, P. I. Krastev, A. Yu. Smirnov, Phys. Rev. D60, 093001, 1999.

[294] G. T. Danby et al., Phys. Rev. Lett. 9, 36, 1962.

[295] Mark II collaboration, G. S. Abrams et al., Phys. Rev. Lett. 63, 2173, 1989.

[296] L3 collaboration, B. Adeva et al., Z. Phys. C51, 179, 1991; OPAL collaboration, G. Alexander et al., Z. Phys. C52, 175, 1991; DELPHI collaboration, P. Abreu et al., Nucl. Phys. B367, 511, 1991; ALEPH collaboration, D. Decamp et al., Z. Phys. C53, 1, 1992.

[297] W. Bothe, H. Becker, Z. Phys. 66, 289, 1930.

[298] I. Curie, F. Joliot, Compt. Ren. 194, 273, 1932; Comt. Ren. 194, 708, 1932; Compt. Ren. 194, 876, 1932.

[299] J. Chadwick, Nature 129, 312, 1932; Proc. Roy. Soc. A136, 692, 1932.

[300] J. M. Robson, Phys. Rev. 77, 747(A), 1950.

[301] C. A. Altshuler, I. E. Tamm, Doklady Akad. Nauk SSSR 8, 455, 1934.

[302] L. W. Alvarez, F. Bloch, Phys. Rev. 57, 111, 1940.

[303] J. Schwinger, Phys. Rev. 52, 1250, 1937.

[304] E. Fermi, Internal report for the Metallurgical Laboratory of the University of Chicago, reprinted in American Journal of Physics 20, 536, 1952.

[305] B. C. Maglic et al., Phys. Rev. Lett. 7, 178, 1961.

[306] V. E. Barnes et al., Phys. Rev. Lett. 12, 204, 1964, Phys. Lett. 12, 134, 1964; G. S. Abrams et al., Phys. Rev. Lett. 13, 670, 1964

[307] I. Firestone et al., Phys. Rev. Lett. 26, 410, 1971.

[308] O. W. Greenberg, Phys. Rev. Lett. 13, 598, 1964.

[309] E. P. Wigner, Nachr. Ges. Wiss. Goett. P 375, 1927.

[310] C. S. Wu, I. Shaknov, Phys. Rev. 77, 136, 1950.

[311] C. S. Wu et al., Phys. Rev. 105, 1413, 1957.

[312] C. Y. Prescott et al., Phys. Lett. B84, 524, 1979.

[313] V. N. Gribov, L. N. Lipatov, Yad. Phys. 15, 781, 1972; Sov. J. Nucl. Phys. 15, 438, 1972; G. Altarelli, G. Parisi, Nucl. Phys. B126, 298, 1977.

[314] R. P. Feynman, Rev. of Mod. Phys. 20, 367, 1948.

[315] R. D. Peccei, H. Quinn, Phys. Rev. D16, 1791, 1977.

[316] L. Bertenza et al., Phys. Rev. Lett. 9, 180, 1962; P. Schlein et al., Phys. Rev. Lett. 10, 368, 1963.

[317] R. G. Astvacaturov et al., Phys. Lett. 27B, 45, 1968.

[318] A. Einstein, Annalen der Physik. Leipzig, 20, 199, 1906.

[319] J. Stark, Phys. Zeitschr. 10, 902, 1909.

[320] C. V. Raman, S. Bhagavantham, Nature 128, 114, 1931.

[321] S. N. Bose, Annalen der Physik. Leipzig, 26, 178, 1924.

[322] C. M. G. Lattes et al., Nature 159, 694, 1947.

[323] A. F. Dunaitsev et al., Phys. Lett. 1, 138, 1962.

[324] R. Durbin, H. Loar, J. Steinberger, Phys. Rev. 83, 646, 1951; D. Clark, A. Roberts, R. Wilson, Phys. Rev. 83, 649, 1951.

[325] W. F. Cartwright et al., Phys. Rev. 81, 652, 1951.

[326] W. K. H. Panofsky et al., Phys. Rev. 81, 565, 1951.

[327] T. B. Day, G. A. Snow, J. Sucher, Phys. Rev. Lett. 3, 61, 1960; M. Leon, H. A. Bethe, Phys. Rev. 127, 636, 1962.

[328] R. G. Glasser, N. Seeman, B. Stiller, Phys. Rev. 123, 1014, 1961.

[329] N. Kroll, W. Wada, Phys. Rev. 98, 1355, 1955.

[330] F. J. Dyson et al., Phys. Rev. 95, 1644, 1954.

[331] J. Hamilton, Theory of Elementary Particles, Clarendon Press, Oxford, U. K. 1959, p. 318.

[332] C. Goebel, Phys. Rev. Lett. 1, 337, 1958.

[333] G. F. Chew, F. E. Low, Phys. Rev. 113, 1640, 1959.

[334] J. Bowcock, W. N. Cottingham, D. Lurie, Nuovo Cim. 19, 142, 1961.

[335] I. Ya. Pomeranchuk, Zh. Eksp. Teor. Fiz. 30, 423, 1956; 34, 725, 1958.

[336] L. B. Okun, I. Ya. Pomeranchuk, Soviet Physics, JETP, 3, 307, 1956.

[337] A. M. Baldin, Nucl. Phys. 18, 310, 1960.

[338] V. I. Goldansky et al., Zh. Eksp. Teor. Fiz. 38, 1693, 1960.

[339] Y. Nambu, Preludes in Theoretical Physics in Honor of V. F. Weisskopf, ed. by A. De-Shalit, H. Feshbach, and L. Van Hove, North-Holland, Amsterdam, 1966, p.133.

[340] O. W. Greenberg, D. Zwanziger, Phys. Rev. 150, 1177, 1966.

[341] H. Fritzsch, M. Gell-Mann, H. Leutwyler, Phys. Lett. 47B, 365, 1973; S. Weinberg, Phys. Rev. Lett. 31, 494, 1973.

[342] D. J. Gross, F. Wilczek, Phys. Rev. D8, 3633, 1973.

[343] R. P. Feynman, Phys. Rev. 74, 1430, 1948; Phys. Rev. 76, 749, 1949; Phys. Rev. 76, 769, 1949; Phys. Rev. 80, 440, 1950.

[344] J. Schwinger, Phys. Rev. 74, 1439, 1948; Phys. Rev. 75, 651, 1949.

[345] S. Tomonaga, Progr. of Theor. Phys. 1, 27, 1946; T. Tati, S. Tomonaga, Prog. of Theor. Phys.3, 391, 1948; Z. Koba, T. Tati, S. Tomonaga, Progr. of Theor. Phys. 2, 101, 1947, and Progr. of Theor. Phys. 2, 198, 1947.

[346] F. J. Dyson, Phys. Rev. 75, 486, 1949, and Phys. Rev. 75, 1736, 1949.

[347] T. Regge, Nuovo Cim. 18, 947, 1960.

[348] G. F. Chew, S. Frautschi, S. Mandelstam, Phys. Rev. 126, 1202, 1962.

[349] P. D. B. Collins, An Introduction to Regge Theory & High Energy Physics, Cambridge University Press, New York, 1977.

[350] W. Pauli, F. Villars, Rev. Mod. Phys. 21, 434, 1949.

[351] G. 't Hooft, M. Veltman, Nucl. Phys. B44, 189, 1972.

[352] E. C. G. Stückelberg, A. Petermann, Helv. Phys. Acta 26, 499, 1953

[353] S. L. Glashow, Nucl. Phys. 22, 579, 1961; S. Weinberg, Phys. Rev. Lett. 19, 1264, 1967; A. Salam, Elementary Particle Theory (1968), reprinted in *New Particles*, J. L. Rosner, Selected Reprints, Stony Brook, Am. Assoc. of Physics Teachers, 1981.

[354] W. Pauli, Phys. Rev. 58, 716, 1940.

[355] G. Carron et al., Phys. Lett. 77B, 353, 1978.

[356] I. M. Ternov, Y. M. Loskutov, L. I. Korovina, Zh. Eksp. teor. Fiz. 41, 1294, 1961; A. A. Sokolov, I. M. Ternov, Doklady Akad. Nauk SSSR 153, 1052, 1963.

[357] R. Dolen, D. Horn, C. Schmid, Phys. Rev. 166, 1768, 1968.

[358] G. Veneziano, Nuovo Cim. 57A, 190, 1968.

[359] J. Scherk, J. H. Schwarz, Nucl. Phys. B81, 118, 1974.

[360] P. Ramond, Phys. Rev. D3, 2415, 1971.

[361] A. Neveu, J. Schwarz, Nucl. Phys. B31, 86, 1071.

[362] M. B. Green, J. Schwarz, Nucl. Phys. B181,502, 1981; Nucl. Phys. B198, 252, 1982; Phys. Lett. B109, 444, 1982.

[363] Th. Kaluza, Sitz. Preuss. Akad. Wiss., K1, 966, 1921; O. Klein, Z. Phys. 37, 895, 1926.

[364] S. Sakata, Prog. Theor. Phys. 16, 686, 1956.

[365] P. Ramond, Phys. Rev. D3, 2415, 1971; A. Neveau, J. H. Schwarz, Nucl. Phys. B31, 86, 1971; J. L. Gervais, B. Sakita, Nucl. Phys. B34, 632, 1971; J. Wess, B. Zumino, Nucl. Phys. B70, 39, 1974, and Phys. Lett. 49B, 52, 1974.

[366] M. L. Perl et al., Phys. Rev. Lett 35, 1489, 1975.

[367] M. L. Perl et al., Phys. Lett. 70B, 487, 1977.

[368] R. Brandelik et al., Phys. Lett. 73B, 109, 1978.

[369] G. Feldman et al., Proceedings of the 19th International Conference on High Energy Physics, Tokyo, p 777, 1978.

[370] L. Susskind, Phys. Rev. D20, 2619, 1979.

[371] S. W. Herb et al., Phys. Rev. Lett. 39, 252, 1977.

[372] R. H. Brown et al., Nature 163, 82, 1949.

[373] M. Y. Han, Y. Nambu, Phys. Rev. 139, B1006, 1965.

[374] E. P. Wigner, Nachr. Akad. Ges. Wiss. Göttingen, 31, 546, 1932.

[375] F. Abe et al., Phys. Rev. Lett., 74, 2626, 1995.

[376] S. Abachi et al., Phys. Rev. Lett., 74, 2632, 1995.

[377] F. Abe et al., Phys. Rev. Lett., 80, 2767, 1998.

[378] B. Abbott et al., Phys. Rev. D58, 052001, 1998; Phys. Rev. D60, 052001, 1999; Phys Rev. Lett. 80, 2063, 1998.

[379] L.D. Landau, Nucl. Phys. 3, 127, 1957; T. D. Lee, C. N. Yang, Phys. Rev. 105,1671, 1957; A. Salam, Nuovo Cim. 5, 299, 1957.

[380] H. Georgi, S. L. Glashow, Phys. Rev. Lett., 32, 438, 1974; H. Fritzsch, P. Minkowski, Ann. Phys. (N.Y.) 93, 193, 1975; F. Gürsey, P. Sikvie, Phys. Rev. Lett., 36, 775, 1976; K. Inoue, A. Kakuto, Y. Nakano, Prog. Theor. Phys. 58, 630, 1977.

[381] A. Sakharov, JETP Lett. 5, 32, 1967.

[382] M. Yoshimura, Phys. Rev. Lett. 41, 281, 1978; S. Dimopoulos, L. Susskind, Phys. Rev. D18, 4500, 1978; D. Toussaint et al., Phys. Rev. D19, 1036, 1979; S. Weinberg, Phys. Rev. Lett. 42, 850, 1979; S. Barr et al., Phys. Rev. D20, 2494, 1979.

[383] C. Berger et al., Phys. Lett. 76B, 243, 1978.

[384] C. W. Darden et al., Phys. Lett 76B, 246, 1978.

[385] J. C. Ward, Phys. Rev. 78, 182, 1950.

[386] UA1 Collaboration, G. Arnison et al., Phys. Lett. 122B, 103, 1983.

[387] UA2 Collaboration, M. Banner et al., Phys. Lett. 122B, 476, 1983.

[388] UA1 Collaboration, G. Arnison et al., Phys. Lett. 126B, 398, 1983.

[389] UA2 Collaboration, P. Bagnaia et al., Phys. Lett. 129B, 130, 1983.